Protein Phosphorylation

The Practical Approach Series

SERIES EDITOR

B. D. HAMES
Department of Biochemistry and Molecular Biology
University of Leeds, Leeds LS2 9JT, UK

See also the Practical Approach web site at **http://www.oup.co.uk/PAS**

★ **indicates new and forthcoming titles**

Affinity Chromatography
★ Affinity Separations
   Anaerobic Microbiology
   Animal Cell Culture
      (2nd edition)
   Animal Virus Pathogenesis
   Antibodies I
   Antibodies II
   Antibody Engineering
   Antisense Technology
★ Apoptosis
   Applied Microbial Physiology
   Basic Cell Culture
   Behavioural Neuroscience
   Bioenergetics
   Biological Data Analysis
   Biomechanics—Materials
   Biomechanics—Structures and
      Systems
   Biosensors
★ Caenorhabditis Elegans
   Carbohydrate Analysis
      (2nd edition)
   Cell-Cell Interactions

   The Cell Cycle
★ Cell Growth, Differentiation
      and Senescence
★ Cell Separation
   Cellular Calcium
   Cellular Interactions in
      Development
   Cellular Neurobiology
   Chromatin
★ Chromosome Structural
      Analysis
   Clinical Immunology
   Complement
★ Crystallization of Nucleic
      Acids and Proteins
      (2nd edition)
   Cytokines (2nd edition)
   The Cytoskeleton
   Diagnostic Molecular
      Pathology I
   Diagnostic Molecular
      Pathology II
   DNA and Protein Sequence
      Analysis
   DNA Cloning 1: Core
      Techniques (2nd edition)

Protein Phosphorylation

Second Edition

A Practical Approach

Edited by

D. G. HARDIE
Department of Biochemistry
University of Dundee, Dundee

OXFORD
UNIVERSITY PRESS

OXFORD

UNIVERSITY PRESS

Great Clarendon Street, Oxford OX2 6DP

Oxford University Press is a department of the University of Oxford
and furthers the University's aim of excellence in research, scholarship,
and education by publishing worldwide in

Oxford New York

Athens Auckland Bangkok Bogotá Buenos Aires Calcutta
Cape Town Chennai Dar es Salaam Delhi Florence Hong Kong Istanbul
Karachi Kuala Lumpur Madrid Melbourne Mexico City Mumbai
Nairobi Paris São Paulo Singapore Taipei Tokyo Toronto Warsaw

and associated companies in Berlin Ibadan

Oxford is a registered trade mark of Oxford University Press

Published in the United States
by Oxford University Press Inc., New York

© Oxford University Press, 1999

A catalogue record for this book is available from the British Library

Library of Congress Cataloging-in-Publication Data
Protein phosphorylation : a practical approach / edited by Grahame
Hardie.—2nd ed.
(The practical approach series ; 211)
Includes bibliographical references and index.
1. Phosphoproteins Laboratory manuals. 2. Phosphorylation
Laboratory manuals. 3. Protein kinases Laboratory manuals.
4. Phosphoprotein phosphatases Laboratory manuals. I. Hardie, D.
G. (D. Grahame) II. Series.
QP552.P5P76 1999 572'.792—dc21 99–32319
ISBN 0 19 963729 6 (Hbk)
0 19 963728 8 (Pbk)

Typeset by Footnote Graphics,
Warminster, Wilts

Preface

Since the publication of the first edition of this book in 1993, the single factor that has had the most dramatic effect on the manner in which we study protein phosphorylation has undoubtedly been the progress of the genome sequencing projects. We now know that *Saccharomyces cerevisiae* has around 120 protein kinases, and *Caenorhabditis elegans* about 420. Within the next 2 or 3 years we will also know how many are present in humans and *Arabidopsis thaliana* (>1000 in each case?). The focus is therefore inevitably shifting from discovery of new protein kinases and phosphatases to working out what the ones that have already been defined are doing. This task is likely to occupy us for many years to come. We can already begin to construct 'cell signalling maps' similar to the metabolic maps that were first produced in the 1960s. However signalling maps are undoubtedly more complex, since the protein kinases and phosphatases in the cell represent a form of 'massively parallel processor' with obvious analogies to a neural network. Following the flow of information down these pathways is not as easy as following the flow of intermediates through metabolism. We have nothing quite equivalent to the use of the radioactively labelled precursors that provided so much information about metabolism. We can of course study protein kinases and phosphatases in cell-free systems just as the metabolic investigators did, but in intact cells the only approach we currently have is to manipulate a component of the signalling network (either upwards or downwards) and examine the consequences. Interpretation is, however, a difficult task because of the very large degree of redundancy and cross-talk which exists between the pathways.

In the light of the new developments, several completely new chapters have been added to the book, while a few of the original chapters have been updated. Jim Garrison (Chapter 1) has reprised his original chapter on studying protein phosphorylation in intact cells, but with Tim Haystead has added to it an exciting new whole-cell 'proteomics' approach to the identification of downstream targets of signalling pathways. Hiroyoshi Hidaka's group (2), and the editor (3), have updated their chapters on the use of pharmacological inhibitors of kinases and phosphatases respectively. An important development in the protein phosphatase inhibitor field was the discovery that the target for the immunosuppressive drugs cyclosporin and FK506 was protein phosphatase-2B. The pharmaceutical companies have also been performing high throughput screens for kinase and phosphatase inhibitors, and the fruits of this effort have already started filtering down (or up?) into the research community. Fears that kinase inhibitors that mimic ATP would inevitably be non-specific have to some extent proved unfounded, as exemplified by SB203580, a highly specific inhibitor of the p38/SAPK2 kinases. Alongside pharmacology, molecular biology provides several possible routes to alter the activity of components of signalling pathways, and Calum Sutherland (4) provides a new chapter on the use of methods such as over-expression of constitutively active or dominant negative mutants, and antisense and ribozyme technology. Targetted gene knock-outs provide another important avenue, although one that is not specific to studies of protein phosphorylation and was felt to be beyond the scope of the book.

With respect to the mapping of phosphorylation sites, Tony Hunter and colleagues (5) have reprised their popular 2-D peptide mapping methods, while Ken Mitchellhill (6) provides a completely new chapter on the mass spectrometric methods that are revolutionizing the identification of phosphorylation sites. Biological mass spectrometry has certainly come of age, and becomes particularly powerful with the availability of the expressed sequence tag and genome databases.

Carol MacKintosh (7, now with Greg Moorhead) has updated her chapter on the assay and purification of phosphatases, including the new microcystin affinity columns. Andrew Garton, Andrew Flint and Nick Tonks (8) describe a powerful new method for identifying substrates for protein-tyrosine phosphatases, previously an extremely difficult task. The remainder of the book mainly concerns methods for studying protein kinases. Chapters by the editor (9), and by Lorenzo Pinna and his colleagues (10), discuss methods for assaying and purifying kinases, and the design of specific peptide substrates. David Carling (11) describes the cloning and expression of protein kinases and their subunits, and explains how to make best use of the EST and genome databases for '*in silico*' cloning, also known as 'cloning by phoning'.

At one time protein kinases were generally identified by their ability to regulate a target, so that identification of the target came first and the protein kinase came later. This is now usually completely reversed, and from the sequence databases we already have numerous protein kinases for which no substrate is known. In most cases we also have little or no information about the regulatory subunits associated with the catalytic subunit that is defined by the DNA sequence. Most of the remaining chapters therefore describe methods for identifying protein kinase targets, and the regulatory subunits with which they associate. Julia Stone and John Walker (12) describe the interaction cloning approach to identify associated subunits, while Rikiro Fukunaga and Tony Hunter (13) present a new expression cloning method to identify downstream targets. Rong Jiang and Marian Carlson (14) discuss the popular yeast two-hybrid approach. With their work on the yeast SNF1 complexes they have shown just how much information this technique can yield, although it can also be misused if applied indiscriminately. Friedrich Herberg and Bastian Zimmermann (15) discuss the application to protein kinases of the other end of the spectrum of methods for studying protein: protein interactions, i.e. the physical methods such as surface plasmon resonance. Finally in this section, Zhou Songyang and Lewis Cantley (16) present their peptide library approach to determine the recognition motif for protein kinases, a method very much in the modern idiom of combinatorial chemistry.

Another important development since the production of the first edition has been the publication of several protein kinase and phosphatase structures determined by X-ray crystallography. A full discussion of crystallography was beyond the scope of this book, and is the subject of a separate volume in the Practical Approach series. However, it was felt that a chapter on the elements of the technique would be useful for protein phosphorylation researchers who might at least wish to initiate a collaboration with a crystallographer. Bruce Kemp (17) was an obvious person to write this because he has successfully entered this field, although most of his career has been spent as a biochemist studying protein kinases.

Many of the new techniques fall into the category of what are sometimes now referred to as 'fishing expeditions'. May I wish you all happy fishing!

Dundee G. H.
June 1999

Contents

Contents

Contents

6. Phosphorylation site analysis by mass spectrometry   127

Ken I. Mitchelhill and Bruce E. Kemp

Contents

14. Analysis of protein kinase interactions using the two-hybrid method   315

Rong Jiang and Marian Carlson

Contents

Contents

17. Crystallization of protein kinases and phosphatases 387

Bostjan Kobe, Thomas Gleichmann, Trazel Teh, Jörg Heierhorst, and Bruce E. Kemp

Contents

Contributors

LEWIS C. CANTLEY
Division of Signal Transduction, Beth Israel Hospital, Harvard Institutes of Medicine, 10th floor, 330 Brookline Avenue, Boston, MA 02215, USA.

DAVID CARLING
Cellular Stress Group, MRC Clinical Sciences Centre, Imperial College School of Medicine, Hammersmith Hospital, Du Cane Road, London W12 0NN, UK.

MARIAN CARLSON
Department of Genetics and Development, College of Physicians and Surgeons, Columbia University, New York, NY 10032, USA.

STEPHEN P. DAVIES
Biochemistry Department, Dundee University, MSI/WTB Complex, Dow Street, Dundee, DD1 5EH, Scotland, UK.

ANDREW FLINT
Demerec Building, Cold Spring Harbor Laboratory, PO Box 100, Cold Spring Harbor, NY 11724-2208, USA.

RIKIRO FUKUNAGA
Department of Genetics, Osaka University Medical School, B-3, 2-2 Yamadaoka, Suita, Osaka 565-0871, Japan.

JAMES C. GARRISON
Department of Pharmacology, University of Virginia, Charlottesvill, VA 22908, USA.

ANDREW GARTON
Demerec Building, Cold Spring Harbor Laboratory, PO Box 100, Cold Spring Harbor, NY 11724-2208, USA.

PETER VAN DER GEER
Molecular Biology and Virology Laboratory, The Salk Institute for Biological Studies, 10010 North Torrey Pines Road, La Jolla, CA 92037, USA.

THOMAS GLEICHMANN
St. Vincent's Institute of Medical Research, 41 Victoria Parade, Fitzroy, Victoria 3065, Australia.

D. GRAHAME HARDIE
Biochemistry Department, Dundee University, MSI/WTB Complex, Dow Street, Dundee, DD1 5EH, Scotland, UK.

Contributors

TIMOTHY A.J. HAYSTEAD
Department of Pharmacology, University of Virginia, Charlottesvill, VA 22908, USA.

JÖRG HEIERHORST
St. Vincent's Institute of Medical Research, 41 Victoria Parade, Fitzroy, Victoria 3065, Australia.

FRIEDRICH W. HERBERG
Institut fur Physiologische Chemie, Ruhr-Universitat Bochum, Abt, Biochemie Supramolekularer Systeme, D-44780 Bochum, Germany.

HIROYOSHI HIDAKA
Department of Pharmacology, Nagoya University School of Medicine, 65 Tsurumai-cho, Nagoya 466-8550, Japan.

TONY HUNTER
Molecular Biology and Virology Laboratory, The Salk Institute for Biological Studies, 10010 North Torrey Pines Road, La Jolla, CA 92037, USA.

RONG JIANG
Department of Genetics and Development, College of Physicians and Surgeons, Columbia University, New York, NY 10032, USA.

BRUCE KEMP
St. Vincent's Institute of Medical Research, 41 Victoria Parade, Fitzroy 3065, Victoria, Australia.

RYOJI KOBAYASHI,
Department of Chemistry, School of Medicine, Kagawa Medical University, 1750-1 Ikenobe, Miki-cho, Kita-gun, Kagawa 761-0793, Japan.

BOSTJAN KOBE
St. Vincent's Institute of Medical Research, 41 Victoria Parade, Fitzroy, Victoria 3065, Australia.

KRUNXIN LUO
The Salk Institute, 10010 North Torrey Pines Road, La Jolla, CA 92037, USA.

CAROL MACKINTOSH
MRC Protein Phosphorylation Unit, Biochemistry Department, Dundee University, MSI/WTB Complex, Dow Street, Dundee, DD1 5EH, Scotland, UK.

KEN MITCHELHILL
St. Vincent's Institute of Medical Research, 41 Victoria Parade, Fitzroy 3065, Victoria, Australia.

GREG MOORHEAD
MRC Protein Phosphorylation Unit, Biochemistry Department, Dundee University, MSI/WTB Complex, Dow Street, Dundee, DD1 5EH, Scotland, UK.

Contributors

YASUHITO NAITO
Department of Pharmacology, Nagoya University School of Medicine, 65 Tsurumai-cho, Nagoya 466-8550, Japan.

LORENZO A. PINNA
Dipartimento di Chimica Biologica, Universita di Padova, and Centro per lo Studio delle Biomembrane del Consiglio Nazionale, delle Ricerche, Viale G. Colombo 3, 35121 Padova, Italy.

MARIA RUZZENE
Dipartimento di Chimica Biologica, Universita di Padova, and Centro per lo Studio delle Biomembrane del Consiglio Nazionale, delle Ricerche, Viale G. Colombo 3, 35121 Padova, Italy.

IAN P. SALT
Biochemistry Department, Dundee University, MSI/WTB Complex, Dow Street, Dundee, DD1 5EH, Scotland, UK.

BARTHOLOMEW M. SEFTON
The Salk Institute, 10010 North Torrey Pines Road, La Jolla, CA 92037, USA.

ZHOU SONGYANG
Division of Signal Transduction, Beth Israel Hospital, Harvard Institutes of Medicine, 10th floor, 330 Brookline Avenue, Boston, MA 02215, USA.

JULIE M. STONE
Department of Molecular Biology, 50 Blossom St., Wellman 10, Massachusetts General Hospital, Boston, MA 02114, USA.

CALUM SUTHERLAND
MRC Protein Phosphorylation Unit, Biochemistry Department, Dundee University, MSI/WTB Complex, Dow Street, Dundee, DD1 5EH, Scotland, UK.

TRAZEL TEH
St. Vincent's Institute of Medical Research, 41 Victoria Parade, Fitzroy, Victoria 3065, Australia.

N.K. TONKS
Demerec Building, Cold Spring Harbor Laboratory, PO Box 100, Cold Spring Harbor, NY 11724-2208, USA.

JOHN C. WALKER
Division of Biological Sciences, Room 308 Tucker Hall, University of Missouri, Columbia, MO 65211, USA.

BASTIAN ZIMMERMANN
Institut fur Physiologische Chemie, Ruhr-Universitat Bochum, Abt, Biochemie Supramolekularer Systeme, D-44780 Bochum, Germany.

Abbreviations

3AT	3-amino-1,2,4-triazole
AET	aminoethanethiol
AMPK	AMP-activated protein kinase
AS	anomalous scattering
ATP	adenosine triphosphate
BSA	bovine serum albumin
CM	carboxymethylated
CMV	cytomegalovirus
CORT	cloning of receptor targets
cpm	counts per minute
DEAE	diethylaminoethyl
DIG	digoxygenin
DMEM	Dulbecco's modification of Eagle's medium
DSC	differential scanning calorimetry
DTT	dithiothreitol
EGF	epidermal growth factor
ESI	electrospray ionization
ESTs	expressed sequence tags
FABMS	fast-atom bombardment mass spectrometry
FCS	fetal calf serum
GFP	green fluorescent protein
GI	gene of interest
GST	glutathione-S-transferase
HBS	HEPES – buffered saline
HEPES	N-[2-hydroxyethyl]piperazine-N'-[2-ethanesulphonic acid]
HPLC	high-pressure liquid chromatography
IDA	iminodiacetic acid
IGF	insulin-like growth factor
IP	immunoprecipitate
IPG	immobilized pH gradient
IPTG	isopropyl β-D-thiogalactopyranoside
ITC	isothermal titration calorimetry
LCMS	liquid chromatography/mass spectrometry
MAD	multi-wavelength anomalous dispersion
MALDITOF	matrix-assisted laser desorption/ionization time-of-flight
MBP	maltose-binding protein
MC	microcystin
Mdha	N-methyl dehydroalanine
MEK	MAP kinase kinase
MES	2-[N-morpholino]ethanesulphonic acid

MIR	multiple isomorphous replacement
MOPS	morpholinopropane sulphonic acid
NGF	nerve growth factor
NHS	*N*-hydroxysuccinimide
NLS	nuclear localization signal
NTA	nitrilotriacetic acid
PAGE	polyacry;amide gel electrophoresis
PBS	phosphate-buffered saline
PCR	polymerase chain reaction
PDEA	2-(2-pyridinyldithio)ethaneamine hydrochloride
PEG	polyethylene glycol
PKA	cyclic AMP-dependent protein kinase
PMSF	phenylmethyl sulfonyl fluoride
*p*NPP	*p*-nitrophenylphosphate
PSK	protein–serine/threonine kinase
PTK	protein–tyrosine kinase
PTP	protein–tyrosine phosphatase
PVP	polyvinylpyrrolidone
SA	streptavidin
SAX	strong anion-exchange
SC	synthetic complete (medium)
SDS	sodium dodecyl sulfate
SMPT	4-succinimidyloxycarbonyl-methyl-α(2-pyridyldithio)toluene
SPDP	3-(2-pyridyldithio)-propionic acid-*N*-succinimidylester
SPR	surface plasmon resonance
TCA	trichloroacetic acid
TEMED	*N,N,N′,N′*-tetramethylethylenediamine
TES	N-tris(hydroxymethyl)methyl-2-aminoethane sulphonic acid
TPA	tetradecanoyl-phorbol-13-acetate
TPCK	*N*-tosyl-L-phenylalanine chloromethyl ketone

1

Study of protein phosphorylation in intact cells

TIMOTHY A.J. HAYSTEAD and JAMES C. GARRISON

1. Introduction

The phosphorylation of proteins on the amino acid residues serine, threonine and tyrosine is generally recognized as a fundamental mechanism by which the regulation of intracellular events is achieved. The process is reversible, enabling cells to respond in a dynamic way to a myriad of signals. The significance of protein phosphorylation is rapidly becoming apparent from the ongoing genome projects, with the discovery that 1–2% of the entire human genome may encode protein kinases (1). Conversely, in the case of protein–tyrosine phosphatases, there may be as many genes encoding these enzymes as there are opposing tyrosine kinases. Although not as readily identified by primary structure predictions alone, there may also be numerous genes encoding regulatory subunits which control the activity of the catalytic subunits of serine/threonine phosphatases (2–5). The finding that nature has put so much evolutionary effort into controlling phosphorylation events in the cell is a testament to its importance, as well as to its complexity. The finding that many protein kinases and phosphatases are localized with their substrates in discrete subcellular compartments, and not floating in the intracellular 'ether' as previously thought, adds additional complexity. Methods that enable phosphorylation events to be examined directly as they happen in the intact cell in response to a particular stimulus are, therefore, highly desirable. This chapter focuses on state of the art, non-invasive, methods that enable changes in intracellular protein phosphorylation to be measured as they occur in intact cells. One of the great advantages of studying protein phosphorylation, as opposed to other regulatory mechanisms such as allosteric regulation, is that protein phosphate esters are stable at physiological pH and can survive homogenization of the cell.

A widely used method for defining a particular phosphorylation event in a cell or tissue is to label the intracellular ATP pools with [^{32}P]orthophosphate, prior to a stimulus expected to change the phosphorylation state of proteins. The phosphorylated proteins are extracted from the cell, resolved on SDS polyacrylamide gels, and visualized using autoradiography or phosphorimaging. Newer, non-radioactive methods offer alternative methods of analysis. Phosphotyrosine phosphorylation events can be studied using anti-phosphotyrosine antibodies (6). Antibodies to phosphoserine or phosphothreonine have not proved as universally successful. However, antibodies that recognize specific peptide sequences within known phosphorylation sites are being developed and are increasingly useful (7). Although not yet widely applied, chemical methods have also been devised that alter phosphoamino acids on proteins and peptides to enable fluors or other probes to be attached. The phosphorylated proteins or peptides are then visualized by their fluorescence (8).

Studying phosphorylation events in intact cells, rather than in cell-free systems, offers a number of advantages:

(a) The approach demonstrates that a phosphorylation reaction studied with purified proteins does indeed occur in the cell. As the protein kinases and substrates are in their native state during the experiment, there can be no question about the physiological relevance of the phosphorylation events.

(b) Changes can be observed in the phosphorylation states of substrates even when the details of the transmembrane signalling system involved are not known. This property has been useful, for example, in studying cellular responses to growth factors, cytokines, insulin, and hormones activating phospholipase C.

(c) The technique can be used to uncover a fairly complete display of the protein phosphorylation events that occur following a given stimulus. This feature allows assessment of the overall response of the cell to a stimulus, and has helped to uncover important new sites of regulation by covalent modification.

(d) Using rapid freezing techniques, time-resolved studies can define important early events in a signalling pathway leading to a particular phosphorylation event. This approach can address, to a certain extent, the cellular architecture of signalling pathways, since one is resolving a pathway both in time and space. The earliest phosphorylation events are likely to be closer to the initiating receptor.

(e) The latest advances in mass spectrometry, or Edman microsequencing, enable picomole or even femtomole amounts of a phosphoprotein resolved on a one-dimensional (1D) or two-dimensional (2D) SDS–polyacrylamide gel electrophoresis (PAGE) to be identified. Indeed, when combined with the various genome projects, these new sequencing technologies may allow examination of the effects of a particular stimulus on the entire 'phosphoprotein proteome' of a cell.

This chapter discusses the latest techniques that enable phosphoproteins to be characterized and identified in the intact cell. The rationale behind each technique, and comments and hints, are provided to help the reader transfer the protocols to almost any type of cell system.

2. Labelling cells with [^{32}P]orthophosphate

2.1 General considerations

For optimal results, the conditions chosen for labelling suspensions of cells or cells in tissue culture need to be adjusted for each different system and experimental objective. However, some guidelines and protocols already proven to be effective are provided below. There are three general considerations important to designing a labelling protocol:

(a) Under most labelling conditions only 10–15% of the added ^{32}P is incorporated into the intracellular ATP pool (9). Therefore, the phosphate content of the incubation buffer must be reduced to obtain the highest possible specific activity of radioactive phosphate during the labelling period. However, eliminating all phosphate can be self-defeating, since this can lead to a depletion of the intracellular phosphate pool below levels where cells can function normally. We suggest that a minimum of 0.2 mM non-radioactive phosphate be included in the medium. Even under these conditions, most experiments require about 0.5–1 mCi of ^{32}P per 20 mg of cell protein to label the cells adequately.

(b) The different pools of phosphate within the cell (phospholipids, nucleotides, nucleic acids, phosphoproteins, etc.) equilibrate with extracellular phosphate at markedly different rates. The length of the labelling time is essentially a compromise between equilibrating the intracellular ATP pool with the incubation medium, and the amount of background labelling of intracellular phosphoproteins. The plasma membranes of certain freshly isolated cells, such as hepatocytes or adipocytes, are fairly permeable to phosphate and the ATP pools may equilibrate within 40–60 minutes (10). However, cells in tissue culture may require several hours to adequately label the ATP pools (the rate of equilibration can be measured, see Section 6). In many instances, the proteins in the cells will not have fully equilibrated with the ^{32}P in the ATP pools during a short labelling period, and will continue to gain phosphate for a number of hours (11). Since many of the phosphoproteins in cells are not responsive to hormones and growth factors, incubations beyond the time necessary to equilibrate the ATP pool increase the total background of radioactive proteins on the autoradiographs used to visualize the labelled proteins. While it is essential that the specific activities of the ATP pools under the control and experimental conditions are identical, it is not necessary to have the ATP pool reach isotopic equilibrium with the extracellular [^{32}P]phosphate for an experiment to be valid. However, the higher the specific activity of the ATP, the more sensitive the detection.

3

(c) Many proteins that are substrates for kinases activated by hormones, neurotransmitters, or growth factors equilibrate rapidly with the ^{32}P in the ATP pool, because their phosphorylation site(s) are subject to continual turnover via the action of kinases and phosphatases. Thus, they will contain small amounts of ^{32}P in the basal state and the phosphorylation level will increase dramatically following stimulation (12). However, it must be recognized that some phosphoproteins contain multiple phosphorylation sites that equilibrate with the ^{32}P in ATP at different rates (via different kinases and/or phosphatases). Thus, the effects of a particular stimulus may be obscured by a high basal phosphorylation of the protein. Adjusting the labelling time may be necessary if the protein of interest falls into this category.

2.2 Labelling cells in suspension

Protocol 1 can be used to label cells maintained in suspension, such as hepatocytes or adipocytes.

Protocol 1. Labelling cells in suspension

Equipment and reagents

- buffer A: Krebs–Ringer bicarbonate buffer (118 mM NaCl, 4.75 mM KCl, 0.2 mM NaH$_2$PO$_4$, 1.2 mmol/MgCl$_2$, 0.26 mM CaCl$_2$, 25 mM Na$_2$HCO$_3$). Maintain pH 7.0–7.4 by equilibrating with CO$_2$.

- buffer B:[a] 125 mM NaCl, 0.2 mM NaH$_2$PO$_4$, 1.2 mM MgCl$_2$, 4.0 mM KCl, 2.6 mM CaCl$_2$, 24 mM ultrapure Hepes (Swarzmann Biotech Inc), 0.5 mM glucose, 1.0% (w/v) fatty acid free bovine serum albumin. Adjust pH to 7.4 with 5 N NaOH.

Method

1. Set up the flask or tubes to be used for equilibrating the cells with ^{32}P behind 1 cm Plexiglas shields. If possible, use disposable plastic tubes for the incubation to simplify clean-up of the radioactive waste.

2. Add nutrients (e.g. glucose, amino acids) needed during the incubation.

3. Suspend the cells at a concentration of about 10–20 mg protein/ml (2–3 × 10^7 cells/ml) in the desired incubation buffer (buffer A or B). If albumin is added to the cell suspension, keep the concentration below 1 mg/ml to avoid distorting the polyacrylamide gels eventually used to resolve the phosphoproteins.[b]

4. Add [^{32}P]phosphate to a concentration that will eventually yield 10–20 μCi/mg cell protein (around 0.5–1.0 mCi/ml). Take care to add exactly the same amount of radioactivity and cells to each tube as this ratio determines the eventual specific radioactivity of the ATP in the cells.

5. Start the incubation by adding the volume of cells needed for the experiment to the tubes prepared above. Shield the tubes during the incubation with sheets of Plexiglas to minimize the radiation flux out of the incubation vessel.

6. Incubate the cells for 45–120 minutes to equilibrate the ATP pools with the extracellular ^{32}P as desired. At the end of the equilibration period, stimulate the cells with the agent(s) of choice for the desired period.

7. Stop the incubations using one of the methods described in Section 3 below.

[a] Buffer B has the advantage of not requiring equilibration with CO_2.
[b] If possible, wash out serum albumin at the end of the experiment by centrifuging (2000 g for 5 minutes), and resuspending cells in Buffer A or B (minus albumin). It will be necessary to include hormones in the resuspension buffer if these are used in the experiment.

2.3 Labelling cells in tissue culture

Protocol 2. Labelling cells growing in tissue culture

Equipment and reagents

- Krebs-Ringer (see *Protocol 1*) containing 0.2 mM phosphate and 20 mM Hepes
- Tissue culture facilities

Method

1. Prepare 60 or 100 mm petri dishes of cells containing about 10^6–10^7 cells each. Each dish should contain approximately the same number of cells.

2. Depending on the purpose of the experiment, the cells may need to be removed from the growth medium and maintained in a defined medium prior to labelling to provide a 'basal' state. For example, this step would be necessary when cells are growing in serum and the serum is known to contain agonists capable of stimulating the cells. If the cells will require a special tissue culture medium during the labelling period, make up the medium with 0.2 mM phosphate and use it in the next two steps. Alternatively, Krebs–Ringer bicarbonate buffer (see *Protocol 1*) containing 0.2 mM phosphate and 20 mM Hepes at pH 7.4 may be suitable.

3. Wash the cells three times in the low phosphate medium, to remove the residual high phosphate from the original medium.

4. Add enough low phosphate medium to the dish to cover the cells (about 1–3 ml) and about 0.5–1 mCi/ml of [^{32}P]phosphate to begin labelling the ATP pools.

Protocol 2. *Continued*

5. Allow ATP pools to label for 1–2 hours in the appropriate atmosphere (5% CO_2/95% O_2 for bicarbonate-containing buffers). For safety reasons you may wish to use a separate incubator dedicated to radioactive experiments.
6. Stimulate the cells for the desired time with agonist or mitogen by adding a concentrated stock of agonist to the cells.
7. After the desired period of stimulation, prepare cell protein extracts for analysis on 1D or 2D gels as described in *Protocol 4* (Section 3.2) below.

3. Extraction of phosphorylated proteins

3.1 General considerations

Once the cells have been stimulated with the appropriate agent, the next step is to extract the relevant proteins, avoiding further modification by proteinases, protein kinases or phosphatases, or other enzymes. The sections below describe protocols for analysing proteins from the whole-cell homogenate or from specific subcellular fractions. The special considerations that apply to each technique are presented in the introductory comments with each protocol.

3.2 Extraction of proteins without cell fractionation

This method of protein extraction has the following advantages:

- speed, simplicity, and breadth of scope, as it extracts almost all of the proteins in the cell;
- it only requires small quantities of cells, and may be the method of choice when one is working with cultured cells or other systems containing small amounts of protein;
- it greatly simplifies handling the large amounts of radioactivity used in the incubation.

Disadvantages include:

- a loss of sensitivity (preparation of subcellular fractions increases the abundance of minor proteins in the sample);
- it provides no information about the intracellular locale of the proteins identified;
- some of the ^{32}P will be incorporated into lipids, RNA, and DNA: if these are not removed prior to SDS–PAGE they tend to cause an increase in background radioactivity, and can obscure hormone-induced changes in minor phosphoproteins, especially in the high molecular mass region of the gel.

Protocols 3 and *4* below are based on the methods developed by Anderson and Anderson (13, 14) and Garrels (15).

Protocol 3. Extracting proteins from cells in suspension for 1D SDS–PAGE

Equipment and reagents
- DNAase/RNAase solution (1 mg/ml DNAase I, 0.5 mg/ml RNAase, 50 mM MgCl$_2$, 50 mM Tris–HCl pH 7.0 at 4°C).

Method

1. Transfer the sample to a microcentrifuge tube with a disposable, plastic transfer pipette.
2. Spin the sample for 30–60 sec in a microcentrifuge ($>$ 2000 g) to pellet the cells. Remove the supernatant medium by aspiration to radioactive waste.
3. Rapidly resuspend the cells in a mixture that will contain final concentrations of the following: 0.5–1 mg protein/ml, 50 mM Tris–HCl pH 6.8, 6.6% (v/v) glycerol, 1.3% (w/v) SDS, 3.3% (v/v) 2-mercapto-ethanol. Boil the mixture in a stoppered Pyrex tube for 5 min.[a]
4. Cool the sample and add 0.1 vol of DNAase/RNAase solution. Shear the DNA by working the sample back and forth through a 20 gauge needle mounted on a 3 ml syringe. This step reduces the DNA to small pieces that can run through the gel.
5. The sample may be stored frozen or prepared immediately for SDS–PAGE as described in Section 4 below.

[a] An alternative to steps 1–3 is to add a concentrated stock solution of glycerol, SDS, and 2-mercaptoethanol directly to the suspension of cells, followed by boiling. This method has the advantage that stopping the reaction is fast, but the disadvantages that all of the [^{32}P]phosphate in the incubation vessel is carried into the sample, and the protein content is more dilute.
[b] Be careful with the RNAase if your laboratory uses molecular biology techniques. Avoid contaminating equipment with this enzyme, and discard all contaminated plasticware.

To improve resolution and reproducibility in two-dimensional SDS–PAGE it is necessary to avoid any salts or buffers when disrupting the cells.

Protocol 4. Extracting proteins from cells in suspension for 2D SDS–PAGE

Equipment and reagents
- homogenization buffer A [8 M urea, 2% Nonidet P40, 1 mM dithiothreitol (DTT)]
- 10 × homogenization buffer B [0.25 M Hepes pH 7.4, 2.5 M sucrose (optional), 10 mM DTT, 15 mM EGTA, 10 mM EDTA, 20% (v/v) Nonidet P40]
- phosphatase inhibitor cocktail (50 mM NaF and 0.15 mM Na vanadate; 10 µM micro-cystin or 10 µM okadaic acid may optionally also be used: these are expensive, but are highly specific inhibitors of protein phos-phatases-1 and -2A, and microcystin-LR is an irreversible inhibitor)

7

Protocol 4. *Continued*

- proteinase inhibitor cocktail [6 μg/ml leupeptin, 1mM benzamidine, 1mM phenyl-methyl sulfonyl fluoride (PMSF)]
- dialysis solution (8M urea, 1 mM dithiothreitol. 2% Nonidet P40)

Method

1. As steps 1 and 2 in *Protocol 3*.

2. After aspiration of the incubation medium to radioactive waste, resuspend pellets in homogenization buffer A. Vortex and sonicate the sample (this step breaks cells, and disrupts DNA and RNA).[a]

3. Centrifuge the homogenate for 30 min at 100 000 *g* at 4°C. Concentrate the sample by placing the supernatant in a dialysis bag (1 × 2 cm) and covering the bag with dry Sepharose beads. Leave at 4°C for 1–2 h (rapidly reduces volume by 5- to 10-fold per h). Then place dialysis bag into dialysis solution and dialyse for 3 h at room temperature. The sample is now ready for 2D gel analysis.

[a] Alternatively, if the cells are in suspension (see footnote to *Protocol 3*) add a 10 × homogenization buffer B.

Protocol 5. Extracting proteins from cells incubated in tissue culture

Equipment and reagents

- centrifugal vacuum concentrator (e.g. Speedvac)
- wash buffer (20 mM Na Hepes pH 7.4, 150 mM NaCl)
- DNAase/RNAase solution (1 mg/ml DNAase I, 0.5 mg/ml RNAase A, 50 mM MgCl₂, 50 mM Tris–HCl pH 7.0)
- lysis buffer [0.3% (w/v) SDS, 65 mM DTT, 1 mM EDTA, 20 mM Tris–HCl, pH 8.0]
- ampholine solution [5.97 g urea, 4 ml of 10% (v/v) Nonidet P40, 0.5 ml of 40% (v/v) ampholines, 154 mg DTT, water to 10 ml; see *Protocol 8* for a description of useful ampholine mixtures]

Method

1. At the end of the incubation period, aspirate the radioactive medium to waste with a plastic transfer pipette, and place the petri dish containing the cells on ice.

2. Rinse cells once with 1–2 ml of ice-cold wash buffer.

3. Lyse cells with 1.0 ml of lysis buffer at 100°C. Keep about 5–7 ml of the buffer in a capped, heat-resistant, plastic 50 ml tube in a boiling water bath during the experiment.

4. Scrape cells to one edge of the plate with a disposable plastic scraper and cool by placing the dish on ice.

5. Add 100 μl of DNAase/RNAase solution. This step reduces the DNA to

small pieces that that can run through the gel (see footnotes to *Protocol 3* for precautions on the use of this solution).

6. Aspirate the sample into a 3 ml syringe equipped with a 20 gauge needle, and expel it into an microcentrifuge tube. Shear the DNA by working the sample back and forth five times through the needle. Be careful not to allow the sample to foam. Finish by using a 26 or 27 gauge needle. Transfer the sample to a clean, 2 ml Sarstedt tube equipped with a cap containing an 'O' ring. It is important to use a tube sealed with an 'O' ring, otherwise tubes will leak during the lyophilization step (step 8).

7. The sample may be snap-frozen in liquid nitrogen if you wish to interrupt the procedure at this point.

8. Change the cap on the Sarstedt tube to one that has been pierced 4 or 5 times with a needle, and lyophilize the sample in a centrifugal vacuum concentrator. Again, the samples can be stored frozen or used immediately, as desired.

9. Resuspend the sample in 1.0 ml (the original sample volume) of ampholine solution.

10. The tube can be centrifuged in a microfuge (13 000 *g*) to remove any insoluble material remaining after the sample is dissolved.

11. Load 30–50 μl of the sample on the focusing gel used in the first dimension of the 2D gel protocol (see *Protocol 8*). The objective is to load equal amounts of protein from each sample on to the focusing gel. However, it is not usually possible to measure protein in samples prepared from only 10^6 cultured cells. Begin with an equal number of cells on each dish and take care that all steps in the protocol are as quantitative as possible.

12. Larger volumes and larger amounts of proteins can be loaded utilizing immobilized ampholine gradient gels (*Protocol 8*).

3.3 Preparing subcellular fractions from radioactively labelled cells

If the phosphorylated proteins in cell fractions such as mitochondria, microsomes, or plasma membranes are of interest, it will be necessary to homogenize the cells and separate the organelles. In this regard, it is important to realize that most subcellular fractionation schemes are performed in a medium of low ionic strength. In contrast, the buffers used to inhibit the actions of phosphatases and kinases following disruption of the cell usually contain 100 mM NaF and 10–15 mM EDTA. Control experiments have demonstrated that hormone-induced changes in the phosphorylation state of cytoplasmic proteins are stable in NaF and EDTA for up to 24 hours at 4 °C, or at least 15–30 minutes at 37 °C (9). However, the inclusion of NaF and EDTA in the buffers may cause particulate material to aggregate and to pellet at lower centrifugal forces than expected. Fortunately, discovery of the potent and selective protein phosphatase inhibitors, e.g. microcystins and okadaic acid, now allows the use of low ionic strength buffers while still inhibiting dephosphorylation of proteins. These compounds inhibit protein phosphatases in the PPP family (e.g. PP1, PP2A, PP4, but not PP2B) at nanomolar concentrations (see Chapter 3). As many phosphoproteins are dephosphorylated by these protein phosphatases, their inclusion in the buffers used to disrupt and fractionate cells should allow the preparation of subcellular organelles and maintain numerous proteins in a phosphorylated state (16). However, these inhibitors do not inhibit protein–tyrosine phosphatases, or protein–serine phosphatases of the PPM family (e.g. protein phosphatase-2C), and thus will not inhibit all potential dephosphorylation events (17). Sodium vanadate (150 μM) inhibits most protein–tyrosine phosphatases. Excluding Mg^{2+} and Ca^{2+} by the use of chelating agents (EDTA, EGTA) is sufficient to prevent dephosphorylation by the PPM and PP2B subfamilies of protein phosphatases.

Major disadvantages of most methods to prepare subcellular organelles are:

(a) Most fractionation procedures require a relatively long time to complete (minutes to hours). This extended period allows more time for proteolysis or dephosphorylation of proteins. By contrast, the procedures described in Protocols 3–5 and 7 require only seconds to prepare the sample prior to boiling in SDS.

(b) By its very nature, the fractionation procedure is likely to spread radioactivity to multiple rooms and pieces of equipment. This will make the containment of the large amounts of ^{32}P used in the experiment more difficult.

Protocol 6. Preparing a homogenate from ^{32}P-labelled cells prior to isolating subcellular fractions

Equipment and reagents
- stop buffer (10 mM Hepes pH 7.4, 220 mM sucrose, 1 μM microcystin-LR, 0.15 mM sodium orthovanadate, 15 mM EDTA, 2 mM EGTA)

Method

1. Incubate labelled cells as described in Protocol 1 or 2.

2. Stop the incubation by adding 10–20 vols of an ice-cold stop buffer. For example, 2–3 ml of cells could be stopped by pouring 35 ml of ice-cold stop buffer on to the cells which are being incubated in a 50 ml

---

plastic centrifuge tube, and placing the tube in an ice bucket. (Cells on petri dishes should be rinsed, scraped into a tube, and pelleted as described in *Protocol 5.*)

3. Pellet the cells at 1000 *g* for 2 min and aspirate the supernatant to the radioactive waste.

4. Homogenize the cells in 0.5–1.5 ml of ice-cold stop buffer using 30–50 strokes of a ground-glass Dounce homogenizer. Rinse the homogenizer three times with stop buffer between each sample and discard the rinses as radioactive waste.

5. Fractionate the homogenate by a method (established in pilot experiments) that will yield the subcellular fraction desired. For example, to examine the phosphorylation of proteins in the cytoplasm, the homogenate can be centrifuged at 100 000 *g* for 1 h to obtain a whole-cell particulate fraction and a cytosolic fraction. The use of protein phosphatase inhibitors also allows the use of column chromatography to enrich the sample in the proteins of interest (see Section 8, *Protocol 12* below).

6. Immediately prepare the fractions for 2D SDS–PAGE as described in Section 4.

3.4 Fractionation of cells with digitonin

If it is desired to extract the cytosolic fraction from cells incubated in suspension, the method developed by Janski and Cornell for use with hepatocytes offers a number of advantages (18). In this procedure, 90–95% of the cytoplasmic proteins are released by lysing the plasma membrane with digitonin, then the rest of the cell structures are removed by rapid centrifugation through a hydrocarbon layer in a microcentrifuge. Overall cell structure is not greatly disrupted by the digitonin (within the time frame of the separation) and the particulate material forms a pellet at the bottom of the hydrocarbon layer, providing a separation of cytoplasmic proteins from particulate fractions within 5–10 sec. The method is advantageous because the proteins can be boiled in SDS within 1 minute of fractionation, minimizing the chances of dephosphorylation or proteolysis. Moreover, clean-up of radioactive waste is simplified. The digitonin causes minimal interference with 1D electrophoresis provided the sample is diluted at least sixfold into the sample buffer. Digitonin does not interfere with 2D separations (12). It is important to realize that the digitonin-induced release of proteins from the cell depends on a number of variables: consult the original paper (18) for further information.

Protocol 7. Digitonin fractionation of cells into cytoplasmic and particulate fractions

Equipment and reagents

- digitonin solution (4 mg/ml digitonin, 10 mM TES, pH 7.4, 50 mM NaF or 1 μM microcystin, 10 mM EDTA, 5 mM EGTA, and 200 mM sucrose)
- hydrocarbon solution [90:110 (v/v) bromodecane:bromododecane, specific gravity = 1.05: other ratios of the two hydrocarbons may be used to vary the density if necessary]

- 1.5 ml microcentrifuge tubes [make new caps from Fisher microfuge caps (catalogue no. 04–978–145) with an X-shaped cut made with a razor: this allows the insertion of the tip of an adjustable micropipette, and the natural resilience of the plastic closes the cap upon withdrawal of the tip]

Method

1. Load the microcentrifuge tubes with 800 μl of digitonin solution. The specific gravity of this medium is very important. It must be light enough to float on the hydrocarbon (see step 3 below) and its osmolarity must be such that it does not cause the cells being fractionated to swell. With cells having densities different from hepatocytes, slight adjustments in the composition of the fractionation medium and/or the hydrocarbon may be needed for the method to work.

2. Layer 150 μl of hydrocarbon mixture under the digitonin fractionation medium using a fresh plastic pipette tip. Since the rate of extraction of protein from the cytoplasm with digitonin is very temperature sensitive, the fractionation tube is maintained at 27 °C in a water bath.

3. Just before the fractionation, the microcentrifuge tube containing medium and hydrocarbon is placed in the microcentrifuge opposite a balance tube.

4. At the end of an incubation as described in *Protocol 1*, remove 200 μl of cell suspension using a variable pipette, insert the tip through the X-shaped cut in the cap of the Eppendorf tube, and inject 200 μl of cell suspension into the fractionation medium, taking care not to disturb the underlying hydrocarbon layer (a pipette tip bent in the shape of a 'J' is useful for this).

5. Allow the fractionation to proceed for 10 sec to ensure release of 90–95% of the cytoplasmic proteins. Run the centrifuge for 30 sec to separate the particulate fraction from the cytoplasm. The particulate fraction should pellet at the bottom of the hydrocarbon layer.

6. Stop the centrifuge and remove the upper layer, containing cytoplasmic proteins, with a plastic transfer pipette. If desired, the proteins in the pellet can be extracted using the procedures described above in *Protocol 3* or *4* for extracting proteins from the whole cell. Discard the tube into radioactive waste.

7. Prepare the sample for SDS–PAGE as described in Section 4.

4. Analysis of ^{32}P-labelled proteins

4.1 Introduction

The most convenient method for observing changes in the phosphorylation state of proteins in ^{32}P-labelled cells is to resolve the proteins by SDS–PAGE and expose the dried gels to X-ray film or a phosphorimager screen. While a detailed description of this powerful separation technique is beyond the scope of this chapter, a brief discussion of the advantages and disadvantages of various gel systems is included below.

One-dimensional SDS–PAGE slab gels run by the method of Laemmli (19) have certain advantages. The theory of this technique has been published (20, 21), and an excellent practical description of the method can be found in a paper by Ames (22). The major advantages of this system are:

- many different samples can be compared in one run
- the technique is simple and in common use
- proteins with a wide range of molecular mass and isoelectric points can be resolved

The major disadvantage of the 1D gel system is that its resolving power is limited compared with 2D gel systems. Two-dimensional gels have the capability to resolve thousands of proteins (23). While this separation method is technically complex, it provides the following advantages:

- the high resolution may allow analysis of phosphorylation changes in major or minor proteins that are not apparent on 1D gels
- recent advances in microsequencing technologies, combined with the excellent resolving power of high resolution 2D gels, may allow the positive identification of the proteins, even in crude cell extracts used without further purification
- when the phosphorylation state of a protein is increased, the protein should focus at a more acidic position in the gel, demonstrating a stoichiometric gain of phosphate (12)
- recent advances in the development of fixed ampholyte gradients enable a high degree of reproducibility between experiments: this feature also facilitates computer analysis, allowing the construction of detailed data-bases that can be accessed over the world-wide web for comparison between laboratories

Apart from its technical complexity, the major disadvantages of the 2D technique are that only one sample can be run per gel, and that quantitation of the resultant autoradiographs or phosphorimages requires more sophisti-cated computer techniques. It also requires two days to complete. The protocols outlined below describe the 2D technique in detail. The samples,

13

the isoelectric focusing gels, and slab gels are prepared on the first day, the samples are focused overnight, and the second dimension slab gel is run on the second day. The protocols described below are for running 2D gels using the immobilized pH gradient (IPG) system, and are adapted from protocols described by Pharmacia. The IPG system is strongly recommended, but protocols describing the preparation of non-fixed pH gradients can be found in the first edition of this book (24).

4.2 Preparation of samples

Protocol 8 can be used to prepare ^{32}P-labelled protein samples prior to 2D electrophoresis. The protocol works equally well with both cytoplasmic and particulate fractions. Samples prepared as described below can be stored frozen and used repeatedly with little deterioration.

Protocol 8. Use of precast immobilized pH gradient (IPG) strips

Equipment and reagents

- pH 4–7 linear IPG strip, 11 cm (Pharmacia Cat no. 18–1016–60) 12/pkg
- pH 3–10 linear IPG strip, 11 cm (Pharmacia Cat no. 18–1016–61) 12/pkg
- pH 4–7 linear IPG strip, 18 cm (Pharmacia Cat no. 17–1233–01) 12/pkg
- pH 3–10 linear IPG strip, 18 cm (Pharmacia Cat no. 17–1234–01) 12/pkg
- pH 3–10 non-linear IPG strip, 18 cm (Pharmacia Cat no. 17–1235–01) 12/pkg
- sample buffer [8M urea, 1 mM DTT, 2%(v/v) Nonidet P40]
- equilibration buffer 1 (12.5 ml 0.5 mM Tris–HCl pH 6.5, 100 mg DTT, make up to 100 ml)
- equilibration buffer 2 [12.5 ml 0.5 mM Tris–HCl pH 6.8, 31 g urea, 30 ml glycerol, 0.1% (w/v) SDS, 5 mg bromophenol blue, 0.45 g iodoacetamide, make up to 100 ml with water]
- agarose solution (0.5g agarose, 12.5 ml 0.5 mM Tris–HCl pH 6.5, 87.5 ml water)
- IEF paper strips (Pharmacia catalogue no. 18–1004–40)
- mineral oil
- flat-bed electrophoresis unit (e.g. Pharmacia Multiphor II)
- programmable power supply (e.g. Pharmacia EPS 3500 XL)
- thermostatic cooling circulator

Method

1. The choice of IPG strips depends upon the nature of the sample. For whole-cell extracts or complex mixtures of proteins, the pH 3–10 non-linear gradient is recommended, since this gives a good spread of cellular proteins across the entire gradient. If the isoelectric point of a protein of interest is in the acidic range, the pH 4–7 linear gradients will provide better resolution.

2. Remove the protective plastic strip. Do not touch the gel surface. Swell the gel strip overnight in 25 ml of sample buffer with gentle rocking.

3. Alternatively, to increase the amount of sample that can be loaded, the gel can be swelled overnight in the presence of protein sample (0.5 ml) prepared either in 8 M urea buffer or dialysed against this

---

buffer (see *Protocol 4*, step 3). A re-swelling tray (Pharmacia) enables the volume of sample to be kept to 0.5 ml. Lay the strip gel side up, and layer on the sample. Cover the sample with 1–2 ml of mineral oil.

4. Add 3–4 ml of mineral oil to the surface of the cooling plate of the flat-bed electrophoresis unit and position the electrode tray on top.

5. Add 4–5 ml mineral oil and place the IPG strip alignment tray on top.

6. Blot the rehydrated strips gently to remove oil or excess buffer then flush the gel with 1–2 ml of water (this procedure prevents urea drying on the surface during subsequent steps). Place the strip, gel side up, on the flat bed with the acidic end (pointed end) towards the anode. The alignment tray will ensure that the strip is straight. Up to 12 strips can be loaded.

7. Cut two IEF paper strips to size and wet with 0.5–1.0 ml of distilled water. Place one strip across the basic end of the gel and the other across the acidic end. Make sure both strips make good contact with the surface of the gel.

8. Gently place the anode and cathode electrodes on to the wetted IEF strip.

9. Proceed to step 10 if the IPG strip was swollen in the presence of sample. If the sample is to be loaded electrophoretically into the gel, attach the sample cup holder ~2 cm from the cathode. Attach the sample cup (Pharmacia catalogue no. 18–1004–35), making sure the bottom of the cup has made contact with the gel surface. Add the sample to the cup (100 μl maximum volume) and cover with mineral oil.

10. Cover the gels with 10–20 ml of mineral oil.

11. Attach the electrophoresis unit lid and turn on the power supply and circulation bath. Set cooling for 22.5°C.

12. Use the voltage/current programs recommended by Pharmacia for the particular IPG strip. The EPS 3500 XL power pack gives excellent separation of phosphoproteins in whole-cell extracts. One of the major advantages of programmable power packs is that they maintain the current at 0.1 μA even after focusing is complete, thus maintaining resolution.

13. During the focusing proceed to *Protocol 8*, Section 4.3 for preparation of the slab gels. (Note the width of the slab gel must be 18 cm to accommodate the IEF strip.)

14. Upon completion of the isofocusing step, remove the strips and blot gently. Place the strips into equilibration buffer 1 for 10 min and then into equilibration buffer 2 for 10 min.

Protocol 8. *Continued*

15. Cut off the plastic tabs, noting which ends are the cathode and anode, respectively. Slide the IPG strip between the glass plates of the slab gel with the plastic backing making contact with the back plate, until the isofocusing gel makes contact with the slab gel surface.[a]

16. Apply melted agarose solution (microwave oven on high for 1–2 min).

17. Proceed to *Protocol 8*, step 8. Apply 40 mA per gel for 3–4 h with cooling.

[a] Pre-wet the top of the gel and plates with equilibration buffer: this makes sliding the strip easier.

4.3 Preparation of slab gels and running the second dimension

The next major step is to prepare the slab gels that will be used in the second dimension separation. *Protocol 9* consistently produces high quality resolving gels.

Protocol 9. Preparation of the second dimension slab gels

Equipment and reagents

- acrylamide solution [e.g. 60 ml for a 10% gel: 24.2 ml of water, 15.0 ml of 1.5 M Tris, pH 8.8, 0.6 ml of 10 % (w/v) SDS and 20.0 ml of acrylamide/bisacrylamide (30:0.8, w/w) and 200 μl of 10% (w/v) ammonium persulfate]
- wetting solution (Kodak Photo-Flo, 5 ml stock into 1 litre of water)
- *N,N,N′,N′*-tetramethylethylenediamine (TEMED)

Methods

1. Wash the glass plates with ethanol, then detergent, then rinse extensively with tap water followed by deionized water. Dip the plates in the wetting agents, allow to drain and air dry.

2. Prepare the acrylamide solution for the concentration of separating gel desired according to standard recipes. Do not add the TEMED at this point. Gels with an acrylamide concentration of 10–12% work well for most experiments. Degas on ice for at least 15 min with occasional swirling. The applied vacuum during degassing should reach at least 500 mm Hg.

3. Add 7.5 μl of fresh TEMED per 30 ml of the acrylamide gel solution and swirl to mix.

4. Pour the separating gel according to the manufacturer's instructions. Bring the gel to a height 1–2 mm above the mark made in step 3 above. The gel will shrink during polymerization.

5. Overlay the separating gel with 0.1% (w/v) SDS using a 24 gauge needle mounted on a 10 ml glass syringe. Apply the SDS to the inside of one glass plate by moving the needle back and forth across the plate allowing the SDS to slowly run down the inside of the plate and gently overlay the acrylamide. Do not disturb the acrylamide.

6. Note the time and allow 30 minutes for polymerization. Polymerization can be observed by the formation of a sharp interface between the acrylamide and the SDS overlay. Polymerization should occur within 30 min or the gel solutions are suspect.

7. Wait another 30 min (1 h from time of TEMED addition) and then wash the top of the gel to remove the unpolymerized acrylamide. Use a 10 ml syringe equipped with a 26–27 gauge needle and filled with 0.1% SDS. Pour off the wash and repeat two times. Finally, add enough 0.1% SDS to fill the entire space between the plates to the top. The separating gel is now ready for use. A stacking gel is not necessary when using the IPG strips. However, if desired a stacking gel can be poured at any time within the next 24 h.

8. When all focusing gels have been transferred to the top of the separating gel and the agarose added, place the slab gels into the apparatus and fill with electrophoresis buffer according to the manufacturer's instructions. Attach leads to the power supply and set the power supply to provide constant current. Run the gels at 40 mA/slab gel (for a 0.75 mm gel; increase the current proportionally for thicker gels).

9. To obtain maximum resolution, the gels should not get warm during the run. Therefore, it is important to cool the gels during the run to about 10–15°C with a cooling system.

10. Following electrophoresis, place the gel into appropriate stain or transfer buffer, noting the orientation of the IPG strip.

5. Autoradiography and phosphorimaging

5.1 Autoradiography

The ^{32}P-labelled proteins resolved in 1D or 2D gels can be detected easily by autoradiography with a number of medical X-ray films. This method is very sensitive (0.5–1 cpm can be observed), provides excellent resolution, and can be quantitative provided care is taken not to saturate the film with over-long exposures. While there is a wide range of X-ray films that are suitable for detection of ^{32}P, it is most convenient to use films that can be developed in automatic film processors, which produce consistent results. The following X-ray films have useful properties for ^{32}P autoradiography: Kodak X-OMAT

17

AR (XAR), Kodak X-OMAT K (XK-1), and Kodak Min R. The choice of film depends upon the degree of resolution required. Generally, a slower film (e.g. Min R) gives the best quality of spot resolution or band resolution, especially if exposures are carried out at room temperature, without the use of intensifying screens in the film cassette. Exposures can take 1 or 2 weeks under these conditions to identify all of the spots that might be detectable by a much shorter exposure on faster film at –70°C with intensifying screens. Kodak XK-1 provides a good compromise between speed and resolution. The density differences on X-ray film can be readily measured with a densitometer, but it should be stressed that the human eye is extremely sensitive to shades of density. It is rare that quantitation of the phosphorylation patterns reveals a density change that is not obvious to visual inspection. The X-ray films listed above respond to ^{32}P with a linear relationship between radioactivity and absorbance up to about 2.5 absorbance units (25). Therefore, it is possible to scan an autoradiogram in a commercial densitometer, or a spectrophotometer fitted with a gel scanner, to quantitate the effects of a stimulus on the phosphorylation of proteins. A recorder trace of the optical density along the film can be used to document the changes in absorbance (i.e. protein phosphorylation) caused by the stimulus, by aligning the valleys on either side of the peak corresponding to the protein of interest.

5.2 Phosphorimaging

As an alternative to film, phosphorimagers (e.g. Molecular Dynamics flat-bed Phosphorimager) have become increasingly popular. Phosphorimagers are estimated to be at least 100 times more sensitive than conventional films in detecting ^{32}P-labelled proteins in polyacrylamide gels. A second advantage is that the results are in digital format, and the response is linear over a wider range. A disadvantage for 2D SDS–PAGE analysis is that the resolution of closely migrating species can be rather poor, and the background is more variable.

6. Ancillary analyses

6.1 Introduction

Two major assumptions implicit in measuring changes in protein phosphorylation in intact cells are that:

- the specific activity of the [^{32}P]ATP in the cells is not changed by the stimulus applied
- the increased density seen on the autoradiograms is due to an increase in the net phosphate content of the protein

Methods for testing the validity of the first assumption are outlined below. The information on charge shifts in Section 7 describes measurements of the gain in net phosphate (see *Protocol 10*).

6.2 Specific radioactivity

The most accurate and convenient method to measure the specific radio-activity of the [^{32}P]ATP pool in cells is to use high-pressure liquid chromato-graphy (HPLC) to resolve the ATP from other ^{32}P-labelled molecules.

Protocol 10. Measurement of the specific radioactivity of ATP

Equipment and reagents

- HPLC system
- anion-exchange column (3.9 × 150 mm)
- 5.0 M K$_2$CO$_3$
- 5 M KH$_2$PO$_4$ buffer pH 4.5

Method

1. This assay does not require high levels of radioactivity. You can incubate the cells with about 100-fold less [^{32}P]phosphate than is used in the protein labelling experiments (about 0.1 μCi per mg of cell protein).

2. At the desired time, stop the reaction by adding ice-cold perchloric acid to a final concentration of 0.5 M (by adding concentrated acid to cell suspensions, or by adding 0.5 M acid directly to cells attached to a petri dish from which the medium has been removed). Add 5 μl of 0.02% (w/v) phenol red to each sample and hold the mixture on ice for 15–30 min.

3. Remove the precipitated protein by centrifugation, place the super-natant in an ice-cold test tube and neutralize it by the addition of 7.0% of the final sample volume of ice-cold 5.0 M K$_2$CO$_3$ (for example add 3.5 μl for a 50 μl sample). Use the phenol red to ensure that neutral-ization is complete.

4. Hold the solution on ice for 30 min, centrifuge it to clarify and remove the supernatant containing the ATP for analysis via HPLC using standard procedures.

5. The [^{32}P]ATP is easily resolved on a Whatman Partisil strong anion-exchange (SAX) column by isocratic elution with 0.5 M KH$_2$PO$_4$ at pH 4.5; however, gradients commonly used to resolve inositol phosphates are also suitable (26).

6. Calibrate the column with adenosine and guanosine nucleotide stan-dards to identify peaks of known retention time. Measure the amount of ATP eluted from the column using its absorbance at 254 nm by comparing it with a standard curve. Collect the peak and determine the ^{32}P content by liquid scintillation counting. The ratio of the radio-activity and the concentration of ATP gives the specific radioactivity.

Protocol 10. *Continued*

7. The method above determines the total radioactivity of the α-, β-, and γ-phosphates. The specific radioactivity of the γ-phosphate can be determined by quantitative enzymatic transfer of the terminal phosphate to glucose with hexokinase (27). Bring the pH of the [^{32}P]ATP purified by HPLC to 8.0 with KOH, add enough of concentrated stock solutions to bring the concentration of ATP to 100 μM, glucose to 10 mM, and MgCl$_2$ to 10 mM.

8. Add enough hexokinase to give an activity of 3 U/ml and allow the reaction mixture to incubate 30 minutes at 25 °C to convert the ATP to glucose-6-phosphate. The extent of conversion can be monitored spectrophotometrically using NADP and glucose-6-phosphate dehydrogenase (27).

9. Separate the radioactive [^{32}P]glucose-6-phosphate from [^{32}P]ATP by adding the reaction mixture to a small column containing 1 ml of BioRad AG-1 ion-exchange resin, and elute the [^{32}P]glucose-6-phosphate with 2 ml of water. The ATP remains on the column: discard it as radioactive waste.

10. The specific radioactivity of the γ-phosphate in the ATP can be determined from the radioactivity in the [^{32}P]glucose-6-phosphate and the concentration of ATP present in the original sample purified by HPLC.

7. Computer analysis of two-dimensional gels

7.1 General considerations

The quantitation of the data obtained from 2D gels is a complex task. Sophisticated computer software is required to:

- match the locations of proteins in one gel with those in another
- create 'master' gels that represent the migration positions of proteins in a given tissue or cell type
- integrate the density information obtained from the spots in each image
- maintain a log of information about each spot
- determine the locations of 'new' spots that appear in a pattern due to charge shifts, or *de novo* synthesis of proteins, following experimental treatments
- create images that represent the protein patterns for comparison with the work of other laboratories

While a full treatment of this subject is beyond the scope of this chapter, a number of vendors have developed instruments and computer software to accomplish these tasks. One property of the 2D gel system useful to investigators studying protein phosphorylation events is that it displays the charge shift that occurs when a protein is phosphorylated. Thus, it is possible to demonstrate increases in protein phosphorylation by changes in the position of the protein in the isoelectric focusing dimension of a 2D gel, because an increase in phosphate content causes the protein to focus at a new, more acidic isoelectric point. Therefore, if a protein moves to a more acidic isoelectric point following a stimulus, its net phosphate content has increased (12). The magnitude of these charge shifts is about 0.3 pH units in the pH 6–7 region of the gel (12). If the protein is abundant enough to be stained with Coomassie Blue or silver, it is possible to demonstrate charge shifts by comparing the stained protein patterns from control and stimulated cells. If higher sensitivity is required, the cells can be labelled with [^{35}S]methionine and the charge shifts determined using autoradiography (28). A disadvantage of this method is that the protein pattern is far more complex because nearly all of the intracellular proteins are displayed.

Computer programs are available for desktop computers that enable the complex patterns produced in 2D gels to be analysed for the identification of proteins whose mobility is altered because of a phosphorylation event. In theory, compilation of these charge shifts will allow the construction of databases which can be used directly by other investigators throughout the world, e.g. over the Internet. In projects in which the entire proteome of a particular cell is being mapped using 2D gels, this type of analysis can be extremely useful. Thus, in the near future it may be possible to run 2D gels containing the phosphorylated proteins in whole-cell extracts, scan the stained gels to create a digital map of the resolved proteins, and compare it against a remote database that has mapped the proteome for that particular cell. Comparison of the maps obtained following different stimuli may identify the proteins modified by the stimulus. The protocol below provides an example of how to analyse 2D gels for charge shifts using one commercially available software package.

Protocol 11. Analysis of complex 2D gel patterns for charge shifts

Equipment and reagents
- flat-bed scanner (e.g. Epson ES1200 C)
- Mellanie II software (BioRad; available in Unix, PC, or Macintosh versions)
- PC or Macintosh with video card
- video cassette recorder

Method

1. Prepare 2D gels of extracts prepared from unstimulated and stimulated cells. Scan the images of 2D gels made from autoradiographs, silver stained, or amido black stained blots at normal resolution.

2. Save the files in TIF format.

Protocol 11. *Continued*

3. Open the Mellanie II program and gel files. Align and overlay the gels according to the software instructions. Use the software to correct for distortions that may have occurred during electrophoresis and overlay the control and experimental images.

4. Toggle back and forth between the overlaid images. This is an extremely effective method for spotting proteins that become modified in response to a stimulus, especially in very complex mixtures containing many hundreds of proteins whose mobility is not affected. Use the software to highlight proteins in the image that shift between the two conditions.

5. Record toggling in real time for 30–60 seconds either as a Quick-time movie or directly to VHS tape (preferable, uses less memory). Playing the video (or movie) back is an effective means for both data analysis as well as presentation of complex 2D maps to audiences.

8. Identification of phosphoproteins in 1D and 2D gels

8.1 Introduction

Full size 2D gels are capable of resolving thousands of proteins. While the power of this purification strategy has been recognized for years, the problem of identifying proteins in the gel pattern has limited the utility of this technique. Methods for identifying the proteins in a 2D gel pattern include:

- running the pure protein through the same gel system to establish its molecular mass and pI
- identification using one of several immunological methods
- direct sequencing of proteins eluted from the gels

While the first two methods provide positive identification of proteins, they have the disadvantage that one must have at least guessed the identity of the protein beforehand. However, the development of extensive DNA and protein databases, combined with major advances in the sensitivity of protein sequencing, now allow investigators to identify proteins of interest directly by the third method. These advances are revolutionizing the way in which investigators can study protein phosphorylation in intact cells, and should allow full utilization of the potential of 2D gel electrophoresis. The paragraphs below provide short outlines of these advances and provide references to the extensive literature on the subject:

(a) The number of sequences in the DNA and protein sequence databases is expanding rapidly. At the time of writing, the complete genomes from yeast, *Caenorhabditis elegans*, and a number of bacteria have been sequenced, and completion of the human genome is predicted to occur by the year 2001. In addition, the number of cDNAs partially or fully identified as expressed sequence tags (ESTs) is expanding daily. These developments make it possible that a protein identified only as a spot on a 2D gel can be partially sequenced and identified by using the information in these databases. Interested readers should refer to refs 4 and 29 for more details.

(b) The development of immobilized ampholytes has greatly improved the reproducibility of the separations obtained with 2D gels. This advance allows different laboratories to compare and identify proteins resolved in the gels. Theoretically, proteome databases from different cells and tissues can be constructed using the contributions of multiple laboratories. These databases will potentially allow individual investigators to identify proteins of interest by comparison of their resolved proteins with master patterns. Examples of such databases already available on the internet can be found at the following URL: http://www.expasy.ch/ch2d/.

(c) Recent advances in protein microsequencing enable picomole or even femtomole amounts of proteins to be extracted from polyacrylamide gels and sequenced. A complete description of these sequencing techniques is beyond the scope of this chapter; however, an introduction and references are provided in Section 8.

These advances have moved the field of study of protein phosphorylation to a new level that allows phosphorylation events to be studied within a physiological context. Conceptually, it is possible that cells could be labelled with [^{32}P]phosphate, stimulated with an agonist, analysed by 2D SDS–PAGE, and every phosphoprotein that is altered in response to the stimulus identified by subsequent microsequencing techniques. Furthermore, with the aid of computer software capable of comparing and matching 2D gel images, it is possible to identify phosphoproteins (and other post-translational modifications) in 2D gels by mobility shift alone. Employing subcellular fractionation techniques (see *Protocols 5* and *6*) and/or affinity methods that enrich samples in certain types of proteins (see *Protocol 12*) and reduce the numbers of proteins loaded on to a 2D gel, increases the power of this approach.

Once a particular protein is identified, many doors are opened to the investigator:

- clones can be obtained for further biochemical, molecular biological, or genetic studies
- antibodies can be obtained enabling immunolocalization studies, immunoprecipitation experiments for detailed phosphorylation site analysis, or binding partner studies by co-immunoprecipitation or far-Western analysis
- time resolution studies can be carried out to dissect the upstream phosphorylation events in a particular pathway leading to the phosphorylation of the identified protein

23

The protocols described in Section 8.3 below outline how to prepare samples for state of the art techniques used to identify proteins in polyacrylamide gels via mixed peptide sequencing or mass spectrometry.

8.2 Identification of the phosphorylated amino acids

If the proteins of interest are labelled with ^{32}P, one question that can be answered relatively easily is which amino acid(s) is(are) phosphorylated following a stimulus. The strategy is to run enough protein into the gel to overload the protein of interest, dry the gel, locate the spot using auto-radiography, and cut out the protein of interest. The protein can be hydrolysed to its amino acids and the phosphoserine, phosphothreonine, and phosphotyrosine resolved via 1D or 2D thin layer chromatography. See ref. 30 and Chapter 5 for methods to resolve phosphoamino acids. Alternatively, phosphotyrosine is relatively easily detected using anti-phosphotyrosine antibodies on Western blots (6).

8.3 Identification of proteins from polyacrylamide gels by direct sequencing

8.3.1 Introduction

The technique of sequencing proteins eluted from polyacrylamide gels has advanced rapidly. In addition, the sensitivity of both the classical Edman and mass spectrometry methods has increased to the extent that it is possible to sequence femtomole (ng) amounts of protein. This sensitivity, combined with the resolving power of a 2D gel, makes it possible to identify a large number of proteins resolved in acrylamide gels. The sections below outline the use of one novel sequencing method, i.e. mixed peptide sequencing, in detail (see *Table 1* and *Figure 1*). This method combines the relative simplicity of the

---

Table 1. Recovery of PTH-amino acids during seven rounds of mixed peptide sequencing

Cycle no.	1st call		2nd call		3rd call	
	AA	pmol	AA	pmol	AA	pmol
	M		M		M	
1	K	37.1	I	27.7	T	10.1
2	D	18.5	N	34.2	(N)	
3	L	24.6	E	16.2	T	23.9
4	D	14.2	G	28.2	V	13.5
5	D	16.0	E	19.1	S	11.1
6	E	18.8	A	13.4	N	8.5
7	D	15.1	E	20.5	L	23.6

AA = PTH-amino acid identified. Data show results for an 18 kDa protein. Methionine (M) is always placed at the top of each string because CNBr was used to cleave the protein. Cycle 2 contained two asparagine residues (N), indicated by peak height and pmol recovery with respect to glutamic acid (D).

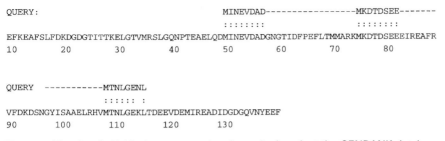

```
QUERY:                                          MINEVDAD----------------MKDTDSEE-------
                                                : : : : : : : :                  : : : : : : : :
EFKEAFSLFDKDGDGTITTKELGTVMRSLGQNPTEAELQDMINEVDADGNGTIDFPEFLTMMARKMKDTDSEEEIREAFR
10        20        30        40        50        60        70        80


QUERY    ----------MTNLGENL
                   : : : : : : :
VFDKDSNGYISAAELRHVMTNLGEKLTDEEVDEMIREADIDGDGQVNYEEF
90       100       110       120       130
```

Figure 1. The data in Table 1 were sorted and matched against the GENBANK database using the FASTF algorithm (ktup 1, protein matrix MD20, expectation (e) 1) and the protein was identified as calmodulin.

Edman method and computer searches of the GENBANK database to identify proteins. The advantages and disadvantages of this method are outlined and compared briefly with mass spectrometry.

8.3.2 Identification of proteins in gels by mixed peptide sequencing

This method enables hundreds of phosphoproteins in complex 2D gels to be identified, provided the protein of interest is present in a database (protein or DNA). Currently, the yeast and *Caenorhabditis elegans* genomes are complete, and at the time of writing the mouse and human genomes are ~60–70% complete. These databases can be accessed at the National Center for Biotechnology Information (NCBI) website (http://www.ncbi.nlm.nih.gov). The technique of mixed peptide sequencing involves transfer of a 1D or 2D gel to a hydrophobic membrane. The membrane is stained and spots/bands excised and treated briefly with cyanogen bromide. The membrane piece is placed directly into an automated Edman sequencer cartridge and 10–12 sequencing cycles carried out. The mixed peptide sequences generated are then used for searching using two computer programs, FASTF (for searching the protein database) or TFASTF (for searching DNA databases), that sort and match the mixed sequence data against the databases. Mixed peptide sequencing has the same sensitivity as fragmentation or time-of-flight mass spectrometry for identifying proteins in gel slices following proteolytic cleavage (4), but has several advantages over these techniques including:

- little purification other than 1D or 2D SDS–PAGE is required;
- only 100–200 fmol (4–8 ng) of protein is required: the technique works well with only 1 pmol (0.5 μg of a 50 kDa protein), but sequencing in the low fmol range is also possible if the sample is kept scrupulously clean;
- low cost;
- lengthy extraction of proteins from gel slices is not required (improving sensitivity);
- standard laboratory equipment is used;

25

- a completed sequence database for the species being analysed is not essential;
- data interpretation is simple.

Disadvantages include:

- the protein of interest must be in the database, although not necessarily from the same species;
- the protein must contain methionine residues [although other cleavage sites may be used, e.g. Skatol (Pierce) cleaves at tryptophan];
- the excised spot or band must be kept scrupulously clear of contamination (airborne or physical) prior to sequencing;
- CNBr is highly toxic.

Protocol 12. Mixed peptide sequencing

Equipment and reagents

- Polyvinyl membrane (Fluortrans, PALL Bio-support, catalog no. PVM020C3R)
- sequencer grade CNBr (Pierce catalog no. 92278; 1 g/ml in 95% formic acid)
- transfer buffer (57g glycine, 12g Tris-base, 0.4g SDS, 800 ml methanol made up to 4 litres with water)
- sequencer grade methanol
- amido black stain (1 mg/ml amido black, 10% methanol, 5% acetic acid)
- destaining solution (10% methanol, 5% acetic acid)
- Western blot transfer apparatus
- Edman automated sequencer (e.g. Applied Biosystems 494)

Method

1. Perform 1D or 2D separation as outlined in *Protocols 7–10*. To increase the amount of protein and improve its resolution from other proteins, the sample should preferably be enriched in the protein prior to SDS–PAGE. One or more of the following steps are recommended: subcellular fractionation; affinity chromatography; precipitation (ammonium sulfate or acetone); anion or cation-exchange chromatography; hydrophobic interaction chromatography.

2. Following electrophoresis, prepare the sample for transfer to the PVM membrane by equilibrating the gel into transfer buffer for 10 min.

3. Assemble the transfer apparatus according to the manufacturer's instructions. On a clean surface, cut PVM membrane to size and wet with sequencer grade methanol (DO NOT TOUCH the membrane). PVDF (Millipore) can also be used, but PVM has a smaller pore size giving better transfer efficiency and improved repetitive yield during Edman sequencing.

4. Transfer proteins to PVM membrane overnight at constant voltage (35 V) following standard transfer procedures.

5. Wash membrane for 5 min with distilled water and repeat two times.

6. Stain with amido black solution for 1 min, then wash for 1 min with destaining solution. Wash with several changes of distilled water until the background is light blue.

7. Air dry the PVM membrane. The background should become white to pale blue, with proteins visible against this background. On PVM membrane, amido black has a similar sensitivity to silver stain. Based on our experience the lowest detectable band or spot is equal to 1–5 ng.

8. Subject the blot to autoradiography (see Section 5) or otherwise analyse (see Section 8) to identify proteins that are of interest. Scan the blot or autoradiogram and print a copy of the gel. Assign a number to each protein and write this next to the spot/band on the printout. This is useful if many phosphoproteins are to be identified from the transfer, as it keeps an accurate record of the proteins that have been sequenced.

9. Excise the stained band/spot with a clean scalpel blade and place in a microcentrifuge tube. Limit the size of the piece excised to within the boundaries of the spot/band of interest. A smaller piece gives a lower background during Edman sequencing.

10. Wash the membrane piece first with 1.0 ml of 50% (v/v) sequencer grade methanol and then 1.0 ml of water. Repeat this wash three times, finishing with 100 µl of sequencer grade methanol. Store at –20 °C.

11. Perform the next two steps in a fume hood (care: CNBr is very toxic!). Remove the methanol and add 30 µl water and 70 µl CNBr solution. Digest for 90 min at room temperature, in the fume hood.

12. Remove the piece of membrane from the CNBr solution and place in a fresh microcentrifuge tube containing 1.0 ml of distilled water. Repeat the methanol/water washing protocol, finishing with 100 µl of methanol.

13. Place the membrane piece into an automated Edman sequencer cartridge and sequence for 10–12 cycles (6 cycles tentatively identifies a protein) using pulse liquid blot chemistry.

The FASTF and TFASTF programs are available from the University of Virginia at ftp://ftp.virginia.edu/pub/fasta, or by contacting the author (TAJH). CNBr treatment of proteins transferred to PVM causes cleavage of the protein on the C-terminal side of methionine, leaving a free N-terminal amino acid. Using this method on over 500–600 proteins, we have found that 3–5 cleavages generally occur in an average protein of molecular mass up to 100 kDa. This relatively low number of cleavages is consistent with the

Timothy A. J. Haystead and James C. Garrison

RESTING PLATELETS   THROMBIN STIMULATED PLATELETS

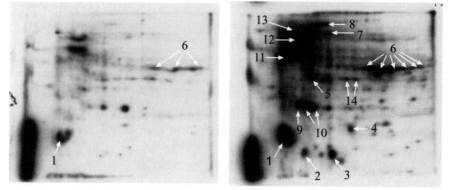

Figure 2. Identification of phosphoproteins in human platelets by mixed peptide sequencing. Washed platelets were prepared from 1.0 ml of human blood and ^{32}P-labelled for 90 min (*Protocol 1*). The platelets were treated±thrombin (0.2 U) for 1 min, then frozen in liquid nitrogen. The cells were homogenized in Hepes buffer (*Protocol 3, Step 6*) and dialysed against 8M urea. The dialysate was characterized on 2D gels following *Protocols 7–9*, and *11*. The separated proteins were transferred to PVM, stained with amido black and autoradiographed. Labelled proteins that corresponded to stained proteins were cut from the PVM, treated with CNBr solution and sequenced by mixed peptide sequencing (*Protocols 12* and *13*). Proteins that were identified ranged from 2 pmol (pleckstrin) to 0.2 pmol (HSP90). Isoelectric values (pI) are theoretical values based on the primary structure of the identified proteins and were determined using the ExPasy program (http://expasy.hcuge.ch/ch2d/pi–tool.html). Key: (1) myosin light chain 20, pI 5.09, 19.8 kDa; (2) rho GDI, pI 5.1, 23.0 kDa; (3) neurogranin, pI 6.53, 7.4 kDa; (4) rap1A, pI 6.39, 20.9 kDa; (5) tubulin, pI 4.75, 49.72 kDa; (6) pleckstrin, pI 8.32, 40.08 kDa; (7) GPIIIα-II, pI 5.3, 84.6 kDa; (8) rho-GAP, pI 5.91, 105 kDa; (9) ESTAA059313; (10) HSP27, pI 5.98, 22.7 kDa; (11) c-raf oncogenic fusion protein, pI 5.15, 47.0 kDa; (12) amphiphysin homologue, pI 4.95, 64.0 kDa; (13) HSP90, pI 4.94, 84.6 kDa; (14) pleckstrin fragments.

frequency of occurrence of methionine in any given protein. Analysis of the current protein databases shows that 4% of all proteins contain no methionines, 9% contain one methionine and 50% contain six methionines. In addition, the number of cleavages may be limited because the rate of cleavage at methionine is dependent on the adjacent amino acid as well as its exposure on the membrane. Figure 2 shows an example of mixed peptide sequencing experiments in which multiple proteins were identified in a 2D gel.

Protocol 13.   Using the FASTF and TFASTF algorithms

Equipment and reagents

- Macintosh or PC computer
- Current web browser program

- Web address: http://fasta.bioch.virginia.edu/

Method

1. Following Edman sequencing, the liberated PTH-amino acids are written in a list in order of abundance. For an example of such a list from a mixed peptide sequencing run see *Table 1*.

2. The called amino acid sequences are then written into the FASTF and TFASTF program dialogue box. For example, for the data shown in *Table 1*: Call 1: MKDLDDED; Call 2: MINEGEAE; Call 3: MTNTVSNL.

3. FASTF searches the protein databases using MD20 as a scoring matrix. TFASTF searches the DNA databases in all three frames and in both directions. MD10 is used as a scoring matrix in this case.

8.4 Identification of phosphoproteins in 1D and 2D gels by mass spectrometry

Alternative means of identifying proteins from polyacrylamide gels is matrix-assisted laser desorption/ionization time-of-flight (MALDITOF) mass spectrometry (MS) and nano-electrospray tandem MS. MALDITOF-MS provides a rapid means of protein identification by measurement of the masses of multiple peptide fragments alone. Peptide masses derived from the analysis of proteolytic fragments are compared against a predicted peptide database. Nano-electrospray MS is a more lengthy procedure, but enables proteins that are not in the database to be identified by peptide sequencing. Although this technique has yet to be applied in a major way to the identification of phosphoproteins in 2D or 1D gels, work by Mann and others on the yeast proteome (29) indicates that this should be a successful approach. Both technologies can identify proteins in the very low femtomole range once peptides are obtained in solution, although it should be borne in mind that extraction of peptide fragments after in-gel digests can result in significant losses of material, reducing the sensitivity of the procedure. The interested reader should consult refs 31–35.

Some advantages of the mass spectrometry approach are:

- little or no purification other than 1D or 2D SDS–PAGE is required;

- high sensitivity (theoretically, proteins can be identified unambiguously at < 1 fmol);

- post-translational modifications, and their positions within a peptide, can be identified without the use of radioactive probes;

- high throughput (many hundreds of proteins can be identified in a short time);

- no toxic chemicals are utilized.

Disadvantages include:

- MALDITOF mass spectrometry requires a complete, species-specific database;
- lengthy gel extraction procedures are required to obtain peptides for sequencing by mass spectrometry;
- instrumentation is expensive;
- mass spectrometry is a rather specialized technique, requiring highly trained personnel.

References

1. Taylor, S.S., Radzio-Andzelm, E. and Hunter, T. Taylor, S.S., Radzio-Andzelm, E., and Hunter, T. (1995) *FASEB J.* **9**, 1255–1266.
2. Campos, M., Fadden, P., Qian, Z., and Haystead, T.A.J. (1996) *J. Biol. Chem.* **271**, 28478–28484.
3. Egloff, M.E., Johnson, D, F., Moorhead, G., Cohen, P.T.W., Cohen, P., and Barford, D. (1997) *EMBO J.* **16**, 1876–1887.
4. Zhao, Y., Boguslawski, G., Zitomer, R.S., DePaoli-Roach, A.A. (1997) *J. Biol. Chem.* **272**, 8256–8262.
5. Damer, C.K., Partridge, J., Pearson, W.R., and Haystead, T.A.J. (1998) *J. Biol. Chem.* **273**, 24396–24405.
6. Donaldson, R. W. and Cohen,S. (1992) *Proc. Natl. Acad. Sci. USA* **89**, 8477–8481.
7. Weng, Q.P., Kozlowski, M., Belham, C., Zhang, A., Comb, M.J., and Avruch, J. (1998) *J. Biol. Chem.* **273**, 16621–16629.
8. Fadden, P and Haystead, T. A.J. (1995) *Anal. Biochem.* 225:81–88.
9. Garrison, J.C. (1978) *J. Biol. Chem.* **253**, 7091–7100.
10. Garrison, J.C., Borland, M.K., Moylan, R.D., and Ballard, B.J. (1981) In *Cold Spring Harbor conferences on cell proliferation* (ed. O. Rosen and E.G. Krebs), Vol. 8, pp. 529–545. Cold Spring Harbor Laboratory, CSH, NY.
11. Avruch, J., Witters, L., Alexander, M.C., and Bush, M.A. (1978) *J. Biol. Chem.* **253**, 4754–4761.
12. Garrison, J.C. and Wagner, J.D. (1982) *J. Biol. Chem.* **257**, 13135–13143.
13. Anderson, N.G. and Anderson, N.L. (1978) *Ann. Biochem.* **85**, 331–340.
14. Anderson, N.G. and Anderson, N.L. (1978) *Ann. Biochem.* **85**, 341–354.
15. James, I. Garrels, J.I. (1979) *J. Biol. Chem.* **254**, 7961–7977.
16. Haystead, T.A., Sim, A.T., Carling, D., Honnor, R.C., Tsukitani,Y., Cohen, P., and Hardie, D.G. (1989) *Nature* **337**, 78–81.
17. Bialojan, C. and Takai, A. (1988) *Biochem. J.* **256**, 283–290.
18. Janski, A. and Cornell, N.W. (1986) *Biochem. J.* **186**, 423–429.
19. Laemmli, U.K. (1970) *Nature* **227**, 680–685.
20. Davis, B.J. (1964) *Annals of N.Y. Acad. Sciences.* **121**, 404–427.
21. Ornstein, L. (1964) *Annals of N.Y. Acad. Sciences.* **121**, 321–349.
22. Ames, F.L.A (1974) *J. Biol. Chem.* **249**, 634–644.
23. O'Farrell, P.J. (1975) *J. Biol. Chem.* **250**, 4007–4021.
24. Garrison, J.C. (1993) In *Protein phosphorylation: a practical approach* (ed. D.G. Hardie), Chapter 1, pp. 1–28. Oxford University Press, Oxford.

25. Garrison, J.C. and Johnson, M.L. (1982) *J. Biol. Chem.* **257**, 13144–13150.
26. Mattingly, R.R., Stephens, L.R.,Irvine, R.F., and Garrison, J.C. (1991) *J. Biol. Chem.* **266**, 15144–15153.
27. Leiter, A.B., Weinberg, A., Isohashi, F., Utter, M.F., and Linn, T. (1978) *J. Biol. Chem.* **253**, 2716–2723.
28. Steinberg, R.A., O'Farrell, P.H., Friedrich, U., and Coffino, P. (1977) *Cell,* **10**, 381–391.
29. Shevchenko, A., Jensen, O.N., Podtelejnikov, A.V., Sagliocco, F., Wilm, M., Vorm, O., Mortensen, P., Shevchenko, A., Boucherie, H., and Mann, M. (1996) *Proc. Natl. Acad. Sci. USA* **93**, 14440–14445.
30. Cooper, J.A., Sefton, B.M., and Hunter, T. (1983) In *Methods in enzymology* (ed. J.D. Corbin and J.G. Hardman), Vol. 99, pp. 387–389. Academic Press, San Diego.
31. Patterson, S.D. and Aebersold R. (1995) *Electrophoresis* **16**, 1791–814.
32. Schwartz, J. C. and Jardine, I. (1996) In *Methods in enzymology*, Vol. 270, pp. 552–586. Academic Press, San Diego.
33. Qin, J., Fenyo, D., Zhao, Y., Hall, W.W., Chao, D.M., Wilson, C.J., Young, R.A., and Chait, B.T. (1997) *Anal. Chem.* **69**, 3995–4001.
34. Cohen, S.L., and Chait. B.T. (1997) *Anal. Biochem.* **247**, 257–267.
35. Larsen, B.S. and McEwen, C.N. (ed) (1998) Mass spectrometry of biological materials, 2nd edn, Chapters 1–17. Marcel Dekker, Inc, New York.

2

Analysis of signal transduction pathways using protein kinase inhibitors and activators

YASUHITO NAITO, RYOJI KOBAYASHI, and
HIROYOSHI HIDAKA

1. Introduction

A complete understanding of the organization and functioning of the protein phosphorylation system requires the pooling of expertise in several different fields, such as molecular pharmacology, genetic manipulation, biochemistry, and cell biology. The advent of a new class of effective pharmacological agents is always an event of considerable interest, in particular for new types of antagonists, i.e. potentially important groups of compounds that act by specifically blocking one or more of the steps in intracellular signalling systems (1). There is now convincing evidence that intracellular protein phosphorylation systems are of fundamental importance in biological regulation. A large number of cellular mechanisms involving protein phosphorylation have been revealed over the past decade. However, interrelationships amongst signal cascades are complex and uncertainties concerning the cellular responses remain. In order to obtain improved understanding, sophisticated methods for estimating changes in the activities of cellular response elements after extracellular stimulation have to be developed. While progress has been made in the biochemistry and molecular biology of protein kinases, it has proved very difficult to elucidate their function in intact cells. For this reason researchers studying second messenger systems and protein phosphorylation have long sought specific and effective inhibitors that would permit definitive determination of physiological roles of protein kinases (2). In order to elucidate the physiological function of individual protein kinases, inhibitors should meet certain criteria:

- direct binding to the protein kinase
- strict specificity
- cell membrane permeability

The availability of such inhibitors should permit understanding of the role of second messengers and second messenger-related protein kinases in cellular responses. At least three different approaches have been followed to identify specific inhibitors of protein kinases. For this endeavour, Hidaka and co-workers have undertaken a systematic programme using synthetic chemistry (1–13), and since 1984 have concentrated attention on H-series inhibitors (naphthalenesulfonamides and isoquinolinesulfonamides). Antibiotic inhibitors such as the staurosporines, and compounds of the K252 class, have also been introduced. While remarkably effective protein kinase inhibitors, these appear to have poor specificity. Natural products other than antibiotics have also been an important source of inhibitors, one of the most interesting examples in this group being sphingosine and its derivatives. These have been used as inhibitors of protein kinase C for studying the physiological roles of this enzyme. However, it has became apparent that these inhibitors also exert other effects. The focus of this chapter will be on synthetic and natural inhibitors, especially those that can be used in intact cells.

2. Overview of synthetic protein kinase inhibitors

In 1977, the calmodulin antagonist, naphthalenesulfonamide, W-7 [N-(6-aminohexyl)-5-chloro-1-naphthalenesulfonamide] (*Figure 1*), which was synthesized in our laboratory, was first introduced (3). We proposed that calmodulin antagonists be defined as agents which bind to calmodulin and selectively inhibit Ca^{2+}/calmodulin-dependent enzymes (4). During the synthesis and selection of novel calmodulin inhibitors among derivatives of naphthalenesulfonamide, we discovered that those with shorter alkyl chains markedly inhibited protein kinases such as myosin light chain kinase, cAMP-dependent protein kinase, and cGMP-dependent protein kinase, through mechanisms differing from those of W-7 (5). Thus a short alkyl chain derivative of W-7, N-(6-aminoethyl)-5-chloro-1-naphthalenesulfonamide (A-3) was found to directly inhibit these protein kinases without any need for calmodulin interaction. Another derivative, 1-(5-chloronaphthalene-1-sulfonyl)-1H-hexahydro-1,4-diazepine (ML-9), proved to be a specific inhibitor of myosin light chain kinase (6, 7), acting competitively with respect to ATP but not with myosin light chain or calmodulin, with a K_i value of 3.8 μM. Accordingly, this compound appears to interact with myosin light chain kinase at or near the binding site for ATP, and has been used to elucidate the physiological roles of this enzyme (8).

At high concentrations, W-7 inhibits Ca^{2+}/phospholipid-dependent protein kinase activity, in a manner competitive with phospholipid. When the naphthalene ring of the naphthalenesulfonamides is replaced by isoquinoline, the derivatives are no longer calmodulin- or phospholipid-interacting agents, but instead directly suppress protein kinase activities. Among them, 1-(5-isoquinolinesulfonyl)-2-methylpiperazine, referred to as H-7, has been found

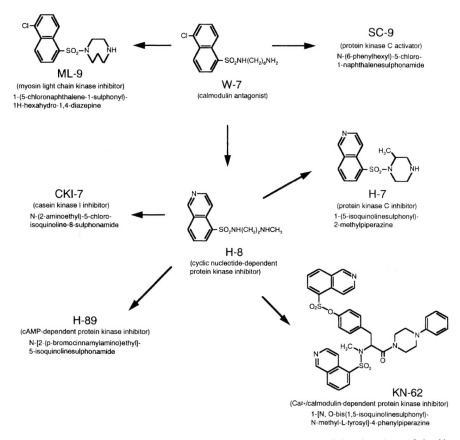

Figure 1. Chemical structures of W-7 and H-8, and the routes of development of the H-series protein kinase inhibitors. The structures of H-89 and CKI-7 are shown in *Figures 2* and *3*, respectively.

to be a specific inhibitor of protein kinase C (9, 10), competitive with respect to ATP and non-competitive with respect to the phosphate acceptor (i.e. substrate proteins). The K_i value against protein kinase C is 6.0 μM. It has now been clearly shown that H-7 directly inhibits protein kinase C, rather than via interaction with Ca^{2+} or phospholipid, and that it has significant effects on various functions by inhibiting protein kinase C-induced phosphorylation (11).

On the other hand, a derivative of naphthalenesulfonamide, *N*-(6-phenylhexyl)-5-chloro-1-naphthalenesulfonamide (SC-9), is a Ca^{2+}-dependent activator with similar effects as phosphoserine on protein kinase C activity (12). Exposure of Swiss 3T3 cells to SC-9 led to increases in hexose uptake, an event also observed when the cells were treated with 12-O-tetradecanoylphorbol-13-acetate (TPA). This activation by SC-9 was inhibited by H-7 (13).

We have established that derivatives of H-7 exhibit a selective inhibition of cyclic nucleotide-dependent protein kinases and that *N*-[2-(methylamino)-ethyl]-5-isoquinolinesulfonamide (H-8, *Figure 1*) is a particularly potent example (9). The inhibition is freely reversible and competitive with respect to ATP. H-8 specifically binds to the ATP-binding site of the catalytic subunit with a stoichiometry of 1:1 and has unique features that differ from the ATP analogues reported by Flockhart *et al.* (14) and others, in that:

- it specifically inhibits cyclic nucleotide-dependent protein kinase
- the binding constant of H-8 is much lower than that of ATP
- the binding of H-8 is independent of magnesium ions
- the binding subsite of H-8 in the active enzyme site differs slightly from that of the ATP (15).

In addition to these synthetic probes, we have synthesized additional compounds that are specific inhibitors of casein kinase I (CKI-7) (16), cAMP-dependent protein kinase (H-89) (17), and Ca^{2+}/calmodulin-dependent protein kinase II (KN-62) (18). The selectivity of these inhibitors is summarized in *Table 1*.

Table 1. Inhibition constants (K_i, μM) for various protein kinases of naphthalenesulfona-mides and isoquinolinesulfonamides

Inhibitors in group I have been used as *in vitro* or *in vivo* inhibitors; those in group II have been used for synthesis of affinity chromatography matrices

Inhibitor	PKA	PKG	CaMKII	MLCK	PKC	CKI	CKII	RoPK
1. Inhibitors for *in vitro/in vivo* use								
A-3	4.3	3.8	–	7.0	47	80	5.1	–
ML-9	32	–	–	3.8	54	–	–	–
H-7	3	5.8	–	97	6	100	780	0.45
H-8	1.2	0.5	–	68	15	133	950	–
H-88	0.4	0.8	7	50	80	60	100	–
H-89	0.05	0.5	3	30	30	40	140	–
KN-62	>100	–	0.9	>100	>100	>100	–	–
CKI-7	550	–	195	–	>1000	9.5	90	–
HA1077	1.6	1.6	–	36	–	–	–	0.33
Y-27632	13	–	–	50	6	–	–	0.14
2. Affinity ligands								
H-9	1.9	0.9	60	70	18	110	>300	
CKI-8	80	260	–	25	>100	–	–	

KEY: PKA, cAMP-dependent protein kinase; PKG, cGMP-dependent protein kinase; CaMKII, calmodulin-dependent protein kinase II; MLCK, myosin light chain kinase; PKC, protein kinase C; CKI, casein kinase I; CKII, casein kinase II; RoPK, Rho-associated protein kinase.

3. Use of protein kinase inhibitors

Detailed protocols are not appropriate for this chapter because they will depend on the experimental system (e.g. cell-free assays, cell suspensions, cell cultures, etc.) used by the experimenter. In brief, inhibitors or activators are dissolved in a suitable aqueous or organic solvent (for example, dimethyl sulfoxide), and diluted in incubation medium before addition to the cells. If an organic solvent is used as the vehicle, an identical volume of solvent should be added to the control incubation: usually the volume added can be kept sufficiently small so that the solvent has no discernible effect on its own. The final concentration of protein kinase inhibitor to be used must be determined by experiment, but concentrations of 10–100 μM are typical. After a suitable period of incubation, experimental parameters expected to be affected by the protein kinase under study may be assessed, as well as parameters not expected to be affected as controls. Another control experiment is to examine the effects of closely related agents that are not selective inhibitors of the protein kinase of interest. Sections 4 and 5 describe the various classes of inhibitors and/or activators available for protein–serine/threonine and protein–tyrosine kinases, respectively. Details of suitable solvents, storage conditions, molecular weights, and sources of supply are presented in *Table 2*.

4. Synthetic serine/threonine protein kinase inhibitors

4.1 Inhibitors of protein kinase C

Since the initial identification of protein kinase C (19), pharmacological approaches have been used to elucidate its physiological roles. Activating compounds include proteases, Ca^{2+}, phospholipids, diacylglycerols, phorbol esters, SC-9, and SC-10 (*Table 3*). These compounds, in particular the phorbol esters, have frequently been used as tools for elucidating the physiological role of protein kinase C in intracellular signal transduction pathways. More pharmacological studies have been carried out on protein kinase C than on any other protein kinase. The interest in synthesizing and using inhibitors for protein kinase C has been in part because of its ability to serve as a receptor for the phorbol ester 12-O-tetradecanoyl phorbol-13-acetate (TPA). In addition, its widespread distribution, the diverse physiological effects of TPA, and the large number of receptors linked to the production of diacylglycerol together suggest that this protein kinase is involved in the regulation of many different cellular responses. However, it is difficult to prove physiological significance without suppressing the enzyme activity using inhibitors. A number of natural or synthesized inhibitors and activators have been prepared. Cal-modulin antagonists, polypeptides, polyamines, local anesthetics, doxorubicin

Table 2. Guidelines for the use of protein-serine/threonine kinase inhibitors

Class	Compound	Stock solution	Storage	Mr	Source
(1) Inhibitors of protein kinase C					
Isoquinolinesulphonamide	H-7	10 mM (water)	4°C	364	Seikagaku
	H-9	10 mM (water)	4°C	348	Seikagaku
(2) Inhibitors of cAMP-dependent protein kinase					
Isoquinolinesulphonamide	H-8	10 mM (water)	4°C	338.3	Seikagaku
	H-88	10 mM (DMSO)	4°C	393.4	Seikagaku
	H-89	10 mM (DMSO)	4°C	446.4	Seikagaku
	H-85	10 mM (DMSO)	4°C	429.5	Seikagaku
(3) Inhibitors of myosin light chain kinase					
Naphthalenesulfonamide	ML-9	10–30 mM (DMSO)	4°C	361.3	Seikagaku
(4) Inhibitor of calmodulin-dependent protein kinase					
Isoquinolinesulfonamide	KN-62	10 mM (DMSO)	4°C or –20°C	721.9	Seikagaku
(5) Inhibitors of casein kinase I and II					
Isoquinolinesulfonamide	CKI-7	10 mM (DMSO)	4°C	285.8	Seikagaku
	CKI-8	10 mM (DMSO)	4°C	311.8	Seikagaku
(6) Antibiotic inhibitors					
Antibiotic	Staurosporine	10 mM (DMSO)	–20°C (in aliquots)	466	Sigma
	K252a	10 mM (DMSO)	4°C or –20°C	312	BIOMOL
	Calphostin C	10 mM (DMSO)	4°C	790	Sigma
(6) Protein–tyrosine kinase inhibitors					
Natural product	Genistein	DMSO[a]	4°C	270	Sigma
	Erbstatin	10 mg/ml (DMSO)	4°C	179	[b]

[a] For use *in vivo* or *in situ* add foroxymithine as an Fe^{3+} chelator.
[b] From Dr K. Umezawa, Yokohama.

Table 3. Activators of protein kinase C

Class	Compound	Stock solution	Storage	Mr	Source
Active phorbol esters[a]	TPA	10 mM (DMSO)	-20°C	617	Sigma
	Phorbol-12,13-didecanoate			673	
	Phorbol-12,13-dibutyrate			505	
	Phorbol-12,13-dibenzoate			573	
Inactive phorbol esters	Phorbol-12-tetradecanoate			575	
	Phorbol-13-acetate			406	
	4α-Phorbol-12,13-didecanoate			673	
Synthetic diacylglycerols	1-Oleyl-2-acetyl-sn-glycerol	100 mg/ml (DMSO)	-20°C[b]	399	Sigma
	1,2-sn-Dihexanoylglycerol			288	
	1,2-sn-Dioctanoylglycerol			344	
	1,2-sn-Didecanoylglycerol			401	
Naphthalenesulfonamide	SC-9	10 mM (DMSO)	4°C	402	Seikagaku
	SC-10			340	
Other tumour promoter[a]	Teleocidin	2 mM (DMSO)	-20°C		[c]
	Mezerein			655	Sigma
	Aplysia toxin				[d]

[a]SAFETY: these compounds are tumour promoters and should be handled with great care: use gloves and take all recommended precautions when working with carcinogens.
[b]If stock solution is diluted in aqueous buffer, sonicate immediately before use.
[c]From Dr H. Fujiki, Japan.
[d]From Dr R. E. Moore, Hawaii.

Table 4. Indirect inhibitors of protein kinase C

Class	Compound	Reference
Calmodulin antagonist	Phenothiazines	65
	W-7	66
	Calmidazolium	67
Polypeptide	Cytotoxins	68
	Polymixin B	69
	Neurotoxins	69
Polyamine	Spermine	70
	1,12-Diaminododecane	70
Local anaesthetic	Dibucaines	71
Other	Adriamycin	65
	Palmitoylcarnitine	61
	Alkyl lysophospholipid	72
	Gangliosides	73
	Sphingosine	48
	Quercetin	53

(adriamycin), and other lipids are indirect inhibitors, being lipophilic and causing inhibition via their interaction with phospholipids. These inhibitors (*Table 4*) are relatively non-selective in their action on phospholipid- and calmodulin-dependent enzymes. Among the isoquinolinesulfonamides, H-7 exhibits a relatively selective inhibition of protein kinase C (*Tables 1 and 2*). Kinetic analysis by double reciprocal plots revealed that this was competitive with respect to ATP. Radiolabelled isoquinolinesulfonamide derivatives are incorporated into cells; evidence for their binding to the ATP-binding site of protein kinases with a binding ratio of 1:1 has been obtained using gel permeation binding assays (unpublished data).

H-7 is now widely used for studies on various biological systems, and has been shown to exert significant effects on various functions by inhibiting protein kinase C in intact cells. For example, in platelets, H-7 enhances the 5-hydroxytryptamine release induced by the phorbol ester TPA or thrombin, in association with inhibition of the protein kinase C-catalysed phosphorylation of the myosin light chain (11). It also attenuates TPA-induced inhibitors of phosphoinositide metabolism and Ca^{2+} mobilization in thrombin-activated human platelets (20). It also prevents TPA-induced inhibition of the NaF-mediated rise in $[Ca^{2+}]$, and thromboxane B_2 generation in platelets (21).

Some selective inhibitors can also be used for affinity chromatography. H-9 possesses a free amino group and can, therefore, be used as an affinity ligand. With H-9–Sepharose gels, we have been able to purify protein kinase C 753-fold compared with the 100 000 *g* supernatant of rabbit brain homogenate. This affinity chromatography method facilitates large scale and rapid preparation (22).

4.2 Inhibitors of cyclic nucleotide-dependent protein kinases

By the time of the discovery of cAMP-dependent protein kinase (23), Sutherland and other researchers had established cAMP to be a second messenger for many extracellular stimuli. Their observations, together with the broad tissue distribution of the kinase, indicated that it was involved in a multitude of responses. While knowledge of its molecular biology and bio-chemistry has progressed rapidly, undoubtedly many aspects of its function remain to be elucidated. Derivatives of W-7 exhibit a relatively selective inhibition toward cAMP- and cGMP-dependent protein kinases, and H-8, *N*-[2-(methylamino)ethyl]-5-isoquinolinesulfonamide, is a potent inhibitor of cyclic nucleotide-dependent protein kinases. A more specific inhibitor of cAMP-dependent protein kinase has now become available with the synthesis of H-89 (*N*-[2-(*p*-bromocinnamylamino)ethyl]-5-isoquinolinesulfonamide) (17).

The structures of H-8, H-88, H-89, and H-85 (a control drug for H-89) are shown in *Figure 2*. As can be seen in *Table 1*, H-88, *N*-(2-cinnamylamino-ethyl)-5-isoquinolinesulfonamide, exerts potent inhibitory effects on both cAMP- and cGMP-dependent protein kinases. H-89, a brominated derivative of H-88, has proven to be the most selective and potent inhibitor of cAMP-dependent protein kinase among the isoquinolinesulfonamide derivatives tested, with a K_i value of 0.048 μM. This contrasts with a 10 times higher K_i for cGMP-dependent protein kinase, showing that the inhibitor is relatively selective. It was also shown to have much less potent effects on other kinases such as protein kinase C, myosin light chain kinase, Ca^{2+}/calmodulin kinase II, and casein kinases I and II. To elucidate the mechanisms involved in their inhibition of kinase activity, both H-88 and H-89 were shown to act competitively with respect to ATP by double-reciprocal plot analyses.

To investigate whether H-89 could serve as a pharmacological probe in intact cells, we studied its effects on forskolin- and nerve growth factor (NGF)-induced phosphorylation in PC12D cells. Pretreatment of the cells

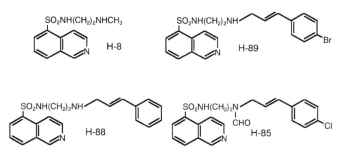

Figure 2. Structures of inhibitors of cyclic nucleotide-dependent protein kinases.

with H-89, 1 hour before the addition of forskolin, markedly inhibited forskolin-induced protein phosphorylation in a dose-dependent fashion, even after 8 hours. Inhibition of NGF-induced protein phosphorylation was not observed in the pretreated PC12D cells, suggesting that H-89 is a useful probe with respect to the selective inhibition of cAMP-dependent protein kinase in intact cells. The addition of H-89 along with forskolin did not affect forskolin-induced increase in cAMP levels in PC12D cells. When examining the effects of H-89 on adenylate cyclase and cyclic nucleotide phosphodiesterase, we found no inhibition up to 100 μM. To elucidate the roles of cAMP-dependent protein kinase in differentiation of PC12D cells, the effects of H-89 on dibutyryl cAMP-, forskolin-, and NGF-induced neurite outgrowth were investigated. When cells were pretreated with H-89 for 30 min before the addition of forskolin, neurite outgrowth was significantly inhibited in a dose-dependent manner from 8 to 48 hours thereafter. Pretreatment with H-89 for 30 min before the addition of NGF, however, exerted no inhibitory action on the neurite outgrowth of PC12D cells. It would thus appear that the cAMP-dependent protein kinase acts as a mediator of forskolin- or dibutyryl cAMP-induced neurite outgrowth, but not of NGF-induced neurite outgrowth. When the cells were treated with ML-9, H-7, CKI-7 (a selective casein kinase I inhibitor, see below) and KN-62 (a specific Ca^{2+}/calmodulin-dependent protein kinase inhibitor, see below), no inhibition was noted. H-85, N-[2-(N-formyl-p-chlorocinnamylamino)ethyl]-5-isoquinolinesulfonamide, was chosen as a control agent for H-88 and H-89 (see *Table 1* for relative potencies). At the concentration of 30 μM this compound did not inhibit the forskolin-induced neurite outgrowth of PC12D cells. There is some discrepancy regarding the effective dose of H-89 for *in vitro* and *in vivo* systems. This can be explained by the permeability of cell membranes to H-89, and the high concentration of ATP in intact cells. The ability of H-89 to inhibit cAMP-dependent protein kinase is 30 times more potent than that of H-8, whereas its ability to inhibit cGMP-dependent protein kinase is 10 times less potent. H-88 and H-89 should serve as useful probes for clarifying the physiological roles of cAMP-dependent protein kinase.

4.3 Inhibitors of myosin light chain kinases

One of the W-7 derivatives, ML-9 [1-(5-chloronaphthalene-1-sulfonyl)-1H-hexahydro-1,4-diazepine], proved to be a specific inhibitor of myosin light chain kinase without being a calmodulin antagonist (6). Myosin light chain kinase can be activated in an irreversible manner by limited proteolysis with trypsin or chymotrypsin (7). The catalytically active fragment produced by this treatment is entirely independent of the Ca^{2+}–calmodulin complex. ML-9 is a potent inhibitor of the Ca^{2+}/calmodulin-dependent form as well as the Ca^{2+}/calmodulin-independent form of myosin light chain kinase of smooth muscle, with the same K_i value of 3.8 μM (6).

As calmodulin activates many enzymes in a Ca^{2+}-dependent manner, it is not possible, using the calmodulin antagonists such as W-7, to evaluate which calmodulin-dependent enzyme is responsible for the calcium-dependent regulatory system in smooth muscle and non-muscle cells. On the other hand, ML-9 has selective inhibitory effects on myosin light chain kinase of smooth muscle and platelets but is less potent in inhibiting skeletal muscle myosin light chain kinase, other calmodulin-dependent enzymes, protein kinase C, and cAMP-dependent protein kinase. The results suggest that ML-9 is a specific inhibitor of myosin light chain kinase, with a K_i value of 3.8 μM (*Table 1*) and hence will be a useful tool for investigating its function in smooth muscle and non-muscle cells. Double-reciprocal plots indicate that ML-9 inhibits competitively with respect to ATP, but not calmodulin or the myosin light chain. ML-9 is not structurally related to ATP nor can it serve as a substrate for the kinase. These results indicate that ML-9 binds at or near the ATP-binding site at the active centre of the kinase, resulting in inhibition of catalytic activity.

To clarify the relationship between the chemical structure and potency of ML-9, derivatives substituted with iodine and bromine for chlorine at the naphthalene ring were examined for their inhibitory effects on myosin light chain kinase. The iodinated compound was found to be the most potent inhibitor, the bromine derivative was intermediate in potency, and the chlorinated form the least potent. These results suggest that the potency of these derivatives may depend on their hydrophobicity. In various types of tissue, ML-9 proved useful for studying the physiological role of myosin light chain kinase-dependent phosphorylation, for example causing the relaxation of vascular strips contracted by high K^+. This was not affected by treatment with adrenergic and cholinergic blocking agents. Thus, the ML-9-induced relaxation is not due to a block of membrane receptor-associated mechanisms. Moreover, ML-9 inhibited the Ca^{2+}-induced contraction in chemically skinned vascular smooth muscle, suggesting that ML-9 is not a Ca^{2+} channel blocker. Increasing concentrations of ML-9 selectively inhibit Ca^{2+}-dependent phosphorylation of the human platelet myosin light chain. The compound also inhibits human platelet aggregation and serotonin secretion induced by collagen and thrombin. These results suggest that myosin light chain kinase is responsible for regulating platelet function.

4.4 Ca^{2+}/calmodulin-dependent protein kinase inhibitor

Calmodulin-dependent protein kinases constitute an important group of enzymes that are involved in many aspects of calcium signalling. One of the Ca^{2+}/calmodulin-dependent enzymes, Ca^{2+}/calmodulin-dependent protein kinase II, is most abundant in the central nervous system and has emerged as a multifunctional, Ca^{2+}-operated switch mechanism. For example, its phosphorylation of tyrosine hydroxylase and tryptophan monooxygenase may alter neurotransmitter synthesis, whereas neurotransmitter release may be facilitated by phosphorylation of synapsin I. In order to clarify the physiological function of calmodulin-dependent protein kinase II, we have developed a specific inhibitor, i.e. KN-62 [1-[*N,O*-bis(1,5-isoquinoline-sulfonyl)-*N*-methyl-L-tyrosyl]-4-phenyl-piperazine] (18).

More than 80% of calmodulin-dependent protein kinase II activity was found to be inhibited by 1 μM KN-62, whereas protein kinase C, cAMP-dependent protein kinase, and myosin light chain kinase were affected only slightly in the presence of higher concentrations (*Table 1*). Double-reciprocal plots revealed that inhibition of calmodulin-dependent protein kinase II by KN-62 was competitive with respect to calmodulin and non-competitive with respect to ATP. Since autophosphorylated calmodulin-dependent protein kinase II is no longer calmodulin dependent, we also examined the effects of KN-62 on the activity of the autophosphorylated enzyme, and demonstrated no influence.

The physiological roles of calmodulin-dependent protein kinase II are postulated to be prolongation of transient Ca^{2+} signals and long-term modulation of synaptic transmission (24–26). Treatment of PC12 cells with extra-cellular signals such as NGF was found to regulate the phosphorylation of several intracellular substrate proteins (24–26). The PC12 cell, therefore, is an appropriate system for observing events related to calmodulin-dependent protein kinase II. PC12D cells were labelled with [^{32}P]phosphate and stimulated by the Ca^{2+} ionophore A23187 in the presence and absence of KN-62. Phosphoproteins immunoprecipitated with an anti-calmodulin-dependent protein kinase II antibody were resolved by SDS-PAGE and visualized by autoradiography. Autophosphorylation of the immunoprecipitated calmodulin-dependent protein kinase II (53 kDa) induced by the ionophore was inhibited in a dose-dependent manner by KN-62 treatment. The results showed that KN-62 enters cells and blocks the calmodulin-dependent protein kinase II activity in PC12D. We have also examined the possible involvement of calmodulin-dependent protein kinase II in long-term potentiation in CA1 cells. When hippocampal slices were treated with KN-62, long-term potentiation was completely suppressed. To define the role of calmodulin-dependent protein kinase II in parietal cell secretion, we examined the effects of KN-62 on acid secretion in isolated parietal cells. Inhibitor pretreatment resulted in decreased [^{14}C]aminopyrine uptake (which is correlated to acid secretion) over a concentration range of 3 to 60 μM. KN-62, however, was also found to inhibit calmodulin-dependent protein kinase IV (27). Care should therefore be used when interpreting its effects.

4.5 Inhibitors of casein kinase I

Casein kinase I has been purified from many tissues, including calf thymus, rabbit reticulocytes, liver, and skeletal muscle. The widespread distribution of the enzyme suggests an importance in cellular function, although its exact roles remain to be clarified. H-9 has a weak inhibitory effect on casein kinase I and II. When the 5-aminoethylsulfonamide chain of H-9 was moved to position 8 on the aromatic ring, the derivative CKI-6 [*N*-(2-aminoethyl)-isoquinoline 8-sulfonamide] demonstrated a more potent inhibition (16).

The structures of CKI-7 and CKI-8 are shown in *Figure 3*. CKI-7 [*N*-(2-aminoethyl)-5-chloroisoquinoline 8-sulfonamide, the chlorinated derivative of CKI-6] exerts potent inhibition with IC_{50} values of 9.5 μM for casein kinase I and 90 μM for casein kinase II, while CKI-7 causes only weak inhibition of protein kinase C at concentrations up to 1 mM. IC_{50} values for CKI-7 are 550 μM for cAMP-dependent protein kinase and 195 μM for Ca^{2+}/calmodulin-dependent protein kinase II activity. Double-reciprocal plots showed that CKI-7 inhibits casein kinase I competitively with respect to ATP. A CKI-7 affinity column can also be used to purify casein kinase I, since casein kinase II does not bind. In contrast, CKI-8, with only one-tenth the inhibitory action of CKI-7 on casein kinase I, binds both casein kinase I and II. However, recovery of casein kinase I from a CKI-7 column is low, suggesting that there is tight binding. Therefore, we have employed a CKI-8 affinity column as the final step for the purification of casein kinase I. Casein kinase I eluted at between 0.53 and 0.8 M L-arginine and 18-fold purification was achieved. Although several casein kinase II inhibitors has been used to investigate the physiological role of the enzyme, a selective casein kinase I inhibitor is not yet available. CKI-7 and CKI-8 should find useful places for determining the physiological role and distribution of casein kinase I in different tissues.

4.6 Rho-associated protein kinase inhibitors

Rho, a Ras homologue and small GTPase, regulates several processes such as cell adhesion, cytokinesis, contractile responses, and cell growth. Rho-associated protein kinases appear to work as Rho effectors, mediating its action in the cytoskeleton. Y-27632 [*trans*-*N*-(4-pyridyl)-4-(1-aminoethyl)-cyclohexanecarboxamide] was reported to have potent inhibitory effects against a Rho-associated protein kinase (28), the affinity being 200–2000 times higher than those for protein kinase C, cAMP-dependent protein kinase, or myosin light chain kinase (*Table 1*). Y-27632 inhibits smooth muscle contraction induced by phenylephrine, but has little effect on

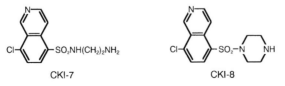

Figure 3. Structures of inhibitors of casein kinases I and II.

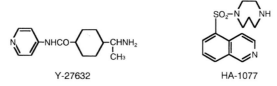

Y-27632   HA-1077

Figure 4. Structures of inhibitors of Rho-associated protein kinase.

contraction induced by KCl. Furthermore, Rho-induced formation of focal adhesions and stress fibres is inhibited by treatment of intact cells with Y-27632 (28). HA-1077 [1-(5-isoqunolinesulfonyl)-homo-pipenazine], originally developed as an arterial vasodilator (29), also inhibits Rho-associated protein kinases competitively with respect to ATP. The structures of Y-27632 and HA1077 are shown in *Figure 4.*

4.7 Antibiotic inhibitors for serine/threonine protein kinases

Another approach for identifying and developing inhibitors of protein kinases has been to screen natural products. A particularly important group of compounds are those that contain the indole carbazole chromophore. Staurosporine (*Figure 5*) was the first compound in this group shown to inhibit protein kinases, being particularly effective against protein kinase C with a K_i value of 1–3 nM (30–32). However, staurosporine also inhibits several other protein–serine/threonine and protein–tyrosine kinases, including the cAMP-dependent protein kinase (33), phosphorylase kinase (34), ribosomal protein S6 kinase (34), pp60^{v-src} (33), and receptor protein–tyrosine kinases (34). The order of potency of staurosporine appears to be roughly: protein kinase C > [cyclic AMP-dependent kinase, Ca^{2+}/calmodulin-dependent protein kinase II, pp60^{v-src}] > insulin receptor kinase > EGF receptor kinase > IGF receptor kinase. The staurosporine binding affinity is greater than that of ATP, and it is thought to target ATP-binding sites.

Another indole carbazole protein kinase inhibitor that has been isolated from microbial organisms is K252a (35, 36). Its structure is similar to that of staurosporine (*Figure 5*) and it also has a broad specificity, although in general its inhibition is somewhat less potent than that by staurosporine (*Table 5*). Several other indole carbazoles have now been isolated or prepared semi-synthetically. While very interesting compounds, they are not specific for protein kinase C and there have been several reports of indole carbazoles paradoxically appearing to mimic the effects of phorbol esters, suggesting actions other than inhibition of protein kinase C in intact cells. Indeed, K252a has been shown to inhibit calmodulin-dependent enzymes and cyclic nucleo-tide phosphodiesterases. Recently, protein kinase C inhibitors that act on the diacylglycerol-binding site have been isolated from a soil fungus (37–39).

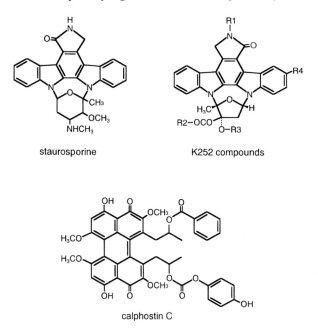

staurosporine   K252 compounds

calphostin C

Figure 5. Structures of antibiotic inhibitors of protein-serine/threonine protein kinases.

These agents have a multi-ring quinone structure and are named calphostins. The most interesting is calphostin C (*Figure 5*), which completely inhibits the binding of [^{3}H]phorbol ester (50 nM) to protein kinase C when present at 1 μM. It inhibits the α-, β-, and γ-isozymes of protein kinase C with equal potency, while having no effect on the activity of either cAMP-dependent protein kinases or pp60^{v-src} at 50 μM. Two other analogues similarly have inhibitory activity against protein kinase C, apparently by binding to the diacylglycerol site. The adriamycin (Fe^{3+}) complex also inhibits protein kinase C with an IC_{50} of 50 μM (40).

4.8 Miscellaneous inhibitors of serine/threonine protein kinases

The narcotic antagonist apomorphine inhibits cAMP-dependent protein kinase and protein kinase C in the μM range. Amiloride (an inhibitor of the Na^+/H^+ exchanger) has been found to inhibit protein kinase C as well as tyrosine protein kinases (41, 42). 5, 6-dichloro-1-(β-D-ribofuranosyl) benzimi-dazole has been reported to be a relatively selective inhibitor of the casein kinases with μM inhibition constants (43, 44). Sangivamycin, an anti-tumour nucleotide analogue, inhibits protein kinase C with a K_i value of 10 μM and cAMP-dependent protein kinase (45–47), as well as the EGF receptor kinase. Many structurally diverse hydrophobic agents are able to compete with

Table 5. Structures and relative potencies of K252 compounds

Compound	Structure				K_i (nM)			
	R1	R2	R3	R4	PKC	PKA	PKG	MLCK
K252a	–H	–CH$_3$	–H	–H	25	16	15	20
K252b	–H	–H	–H	–H	20	90	100	147
KT5720	–H	*n*-hexyl	–H	–H	>2000	56	>2000	>2000
KT5823	–CH$_3$	–CH$_3$	–CH$_3$	–H	4000	>104	234	>104
KT5926	–H	–CH$_3$	–H	O-*n*-propyl	723	1200	158	18

KEY: PKC, protein kinase C; PKA, cAMP-dependent protein kinase; PKG, cGMP-dependent protein kinase; MLCK, myosin light chain kinase.

protein kinase C for binding phosphatidylserine and thus inhibit kinase activity. However, many of these are non-specific and effective only at high concentration. Sphingosine, an 18-carbon chain lipid base, inhibits protein kinase C competitively with respect to diacylglycerol, Ca^{2+}, and phosphatidylserine (48) (*Table 4*), and is thought to be a natural negative regulator in intact cells. However, it has become apparent that it also exerts other effects, for example, as an antagonist of calmodulin (49), activating EGF receptor kinase (50, 51), inhibiting pp60$^{v–src}$ activity (51), and affecting biological processes in a manner that is independent of its ability to inhibit protein kinase C (52). Thus sphingosine is not suitable for studies of protein kinases in intact cells.

5. Inhibitors of protein–tyrosine kinases

As in the case of protein–serine/threonine kinases, natural products have been an important source of protein–tyrosine kinase inhibitors, e.g. quercetin, a flavonoid compound, (53, 54). In 1986, Ogawara *et al.* (55), isolated genistein (*Figure 6*), an isoflavonoid from a strain of *Pseudomonas*, as an EGF receptor/kinase inhibitor (55). Both genistein and quercetin inhibit EGF receptor kinase and pp60$^{v–src}$ with K_i values in the μM range, and kinetic analyses have shown that in the former case this is competitive with respect to ATP (56). However, quercetin can also inhibit protein kinase C and phosphorylase kinase, while genistein has little or no effect on these enzymes. Genistein also did not inhibit a protein–tyrosine kinase purified from thymus, and thus the compound is not effective against all protein–tyrosine kinases. Most of the work with genistein has been carried out with intact cells such as T cells, keratinocytes, and platelets (57). Another inhibitor of protein–tyrosine kinases was isolated from the culture medium of a *Streptomyces* strain (58). The compound, a tertiary amine substituted with three phenyl groups and named lavendustin A, inhibits the EGF receptor kinase activity with a K_i of 12 nM, and is competitive with respect to ATP. It has no effects

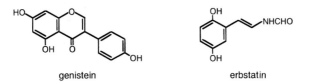

genistein   erbstatin

Figure 6. Structures of inhibitors of protein-tyrosine kinases.

on protein kinase C, and only weakly inhibits cAMP-dependent protein kinase.

In 1989 Umezawa *et al.* reported the isolation of an EGF receptor kinase inhibitor from another strain of *Streptomyces* (59). The compound, named erbstatin (*Figure 6*), was found to inhibit EGF receptor kinase at μM concentrations, but to have almost no effect on cAMP-dependent protein kinase. The same group further reported that the two hydroxyl groups on the ring of erbstatin are necessary for its inhibitory activity (60, 61), and that it is competitive with respect to peptide substrate (62). The compound is relatively small and uncharged, and can thus penetrate intact cells.

6. Concluding remarks

The focus of this chapter has been on synthetic and natural protein kinase inhibitors. To clarify the physiological role and molecular mechanisms of protein phosphorylation, selective protein kinase inhibitors are clearly essential as pharmacological probes (63). However, since more and more novel protein kinases are being identified, lack of specificity remains a problem. One way around this is to employ different types of inhibitors, which work by different mechanisms, in the same study. Such 'combination' experiments make the results more reliable. Nevertheless, more specific inhibitors are still needed. Recently, crystal structures of cAMP-dependent protein kinase in complex with H-7, H-8, and H-89 have been described (64), and this approach may facilitate the design of more potent and specific inhibitors. In fact, potent calmodulin antagonists have already been designed on the basis of the three-dimensional structure of Ca^{2+}/calmodulin complexed with W-7. To conclude, protein kinase inhibitors are very useful and informative tools, but care should always be taken in drawing the conclusions from inhibition experiments.

References

1. Hidaka, H., Tanaka, T., Saitoh, M., and Matsushima, M. (1987). In *Calcium-binding proteins in health and disease* (ed. A. W. Norman, T. C. Vanaman, and A. R. Means), p. 170. Academic Press, New York.

2. Hidaka, H. and Tanaka, T. (1988). In *Methods in enzymology* (ed. A. R. Means and P. M. Conn), Vol. 139, p.153. Academic Press, London.

3. Hidaka, H., Asano, T., Iwadare, S., Matsumoto, I., Totsuka, T., and Aoki, N. (1978). *J. Pharmacol. Exp. Ther.*, **207**, 8.

4. Hidaka, H., Yamaki, T., Naka, M., Tanaka, T., Hayashi, H., and Kobayashi, R. (1980). *Mol. Pharmacol.*, **17**, 66.

5. Inagaki, M., Kawamoto, S., Itoh, H., Saitoh, M., Hagiwara, M., Takahashi, J., and Hidaka, H. (1986). *Mol. Pharmacol.*, **29**, 571.

6. Saitoh, M., Ishikawa, T., Matsushima, S., Naka, M., and Hidaka, H. (1987). *J. Biol. Chem.*, **262**, 7796.

7. Saitoh, M., Naka, M., and Hidaka, H. (1987). *Biochem. Biophys. Res. Commun.*, **140**, 280.

8. Ishikawa, T., Chijiwa, T., Hagiwara, M., Mamiya, S., Saitoh, M., and Hidaka, H. (1988). *Mol. Pharmacol.*, **33**, 598.

9. Hidaka, H., Inagaki, M., Kawamoto, S., and Sasaki, Y. (1984). *Biochemistry*, **23**, 5036.

10. Kawamoto, S., and Hidaka, H. (1984). *Biochem. Biophys. Res. Commun.*, **125**, 258.

11. Inagaki, M., Kawamoto, S., and Hidaka, H. (1984). *J. Biol. Chem.*, **259**, 14321.

12. Ito, M., Tanaka, T., Inagaki, M., Nakanishi, K., and Hidaka, H. (1986). *Biochemistry*, **25**, 4179.

13. Nishino, H., Kitagawa, K., Iwashima, A., Ito, M., Tanaka, T., and Hidaka, H. (1986). *Biochem. Biophys. Acta*, **889**, 236.

14. Flockhart, D. A., Freist, W., Hoppe, J., Lincoln, T. M., and Corbin, J. D. (1984). *Eur. J. Pharmacol.*, **140**, 289.

15. Hagiwara, M., Inagaki, M., and Hidaka, H. (1987). *Mol. Pharmacol.*, **31**, 523.

16. Chijiwa, T., Hagiwara, M., and Hidaka, H. (1989). *J. Biol. Chem.*, **264**, 4924.

17. Chijiwa, T., Mishima, A., Hagiwara, M., Sano, M., Hayashi, K., Inoue, T., Naito, K., Toshioka, T., and Hidaka, H. (1990). *J. Biol. Chem.*, **265**, 5267.

18. Tokumitsu, H., Chijiwa, T., Hagiwara, M., Mizutani, A., Terasawa, M., and Hidaka, H. (1990). *J. Biol. Chem.*, **265**, 4315.

19. Nishizuka, Y. (1984). *Nature*, **308**, 693.

20. Tohmatsu, T., Hattori, H., Nagao, S., Ohki, K., and Nozawa, Y. (1986). *Biochem. Biophys. Res. Commun.*, **134**, 868.

21. Poll. C., Kyrle, P., and Westwick, J. (1986). *Biochem. Biophys. Res. Commun.*, **136**, 381.

22. Ohno, S., Kawasaki, H., Imajoh, S., Suzuki, K., Inagaki, M., Yokokura, H., Sakoh, T., and Hidaka, H. (1987). *Nature*, **325**, 161.

23. Walsh, D. A., Perkins, J. P., and Krebs., E. G. (1968). *J. Biol. Chem.*, **243**, 3763.

24. Halegoua, S. and Patrick, J. (1980). *Cell*, **22**, 571.

25. Greene, L. A., Liem, R. K. H., and Shelanski, M. L. (1983). *J. Cell Biol.*, **96**, 76.

26. Nairn, A. C., Nichols, R. A., Brady, M. J., and Palfrey, H. C. (1987). *J. Biol. Chem.*, **262**, 14265

27. Enslen, H., Sun, P., Brickey, D., Soderling, S. H., Klamo, E., Soderling, T. R. (1994). *J. Biol. Chem.*, **269**, 15520.

28. Uehata, M., Ishizaki, T., Satoh, H., Ono, T., Kawahara, T., Morishita, T., Tamakawa, H., Yamagami, K., Inui, J., Maekawa, M., and Narumiya, S. (1997). *Nature*, **389**, 990.

29. Asano, T., Suzuki, T., Tsuchiya, M., Satoh, S., Ikegaki, I., Shibuya, M., Suzuki, Y., and Hidaka, H. (1989). *Br. J. Pharmacol.*, **98**, 1091.
30. Tamaoki, T., Nomoto, H., Takahashi, I., Kato, Y., Morimoto, M., and Tomita, F. (1986). *Biochem. Biophys. Res. Commun.*, **135**, 397.
31. Gross, J. L., Herblin, W. F., Do, U. H., Pounds, J. S., Buenaga, L. J., and Stephens, L. E. (1990). *Biochem. Pharmacol.*, **40**, 343.
32. Herbert, J. M., Seban, E., and Maffrand, J. P. (1990). *Biochem. Biophys. Res. Commun.*, **171**, 189.
33. Nakano, H., Kobayashi, E., Takahashi, I., Tamaoki, T., Kuzuu, Y., and Iba, H. (1987). *J. Antibiot.*, **40**, 706.
34. Meyer, T., Regenass, U., Fabbro, D., Alteri, E., Rösel, J., Müler, M., Garavatti, G., and Matter, A. (1989). *Int. J. Cancer*, **43**, 851.
35. Kase, H., Iwahashi, K., and Matsuda, Y. (1986). *J. Antibiot.*, **39**, 1059.
36. Kase, H., Iwahashi, K., Nakanishi, S., Matsuda, Y., Yamada, K., Takahashi, M., Murakata, C., Sato, A., and Kaneko, M. (1987). *Biochem. Biophys. Res. Commun.*, **142**, 436.
37. Iida, T., Kobaysh, E., Yoshida, M., and Sano, H. (1989). *J. Antibiot.*, **42**, 1475.
38. Kobayashi, E., Ando, K., Nakano, H., Iida, T., Ohno, H., Morimoto, M., and Tamaoki, T. (1989). *J. Antibiot.*, **42**, 153.
39. Kobayashi, E., Ando, K., Nakano, H., Iida, T., Ohno, H., Morimoto, M., and Tamaoki, T. (1989). *J. Antibiot.*, **42**, 1470.
40. Hannun, Y. A., Foglesong, R. J., and Bell. R. M. (1989). *J. Biol. Chem.*, **264**, 9960.
41. Besterman, J. M., May, W. S. J., LeVine, H. D., Cragoe, E. J. J., and Cuatrecasas, P. (1985). *J. Biol. Chem.*, **260**, 1155.
42. Davis, R. J., and Czech, M. P. (1985). *J. Biol. Chem.*, **260**, 2543.
43. Zandomeni, R., Zandomeni, M. C., Shugar, D., and Weinmann, R. (1986). *J. Biol. Chem.*, **261**, 3414.
44. Meggio, F., Shugar, D., and Pinna, L. A. (1990). *Eur. J. Biochem.*, **187**, 89.
45. Osada, H., Sonoda, T., Tsunoda, K., and Isono, K. (1989). *J. Antibiot.*, **42**, 102.
46. Osada, H., Takahash, H., Tsunoda, K., Kusakabe, H., and Isono, K. (1990). *J. Antibiot.*, **43**, 163.
47. Loomis, C. R. and Bell, R. M. (1988). *J. Biol. Chem.*, **263**, 1682.
48. Hannun, Y. A., Loomis, C. R., Merrill, A. H. J., and Bell, R. M. (1986). *J. Biol. Chem.*, **261**, 12604.
49. Jefferson, A. B., and Schulman, H. (1988). *J. Biol. Chem.*, **263**, 15241.
50. Faucher, M., Gironès, Hannun, Y. A., Bell, R. M., and Davis, R. J. (1988). *J. Biol. Chem.*, **263**, 5319.
51. Davis, P. D., Hill, C. H., Keech, E., Lawton, G., Nixon, J. S., Sedgwick, A. D., Wadsworth, J., Westmacott, D., and Wilkinson, S. E. (1989). *FEBS Lett.*, **259**, 61.
52. Igarashi, Y., Hakomori, S., Toyokuni, T., Dean, B., Fujita, S., Sugimoto, M., Ogawa, T., el-Ghendy, K., and Racker, E. (1989). *Biochemistry*, **28**, 6796.
53. Cochet, C., Feige, J. J., Pirollet, F., Keramidas, M., and Chambaz, E. M. (1982). *Biochem. Pharmacol.*, **31**, 1357.
54. Hagiwara, M., Inoue, S., Tanaka, T., Nunoki, K., Ito, M., and Hidaka, H. (1988). *Biochem. Pharmacol.*, **37**, 2987.
55. Ogawara, H., Akiyama, T., Ishida, J., Watanabe, S. and Suzuki, K. (1986). *J. Antibiot.*, **39**, 606.

56. Akiyama, T., Ishida, J., Nakagawa, S., Ogawara, H., Watanabe, S., Itoh, N., Shibuya, M., and Fukami, Y. (1987). *J. Biol. Chem.*, **262**, 5592.

57. Ogawara, H., Akiyama, T., Watanabe, S., Ito, N., Kobori, M., and Seoda, Y. (1989). *J. Antibiot.*, **42**, 340.

58. Onoda, T., Iinuma, H., Sasaki, Y., Hamada, M., Isshiki, K., Naganawa, H., Takeuchi, T., Tatsuta, K., and Umezawa, K. (1989). *J. Nat. Prod.*, **52**, 1252.

59. Umezawa, H., Imoto, M., Sawa, T., Isshiki, K., Matsuda, N., Uchida, T., Iinuma, H., Hamada, M., and Takeuchi, T. (1986). *J. Antibiot.*, **39**, 170.

60. Isshiki, K., Imoto, M., Takeuchi, T., Umezawa, H., Tsuchida, T., Yoshioka, T., and Tatsuta, K. (1987). *J. Antibiot.*, **40**, 1207.

61. Isshiki, K., Imoto, M., Sawa, T., Umezawa, K., Takeuchi, T., Umezawa, H., Tsuchida, T., Yoshioka, T., and Tatsuta, K. (1987). *J. Antibiot.*, **40**, 1209.

62. Imoto, M., Umezawa, K., Isshiki, K., Kunimoto, S., Sawa, T., Takeuchi, T., and Umezawa, H. (1987). *J. Antibiot.*, **40**, 1471.

63. Hidaka, H., Watanabe, M., and Kobayash, R. (1991). In *Methods in enzymology* (ed. T. Hunter, and B. M. Sefton), Vol. 201, p. 328. Academic Press, London.

64. Engh, R. A., Girod, A., Kinzel, V., Huber, R., Bossemeyer, D. (1996). *J. Biol. Chem.*, **271**, 26157.

65. Wise, M. B. C. and Kuo, J. F. (1983). *Biochem. Pharmacol.*, **32**, 1259.

66. Tanaka, T., Ohmura, T., Yamakado, T., and Hidaka, H. (1982). *Mol. Pharmacol.*, **22**, 408.

67. Mazzei, G. J., Schatzman, R. C., Turner, R. S., Volger, W. R., and Kuo, J. F. (1984). *Biochem. Pharmacol.*, **33**, 125.

68. Kuo, J. F., Raynor, R. L., Mazzei, G. J., Schatzman, R. C., Turner, R. S., and Kem, W. R. (1983). *FEBS Lett.*, **153**, 183.

65. Mazzei, G. J., Katoh, N., and Kuo, J. F. (1982). *Biochem. Biophys. Res. Commun.*, **109**, 1129.

69. Qi, D. F., Schatzman, R. C., Mazzei, G. J., Turner, R. S., Raynor, T. L., Liano, S., and Kuo, J. F. (1983). *Biochem. J.*, **213**, 281.

70. Mori, T., Takai, Y., Minakuchi, R., Yu, B., and Nishizuka, Y. (1980). *J. Biol. Chem.*, **255**, 8378.

71. Helfman, D. M., Darnes, K. C., Kinkade, J. M., Jr, Volger, W. R., Shoji, M., and Kuo, J. F. (1983). *Cancer Res.*, **43**, 2955.

72. Kim, J. Y. H., Goldenring, J. R., DeLorenzo, R. J., and Yu, R. K. (1986). *J. Neurosci. Res.*, **15**, 159.

3

Analysis of signal transduction pathways using protein–serine/threonine phosphatase inhibitors

D. GRAHAME HARDIE

1. Introduction

A widely used approach for studying protein phosphorylation is the pharmacological one, i.e. to study the effects of cell-permeable inhibitors or activators of the enzymes which catalyse changes in phosphorylation. In Chapter 2 the use of protein kinase activators and inhibitors is discussed. Over the past 10 years, a number of potent, cell-permeable protein phosphatase inhibitors have also become available, and their impact has been considerable. Their applications may be summarized as follows:

(a) Improving the specificity of protein phosphatase assays in cell-free extracts. This approach is discussed in Chapter 7, and the present chapter will concentrate instead on their uses in intact cell studies.

(b) Provision of initial evidence that a physiological system is regulated by protein phosphorylation. This is particularly valuable in the case of a system where the major components are not well characterized, in which case more direct methods for the study of protein phosphorylation cannot be utilized.

(c) In cases where the role of protein phosphorylation is already established, the inhibitors may in some cases be used to identify the protein phosphatase(s) active against particular substrates in intact cells.

2. The protein–serine/threonine phosphatases

Two distinct families of serine/threonine-specific protein phosphatases have been found in mammalian cells. The PPP family (1) includes protein phosphatases-1, -2A, -2B, -4, -5, and -6 (PP1, PP2A, PP2B, PP4, PP5, PP6)

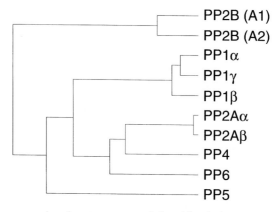

Figure 1. Dendrogram showing sequence relationships between members of the PPP protein phosphatase family. The dendrogram was produced using the program PILEUP (52) using human amino acid sequences derived from the EMBL database.

whereas the unrelated PPM family include various forms of protein phosphatase-2C (PP2C) and the mitochondrial enzyme pyruvate dehydrogenase phosphatase (1). A dendrogram showing the sequence relationships between the catalytic subunits of members of the PPP family is shown in *Figure 1*. The inhibitors discussed in this chapter usually act on several PPP family members, but with a certain degree of selectivity. Familiarity with the sequence relationships shown in *Figure 1* is helpful when discussing this selectivity. The PPM family is not sensitive to any of the inhibitors described below, and as yet there are unfortunately no specific inhibitors available for this family.

Figure 1 shows that PP1 and PP2A isoforms cluster into two distinct subfamilies, with PP4 and PP6 being closer to the PP2A subfamily. PP5 and the PP2B isoforms lie on their own distinct branches. Most of the PPP phosphatases exist as complexes between the catalytic subunits represented in *Figure 1* and non-catalytic regulatory subunits. For example, the PP2B subfamily (also known as calcineurins) are Ca^{2+}-dependent phosphatases and exist as complexes between catalytic (A) subunits and Ca^{2+}-binding (B) subunits.

3. Structure and origin of protein phosphatase inhibitors

3.1 Inhibitors of the PP1/PP2A/PP4/PP6 subfamilies

3.1.1 Okadaic acid

The first potent and selective inhibitor of the PPP family to be described, okadaic acid, is a complex fatty acid derivative containing many cyclic polyether linkages. Okadaic acid and its close relatives, dinophysistoxin-1 and

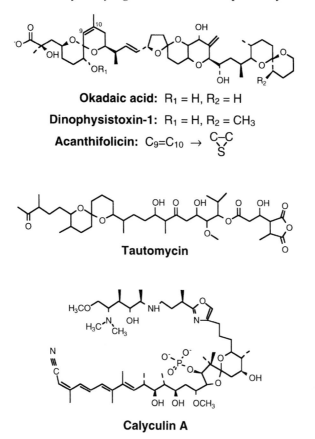

Okadaic acid: $R_1 = H$, $R_2 = H$

Dinophysistoxin-1: $R_1 = H$, $R_2 = CH_3$

Acanthifolicin: $C_9=C_{10} \rightarrow$

Tautomycin

Calyculin A

Figure 2. Structures of okadaic acid (and its relatives dinophysistoxin-1 and acanthifolicin), tautomycin, and calyculin A.

acanthifolicin (*Figure 2*), are marine toxins produced by genera of the dinoflagellates such as *Dynophysis*. Okadaic acid accumulates in filter feeders such as shellfish and the sponge *Halichondria okadai* (from which it was first isolated and from which it obtained its name). In this way it enters the food chain, and okadaic acid is believed to be the major causative agent of diarrhetic shellfish poisoning. In 1982 it was shown to cause contraction of isolated smooth muscle (2). Since contraction was known to be initiated by phosphorylation of one of the myosin light chains, this suggested that it either activated myosin light chain kinase or inhibited the relevant protein phosphatases. The latter turned out to be correct, and okadaic acid was shown to be an extremely potent inhibitor of PP1 and PP2A (3). It was subsequently shown that it caused dramatic increases in phosphorylation of numerous proteins in isolated hepatocytes and adipocytes (4).

3.1.2 Tautomycin

Tautomycin (*Figure 2*) bears an obvious structural relationship to okadaic acid, but was isolated from a completely different source, i.e. culture filtrates of the soil bacteria *Streptomyces griseochromogenes* and *S. verticillatus*. Its relative potency with different protein phosphatases is similar to that of okadaic acid (5).

3.1.3 Calyculin A

Calyculin A (*Figure 2*) was, like okadaic acid, originally derived from a marine sponge, *Discodermia calyx*. Although not obviously related in structure to okadaic acid, it also causes contraction of isolated smooth muscle (6) and is a potent inhibitor of PP1 and PP2A in cell-free assays (7).

3.1.4 Microcystins and nodularins

The microcystins and nodularins (*Figure 3*) are peptide toxins produced by blue-green algae of the genera *Microcystis*, *Anabaena*, and *Nodularia*. Blooms of these algae in freshwater ponds and reservoirs have caused poisoning of humans as well as farm and domestic animals. The microcystins are cyclic heptapeptides while the nodularins are cyclic pentapeptides, and although both contain hydrophobic side chains they are otherwise unrelated in structure to okadaic acid. Nevertheless, like the latter, they are extremely potent inhibitors of PP1 and PP2A (8, 9). Microcystins, but not nodularins, are unusual among protein phosphatase inhibitors in that after binding through non-covalent interactions there is a secondary covalent reaction, involving

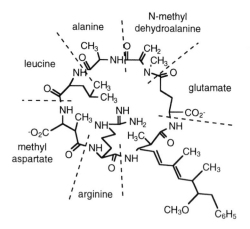

microcystin-LR

Figure 3. Structure of microcystin-LR. Members of the microcystin family differ in the amino acids present at two variable positions (the LR variant having leucine and arginine) and in the presence or absence of certain methyl groups.

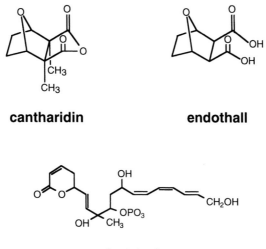

cantharidin   **endothall**

fostriecin

Figure 4. Structures of cantharidin, endothall, and fostriecin.

cysteine residues on the protein phosphatase and the dehydroalanine residue on the inhibitor, such that the inhibition is irreversible (10). The structure of the complex between microcystin-LR and the catalytic subunit of PP1 has been determined by X-ray crystallography (11).

3.1.5 Cantharidin and endothall

Cantharidin ('Spanish fly') is derived from blister beetles. Although reputed to be an aphrodisiac, it is extremely toxic (12) and not to be recommended for this purpose! In 1992 it was shown that the major binding protein for cantharidin in mouse liver extracts was PP2A (13). Cantharidin and related compounds, such as the herbicide endothall (*Figure* 4), were subsequently shown to be potent inhibitors of both PP1 and PP2A (14, 15).

3.1.6 Fostriecin

Fostriecin is an antibiotic isolated from *Streptomyces pulveraceus* (subspecies *fostreus*) (16) that is under development as an anti-cancer drug (17). The compound is an extremely potent inhibitor of PP2A and PP4 (*Table 1*), but of particular interest is the finding that inhibition of PP1 required >10 000-fold higher concentrations (18).

3.2 Inhibitors of the PP2B (calcineurin) subfamily

Cyclosporin A (CsA) and FK506 (also known as tacrolimus) are immuno-suppressant drugs that inhibit T-cell responses to activation of T-cell receptors, such as the induction of the cytokine interleukin-2 (19) (*Figure 5*).

Table 1. Specificity of inhibitors for different members of the PPP superfamily

Inhibitor	PP1	PP2A	PP4	PP5	PP2B	References
			IC_{50} values^a (nM)			
Okadaic acid	20	0.1	0.1	1.4	4000	25, 26
Acanthifolicin	20	2				35
Dinophysistoxin-1	55	0.6			35	
Calyculin A	2	1	0.2			7, 25
Tautomycin	3	30	0.4		>10 000	5, 25
Microcystin-LR	0.1	0.1	0.15	15		25, 26
Cantharidin	470	40	50		>30 000	14, 25
Palasonin	660	120				14
Cantharidic acid	560	53			14	
Endothall	5000	970			>60 000	14
Fostriecin	45 000	1.5	3		No effect	18, 25

^{a} Because of the tight binding of many of these inhibitors, IC_{50} values may depend on the concentration of the protein phosphatase used in the assay, and cannot be directly equated with dissociation constants.

Cyclosporin is a hydrophobic cyclic peptide isolated from the fungus *Tolypocladium inflatum*, while FK506 is a macrolide isolated from *Streptomyces tsukubaensis*. They act by forming complexes with distinct binding proteins [cyclophilins and FK506-binding proteins (FKBPs), respectively], which are known collectively as *immunophilins*. Surprisingly, although the immunophilins all carry a peptidyl–prolyl *cis–trans* isomerase activity, this is not required for their immunosuppressive function. Instead, the CsA–cyclophilin and FK506–FKBP complexes appear to act by binding to, and inhibiting, PP2B (calcineurin) (20). This hypothesis was strongly supported by observations that overexpression of PP2B in Jurkat T cells rendered them more resistant to the actions of CsA and FK506 (21). The CsA–cyclophilin and FK506–FKBP complexes inhibit PP2B with half-maximal effects at around 30 nM. Inhibition requires the presence of the Ca^{2+}-binding B subunit as well as the catalytic A subunit of PP2B (22, 23). Since none of the other members of the PPP superfamily bind the B subunit, this explains why inhibition appears to be specific for the PP2B subfamily (24). Although CsA and FK506 are utilized medically for their immunosuppressive effects on T cells, the immunophilins are abundant proteins that are ubiquitously distributed, so that the drugs can be utilized to assess the roles of PP2B in almost any cell type.

4. Selectivity of protein phosphatase inhibitors

Table 1 provides a summary of the selectivity of inhibitors of the PP1 and PP2A subfamilies. Okadaic acid and its relatives (e.g. acanthifolicin, dinophysistoxin-1) inhibit PP2A at lower concentrations than PP1. In cell-free

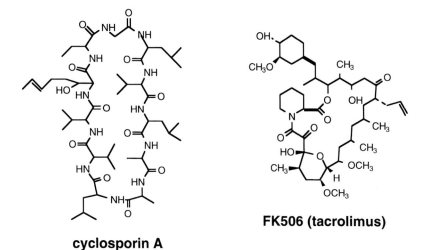

cyclosporin A

FK506 (tacrolimus)

Figure 5. Structures of cyclosporin A and FK506.

assays where the protein phosphatase is diluted to concentrations at or below 1 nM, PP2A activity is completely abolished by 1 nM okadaic acid, whereas PP1 activity is scarcely affected. This provides the basis of cell-free assays that distinguish PP1 and PP2A (see Chapter 7). However, the concentrations of PP1 and PP2A in *intact* cells are usually between 0.1 and 1 μM, so that concentrations of up to 1 μM okadaic acid will be required to completely titrate out either PP1 or PP2A. In intact cells these inhibitors do not therefore distinguish between the effects of PP1 and PP2A. With the possible exception of fostriecin, similar considerations apply to all of the other inhibitors in *Table 1*. Calyculin A and microcystin-LR have similar potencies with PP1 and PP2A, whereas tautomycin is a rather more potent inhibitor of PP1 than of PP2A. Cantharidin and its relatives (palasonin, cantharidic acid, and endo-thall) are somewhat more potent inhibitors of PP2A than of PP1. The only case where there is a marked difference in potency is fostriecin, which is 30 000-fold less potent in inhibition of PP1 than of PP2A. Although this does not yet appear to have been rigorously tested, it therefore seems likely that fostriecin, unlike the other inhibitors, could be used to distinguish between the effects of the PP1 and PP2A subfamilies in intact cells.

In general, the spectrum of inhibition of PP4 by different toxins appears to be similar to that of the PP2A subfamily (25), consistent with their close sequence relationship (*Figure 1*). There are, therefore, no inhibitors currently available that can distinguish between the effects of PP2A and PP4. PP5 is potently inhibited by okadaic acid and a little less potently by microcystin-LR (26): the effects of the other toxins do not yet appear to have been tested. It has not proved possible to express PP6 in active form, so nothing is known of its inhibitor sensitivity.

The PP2B subfamily is rather distantly related to PP1 and PP2A (*Figure 1*), and in general the toxins, which are potent inhibitors of the latter, are very poor inhibitors of PP2B, requiring >1 µM concentrations to achieve half-maximal inhibition (*Table 1*). If used at concentrations <1 µM in intact cell experiments, it can be safely assumed that these inhibitors are not inhibiting PP2B. The role of the PP2B subfamily can, of course, be addressed using CsA or FK506. The CsA–cyclophilin and FK506–FKBP complexes inhibit PP2B with half-maximal effects at around 30 nM. As discussed in Section 3.2, this requires the presence of the Ca^{2+}-binding B subunit of PP2B, and these complexes have no effect on PP1, PP2A, PP2C, or PP2B. The PPM protein phosphatase family (including PP2C) is unrelated to the PPP family, and none of the inhibitors in *Table 1* inhibit PP2C. Unfortunately, to our knowledge, selective inhibitors of PP2C have not yet been described.

Most of the inhibitors discussed above act at nM concentrations, and if used in intact cell experiments at 1 µM or below, all of their effects can probably be accounted for by protein phosphatase inhibition rather than other, non-specific effects. In short-term (15 minute) incubations in isolated hepatocytes 1 µM okadaic acid did not affect cell viability as judged by Trypan Blue exclusion or ATP/ADP levels (4). However since these inhibitors appears to totally prevent dephosphorylation of most proteins (4), and given the central importance of protein phosphorylation in cellular regulation, it seems inevitable that they will have many secondary effects in longer incubations. For example, it has been reported that incubation of hepatocytes in 1 µM okadaic acid for several hours results in morphological changes characteristic of apoptosis (27). Cultured cells have been incubated in okadaic acid for up to 10 days, and cause reversion of a transformed phenotype (28). However, this study utilized much lower concentrations of the toxin (10 nM), which probably only inhibit dephosphorylation partially.

5. Guidelines for the use of the inhibitors in intact cells

The choice of inhibitor will depend on both availability and efficacy. Microcystin-LR is readily available and cheap, but may not be readily permeable to all cells. The microcystins are liver toxins *in vivo*, and the basis of this appears to be that they are taken up into liver cells by specific mechanisms, probably via a bile salt transport mechanism (29). Microcystins do appear to be able to permeate plant cells (30). Okadaic acid, calyculin A, and cantharidin are all commercially available, and being lipophilic they appear to permeate all cells. Since their structures are very different, the demonstration that a functional effect is produced by all three will lend weight to the conclusion that this is due to the inhibition of protein phosphatases. For reasons discussed in Section 4, these inhibitors do not in general distinguish between effects of the PP1 and PP2A/PP4 subfamilies in intact cells. However, fostriecin, which is not yet commercially available, may allow such a distinction to be made.

Given the central importance of protein phosphorylation in cellular regulation, and the widespread functions of the PPP family, it is unlikely that the use of these toxins *in vivo* would be very informative, although the tumour-promoting activity of okadaic acid was originally tested in the mouse skin *in vivo* model (31). Their main use is in the testing of roles for protein phosphorylation in some biochemical or physiological parameter which can be measured in short-term (say, 15 minute) experiments in isolated or cultured cells. The methods given as *Protocols 1* and *2* were developed for studies using freshly isolated rat hepatocytes or adipocytes in suspension, but can be readily adapted for any isolated or cultured cell.

Protocol 1.   Use of protein phosphatase inhibitors in isolated cells

Equipment and reagents

- okadaic acid, calyculin A, or tautomycin may be dissolved to 5 mM in DMSO, and may be stored indefinitely in this form at −20°C
- microcystins are soluble in water and can be stored in this form at −20°C

Method

1. Cell suspensions or monolayers are prepared as appropriate. If one wishes to study the effect of the toxin on protein phosphorylation, cells should be prelabelled with [^{32}P]phosphate. The methodology for this is discussed in *Chapter 1*.

2. Toxin is added, typically at a concentration of 1 μM to cause a total block of dephosphorylation, and incubation continued for a short period (e.g. 15 min). In isolated rat adipocytes, inhibition of protein dephosphorylation appears to be maximal within 1 min. For the lipophilic toxins, an equivalent amount of solvent should be added to controls, although the amounts of DMSO involved are unlikely to have any effects on their own.

3. Some biochemical or physiological parameters may be measurable in the intact cells. For effects which can only be measured in cell-free extracts, cells should be homogenized in the presence of additional protein phosphatase inhibitors, e.g. NaF (50 mM) and/or Na pyrophosphate (5 mM) (while the toxin added may be sufficient to block PP1, PP2A, or PP4, homogenization may expose substrates to toxin-insensitive protein phosphatases, e.g. PP2C). EDTA and EGTA should also be added to chelate Mg^{2+} and Ca^{2+} and thus block protein kinases and Ca^{2+}-dependent protein phosphatases, i.e. PP2B. A typical method is given in *Protocol 2*.

Protocol 2. Homogenization of toxin-treated, ^{32}P-labelled hepatocytes/adipocytes

Equipment and reagents
- homogenization medium (0.25 M mannitol, 2 mM EDTA, 1 mM EGTA, 50 mM NaF, 50 mM Tris–HCl, pH 7.4)
- bench-top centrifuge and ultracentrifuge

Method

1. Transfer cells to a plastic centrifuge tube and centrifuge for a few seconds at 3000 rpm in a bench-top centrifuge. Hepatocytes will form a cell pellet from which the medium may be poured off. Adipocytes will float to the top of the tube: remove the medium by aspiration using a piece of fine plastic tubing attached to a syringe. This can be done without disturbing the cell layer if the plastic tubing was placed into the cell suspension before centrifugation.

2. Resuspend the cells in 2.5 ml/g of homogenization medium. For hepatocytes, homogenize using 40 strokes of a tightly fitting ground glass (Dounce) homogenizer. For adipocytes, homogenize using a Polytron-type homogenizer (e.g. an Ystral T1500) at medium speed for 3–5 sec.

3. Preparation of cell fractions: prepare a post-mitochondrial supernatant by centrifugation (10 000 g; 15 min; 4°C). In the case of hepatocytes, this can be further fractionated by centrifugation (100 000 g; 60 min; 4°C). The supernatant represents the soluble fraction: the pellet may be separated into a lower, hard, white fraction (glycogen pellet) and an upper, buff-coloured, fluffy fraction (microsomal fraction). The microsomal fraction may be resuspended by gentle swirling in homogenization buffer without disturbing the glycogen pellet. The latter may then be resuspended by scraping it off and rehomogenizing in a Dounce homogenizer. These fractions can be washed by recentrifugation and resuspension.

6. Effects of inhibitors on protein phosphorylation in intact cells

The earliest studies of the effects of protein phosphatase inhibitors on protein phosphorylation were conducted using okadaic acid in ^{32}P-labelled, isolated rat hepatocytes and adipocytes (4), bovine chromaffin cells (32), and human fibroblasts (33). Hepatocytes were prelabelled with [^{32}P]phosphate for 60 minutes, treated with okadaic acid for 15 minutes, and cell fractions prepared

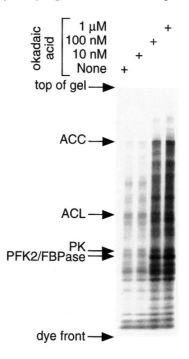

Figure 6. Phosphorylation of soluble proteins in response to different concentrations of okadaic acid in ^{32}P-labelled hepatocytes. The photograph shows an autoradiogram of proteins separated by SDS–PAGE. Phosphorylated polypeptides identified were: ACC, acetyl–CoA carboxylase; ACL, ATP–citrate lyase; PK, pyruvate kinase; PFK2/FBPase, 6-phosphofructo-2-kinase/fructose-2,6-bisphosphatase.

as described in *Protocols 1* and *2*. *Figure 6* shows analysis of the soluble fraction of hepatocytes by SDS–PAGE and autoradiography. This reveals that okadaic acid increases the phosphorylation of essentially all soluble phosphoproteins, an effect that was almost saturated at 100 nM okadaic acid. Analysis of ^{32}P-labelling of total protein in the soluble fraction, by precipitation with trichloroacetic acid (5%, w/v), indicates that okadaic acid increases phosphorylation 2.7-fold, with a half-maximal effect at ~200 nM and a maximal effect at ~1 μM. Similar results were obtained using isolated adipocytes, in which time courses also revealed that the effect of 1 μM okadaic acid was maximal within 1 minute (4). In isolated bovine chromaffin cells, 1 μM okadaic acid increased ^{32}P-labelling of total protein 2.7-fold, and phosphorylation of tyrosine hydroxylase 3.1-fold (32). Using lower concentrations of the toxin and longer incubation times (93 nM for 90 min), okadaic acid was shown to stimulate phosphorylation of a nucleolar protein in primary human fibroblasts (33).

Table 2. Examples of use of protein phosphatase inhibitors for functional studies

Cell type	Inhibitor	Effect	Reference
Smooth muscle	Okadaic acid	Contraction	2
Mouse skin	Okadaic acid	Tumour promotion	31
Rat hepatocytes	Okadaic acid	Stimulates gluconeogenesis	4
Rat adipocytes	Okadaic acid	Stimulates glucose uptake, lipolysis	4
Rat adipocytes	Okadaic acid	Inhibits fatty acid synthesis	4
Paramecium tetraurelia	Okadaic acid	Prolongs backward swimming	36
NIH-3T3 cells	Okadaic acid	Reversion of transformed phenotype	28
Spinach leaf	Okadaic acid	Inhibits sucrose phosphate synthase	30
Smooth muscle	Calyculin A	Contraction	6
Mouse 3T3 cells	Calyculin A	Shape/cytoskeleton changes	37
Sea urchin eggs	Calyculin A	Chromosome condensation	38
Rat macrophages	Dinophysistoxin-1	Production of prostaglandin E_2	39
Smooth muscle	Tautomycin	Contraction	40
Spinach leaf	Microcystin-LR	Inhibits sucrose phosphate synthase	30
Hepatocytes	Microcystin-LR	Shape/cytoskeleton changes	41
Aplysia sensory neurones	Microcystin-LR	Prolongs cAMP-induced currents	42
Guinea pig cardiac muscle	Cantharidin	Increased contraction	43
Mouse 3T3 cells	Endothall	Shape/cytoskeleton changes	44
CHO cella	Fostriecin	Cell cycle arrest	45
BHK cells	Fostriecin	Shape/cytoskeleton changes	46
Jurkat T cells	CsA/FK506	Inhibit interleukin-2 production	21
Mouse AtT20 pituitary cells	CsA/FK506	Agonist-induced cAMP production	47
Cytotoxic T lymphocytes	CsA/FK506	Inhibit degranulation	48
PC12 cells	CsA/FK506	Changes in gene expression	49
L929 cells	CsA	Inhibits TNF-induced NO production	50
Saccharomyces cerevisiae	CsA/FK506	Inhibit mating factor arrest recovery	51

7. Effects of inhibitors on physiological function in intact cells

Since the original observation that okadaic acid caused contraction of smooth muscle (2), protein phosphatase inhibitors have been shown to dramatically modulate many functional parameters in intact cells. This confirms that the elevation in ^{32}P-labelling discussed in Section 6 was due to real increases in protein phosphorylation, rather than being merely due to effects on phosphate turnover. Okadaic acid stimulates glucose output and gluconeogenesis in hepatocytes, and stimulates lipolysis and inhibits fatty acid synthesis in adipocytes (4). These are exactly the effects predicted if the toxin acts exclusively to increase protein phosphorylation, strengthening the view that it has no non-specific toxic effects, at least in these short-term incubations (15 minutes). Given the dramatic effects of okadaic acid on protein phosphorylation, it might be expected that all manner of toxic effects secondary to increases in protein phosphorylation would occur in long-term incubations. Okadaic acid has been incubated with NIH-3T3 cells at 10 nM for up to 10 days, and caused reversion from a transformed phenotype, but at higher concentrations it killed the cells (28).

Clearly the most exciting use of the protein phosphatase inhibitors is in obtaining evidence for novel roles for protein phosphorylation, and many such studies have now been performed. A small selection of published studies is presented in *Table 2*, which have been chosen to illustrate the range of inhibitors used and functional effects studied.

One point to bear in mind in this type of study is that okadaic acid can only be effective if the kinase whose effect it reverses is at least partially active. This is illustrated by the fact that inhibition of protein synthesis by okadaic acid in reticulocyte lysates only occurs in the presence of Ca^{2+}, because elongation factor-2 kinase is Ca^{2+}/calmodulin-dependent (34). Another example may be the effect of the toxin on glucose transport in adipocytes. Although okadaic acid quantitatively mimics the effect of insulin, it only does so after a lag of several minutes (4). Since inhibition of protein phosphatases occurs within 1 minute (see above), this time lag may indicate that the kinase(s) responsible for activating glucose transport is almost inactive in the absence of insulin, so that it takes some time for the phosphorylation to reach a new steady-state level.

References

1. Cohen, P.T. (1997). *Trends Biochem. Sci.* **22**, 245.
2. Shibata, S., Ishida, Y., Kitano, H., Ohizumi, Y., Habon, J., Tsukiyani, Y., and Kikuchi, H. (1982). *J. Pharmacol. Exp. Ther.* **223**, 135.
3. Bialojan, C. and Takai, A. (1988). *Biochem. J.* 256, 283.

4. Haystead, T.A.J., Sim, A.T.R., Carling, D., Honnor, R.C., Tsukitani, Y., Cohen, P., and Hardie, D.G. (1989). *Nature* **337**, 78.
5. MacKintosh, C. and Klumpp, S. (1990). *FEBS Lett.* **277**, 137.
6. Ishihara, H. *et al.* (1989). *J. Pharmacol. Exp. Ther.* **250**, 388.
7. Ishihara, H. *et al.* (1989). *Biochem. Biophys. Res. Commun.* **159**, 871.
8. MacKintosh, C., Beattie, K.A.,Klumpp, S., Cohen, P., and Codd, G.A. (1990). *FEBS Lett.* **264**, 187.
9. Yoshizawa, S., Matsushima, R., Watanabe, M.F., Harada, K., Ichihara, A., Carmichael, W.W., and Fujiki, H. (1990). *J. Cancer Res. Clin. Oncol.* **116**, 609.
10. MacKintosh, R.W., Dalby, K.N., Campbell, D.G., Cohen, P.T., Cohen, P., and MacKintosh, C. (1995). *FEBS Lett.* **371**, 236.
11. Goldberg, J., Huang, H.B., Kwon, Y.G., Greengard, P., Nairn, A.C., and Kuriyan, J. (1995). *Nature* **376**, 745.
12. Karras, D.J., Farrell, S.E., Harrigan, R.A., Henretig, F.M., and Gealt, L. (1996). *Am. J. Emerg. Med.* **14**, 478.
13. Li, Y.M. and Casida, J.E. (1992). *Proc. Natl. Acad. Sci. USA* **89**, 11867.
14. Li, Y.M., MacKintosh, C., and Casida, J.E. (1993). *Biochem. Pharmacol.* **46**, 1435.
15. Honkanen, R.E. (1993). *FEBS Lett.* **330**, 283.
16. Stampwalla, S.S., Bunge, R.H., Hurley, T.R., Willmer, N.E., Brankiewicz, A.J. Steinman, C. E., Smitka, T.A., and French, J.C. (1983). *J. Antiobiot.* (*Tokyo*) **36**, 1601.
17. de Jong, R.S., de Vrieas, E.G., and Mulder, N.H. (1997). *Anticancer Drugs* **8** 413.
18. Walsh, A.H. Cheng, A., and Honkanen, R.E. (1997). *FEBS Lett.* **416**, 230.
19. Braun, W., Kallen, J., Mikol, V., Walkinshaw, M.D., and Wuthrich, K. (1995). *FASEB J.* **9** 63.
20. Liu, J., Farmer, J.D., Jr, Lane, W.S., Friedman, J., Weissman, I., and Schreiber, S.L. (1991). *Cell* **66** 807.
21. Clipstone, N.A. and Crabtree, G.R. (1992). *Nature* **357**, 695.
22. Li, W and Handschumacher, R.E. (1993). *J. Biol. Chem.* **269**, 14040.
23. Husi, H., Luytel, M.A., and Zurini, M.G. (1994). *J. Biol. Chem.* **269**, 14199
24. Liu, J. *et al.* (1992). *Biochemistry* **31**, 3896.
25. Hastie, C.J. and Cohen, P.T. (1998). *FEBS Lett.* **431**, 357.
26. Chen, M.X., McPartlin, A.E., Brown, L., Chen, Y.H., Barker, H.M., and Cohen, P.T. (1994). *EMBO J.* **13**, 4278.
27. Boe, R., Gjertsen, B.T., Vintermyr, O.K., Houge, G., Lanotte, M., and Doskeland, S.O. (1991). *Exp. Cell Res.* **195**, 237.
28. Sakai, R., Ikeda, I., Kitani, H., Fujiki, H., Takaku, F., Rapp, U., Sugimura, T., and Nagao, M. (1989). *Proc. Natl. Acad. Sci. USA* **86**, 9946.
29. Eriksson, J.E., Gronberg, L., Nygard, S., Slotte, J.P., and Meriluoto, J.A. (1990). *Biochim. Biophys. Acta* **1025**, 60.
30. Siegl, G., MacKintosh, C., and Stitt, M. (1990). *FEBS Lett.* **270**, 198.
31. Suganuma, M. *et al.* (1988). *Proc. Natl. Acad. Sci. USA* **85**, 1768.
32. Haavik, J., Schelling, D.L., Campbell, D.G., Andersson, K.K., Flatmark, T., and Cohen, P. (1989). *FEBS Lett.* **251**, 36.
33. Issinger, O.G., Martin, T., Richyer, W.W., Olson, M., and Fujiki, H. (1988). *EMBO J.* **7**, 1621.
34. Redpath, N.T. and Proud, C.G. (1989). *Biochem. J.* **262**, 69.
35. Holmes, C.F., Luu, H.A., Carrier, F., and Schmitz, F.J. (1990). *FEBS Lett.* **270**, 216.

36. Klumpp, S., Cohen, P., and Schultz, J.E. (1990). *EMBO J.* **9**, 685.
37. Chartier, L., Rankin, L.L., Allen, R.E., Kato, Y., Fusetani, N., Karaki, H., Watabe, S., and Hartshporne, D.J. (1991). *Cell. Motil. Cytoskel.* **18**, 26.
38. Tosuli, H., Mabuchi, L., Fusetani, N., and Nakazawa, T (1992). *Proc. Natl. Acad. Sci. USA* **89**, 10613.
39. Ohuchi, K., Tamura, T., Ohashi, M. Watanabe, M., Hirasawa, N., Tsurufuji, S., and Fujiki, H. (1989). *Biochim. Biophys. Acta* **1013**, 86.
40. Hori, M., Magae, J., Han, Y.G., Hartshorne, D.J., and Jaraji, H. (1991). *FEBS Lett.* **285**, 145.
41. Eriksson, J.E., Toivola, D., Meriluoto, J.A., Karaki, H., Han Y.G., and Hartshorne, D. (1990). *Biochem. Biophys. Res. Commun.* **173**, 1347.
42. Ichinose, M., Endo, S., Critz, S.D., Shenolikar, S., and Byrne, J.H. (1990). *Brain Res.* **533**, 137.
43. Neumann, J., Herzig, S., Boknik, P., Apel, M., Kaspareit, G., Schmitz, W., Scholz, H., Tepel., M., and Zimmermann, N. (1995). *J. Pharmacol. Exp. Ther.* **274**, 530.
44. Erdodi, F., Toth, B., Hirano, K., Hirano, M., Hartshorne, D.J., and Gergely, P. (1995). *Am. J. Physiol.* **269**, C1176.
45. Cheng, A., Balczon, R., Zuo, Z., Koons, J.S., Walsh, A.H., and Honkanen, R.E. (1998). *Cancer Res.* **58**, 3611.
46. Ho, D.T. and Roberge, M. (1996). *Carcinogenesis* **17**, 967.
47. Antoni, F.A., Barnard, R.J., Shipston, M.J., Smith, S.M., Simpson, J., and Paterson, J.M. (1995). *J. Chem.* **270**, 28055.
48. Dutz, J.P., Fruman, D.A., Burakoff, S.J., and Bierer, B.E. (1993). *J. Immunol.* **150**, 2591.
49. Enslen, H. and Soderling, T.R. (1994). *J. Biol. Chem.* **269**, 20872.
50. Fast, D.J., Lynch, R.C., and Leu, R.W. (1993). *J. Interferon Res.* **13**, 235.
51. Foor, F., Parent, S.A., Morin, N., Dahl, A.M., Ramadan, N. Chrebet, G., Bostian, K.A., and Nielsen, J.B. (1992). *Nature* **360**, 682.
52. Devereux, J., Haeberli, P., and Smithies, O. (1984). *Nucl. Acids Res.* **12**, 387.

4

Analysis of signal transduction pathways using molecular biological manipulations

CALUM SUTHERLAND

1. Introduction

This chapter describes the approaches currently available to generate molecular biological tools for the dissection of intracellular signalling pathways, together with protocols required for their application. Many standard techniques of molecular biology, which have been previously documented, are required. In such cases, specific examples are provided with reference to alternative protocols.

Signalling pathways require three general components: a receptor system, an information interpretation and transfer system, and an ultimate acceptor component (*Figure 1*). Most receptor complexes lie at the cell surface, enabling the cell to respond to factors in its extracellular environment (e.g. hormones, nutrients, cell–cell contacts). However, there are also intracellular receptors, such as steroid hormone receptors, that can bind to agents that can permeate the plasma membrane. The molecular mechanisms underlying the interpretation and transfer of signals are becoming clearer, but much remains to be elucidated. Molecules accepted to have major roles in signal transduction include second messengers (e.g. cAMP, cGMP, diacylglycerol, ceramide), protein kinases (e.g. MAP kinases, cAMP-dependent protein kinase), protein phosphatases (e.g. protein phosphatases-1, -2A, calcineurin), small G-proteins (e.g. p21Ras, Rac, Rho), heterotrimeric G-proteins, adaptor molecules (e.g. Grb2, Shc, IRS1) and lipid kinases (e.g. phosphatidylinositol 3-kinase). Finally, the ultimate acceptor for a given signalling pathway must interpret the signal and adapt its function accordingly. For example, a cytoplasmic substrate for a regulatable protein kinase may translocate to the nucleus upon stimulation, resulting in a shift in the spectrum of action of that target protein.

The flow of information through such a system has generally been regarded to be a simple linear series of events, but it is becoming clear that signalling *in*

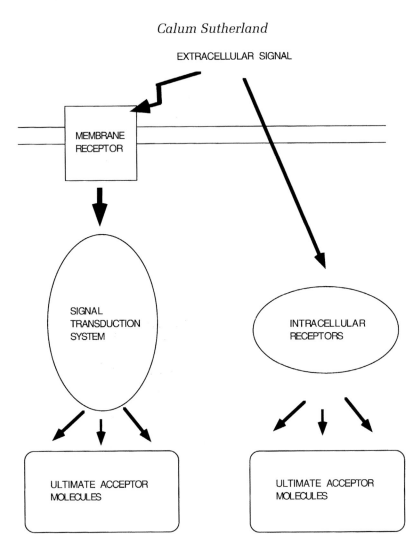

Figure 1. Schematic representation of typical signalling cascades.

vivo is much more complex than this, which must be kept in mind when designing tools for dissection of signalling pathways. For example, multiple distinct signalling systems may utilize common downstream signalling molecules (*convergence*). In addition, downstream components of the pathway can affect the activity of upstream regulators (*feedback*). Similarly, the activation of one signalling pathway can affect the ability of a second pathway to respond to a distinct signal (*cross-talk*, e.g. between the cAMP and MAP kinase pathways). A further level of complexity is created through the existence of multiple isoforms of signalling molecules. This can give rise to the phenomenon of *redundancy*; i.e. the ability of a related molecule to compensate for the loss of a given molecule. The existence of redundancy can reduce the success of experiments designed to deplete cells of a specific signal-ling component. In addition, many protein kinases and protein phosphatases show a high level of substrate promiscuity in cell-free assays, and when over-expressed in cells. Thus, care must be taken when interpreting such experi-ments. In order to understand fully the role of a given molecule or pathway, the employment of several distinct approaches is mandatory.

In the remainder of this chapter, techniques for designing molecular tools that can be used to determine the role of a given molecule in a signalling system will be discussed, as will the methods required for the application of these tools.

2. Recombinant DNA constructs for manipulation of intracellular signalling

2.1 General considerations

There are two general approaches taken to investigate the role of a protein in a signalling pathway:

- removal of the function of the signalling protein, to establish whether it is *necessary* in the signalling pathway
- artificial induction of the function of the protein, to determine whether it can replace upstream components, and to determine if its activation is *sufficient* to switch on the pathway

The activity of a molecule can be interfered with by expression of a dominant negative mutant, by introduction of specific inhibitors or inhibitory proteins, or by 'knocking out' the expression of the molecule. Conversely, the function of a molecule can be induced artificially by overexpressing the molecule, an activator of the molecule, a mutant form of the molecule exhibit-ing constitutive activity, or by depleting the cell of an inhibitory factor.

2.2 Inhibition of signalling pathways

The generation of dominant negative mutants involves the design of an inactive form of the protein that can 'sequester' interacting proteins, thus preventing these proteins from binding the endogenous protein (*Table 1*). Some knowledge of the mechanism of regulation or function of the protein of interest is helpful when designing these molecules. However, there are some general approaches that have produced effective dominant negative mutants (1–7). For example, the activity of protein kinases requires the binding of ATP to a conserved protein motif (*Figure 2*) which can be blocked by mutation of a single amino acid (*Table 1*). Overexpression of an inactive form of the kinase (unable to bind ATP) may act as a dominant negative mutant by sequestering interacting proteins and thus inhibiting the activity of the endogenous wild-type kinase.

Table 1. Different types of dominant negative forms of signalling molecules

Mutation	Example	Reference
Point mutation	Kinase dead: K, E, or D required for ATP binding or catalytic activity mutated to A	1 and *Fig. 2*
	Inactivatable kinase: mutation of phosphorylation site required for activity, e.g. S/T to A in PKA, GSK3	10, 11
	Mutation causes conformational change, e.g. Ser to Asn in p21 ras (N17)	27
Truncation	e.g. N-terminal fragment of PI 3-kinase	3
(lacking full active site)	DNA-binding domain of transcription factors, e.g. c-*jun*	2, 4
Overexpression of inhibitory protein/subunit	e.g. truncated IκB subunit: inhibits activation of NFκB	8
Overexpression of pseudosubstrate	e.g. protein kinase C	5–7

ATP-binding region of protein kinases

	subdomain II	subdomain III	subdomain VIB
PKA-Cα	TGNHYAM**K**ILDKQKVVKLK	QIEHTLN**E**KRILQAV	YR**D**LKPENLL
PKCα	TEELYAI**K**ILKKDVVIQDD	DVECTMV**E**KRVLALLD	YR**D**LKPENLI
GSK3α	TRELVAI**K**KVLQ	DKRFKNR**E**LQIMRKL	HR**D**IKPQNLL
CONSENSUS	----oAoK-o-------	-----E--oo---	-RDoKP-Noo

Figure 2. ATP-binding domains of protein kinases. Primary sequences of subdomains II, III, and VIB are presented, showing (in bold) the invariant residues found in all protein kinases. These residues are equivalent to Lys-72, Glu-91, and Asp-166 in cAMP-dependent protein kinase-Cα. These are key residues in the proper binding and positioning of ATP (Lys-72, Glu-91) and in catalysis (Asp-166). Mutation of any of these residues produces protein with very little, if any, kinase activity. The consensus sequence for these subdomains is also shown: o = non-polar residue.

Many signalling molecules are present in cells in an inactive form, becoming active upon stimulation of the appropriate signalling pathway. Inhibition is often achieved by interaction of the molecule with a regulatory protein or subunit, or an inhibitory domain within the same polypeptide (e.g. the pseudosubstrate sequences found in some protein kinases). Thus, in some cases, overexpression of an inhibitory domain or subunit can reduce or inhibit the ability of the pathway to stimulate the endogenous protein. For example, overexpression of IκB, a regulatory subunit of NFκB, severely impedes the translocation of this transcription factor to the nucleus (8).

Similarly, overexpression of a substrate analogue (*pseudosubstrate*) that still binds the enzyme but cannot be converted to product, can often result in inhibition of signalling, as it will compete against the endogenous substrate (*Table 1*). Also, single amino acid mutations of signalling molecules can cause conformational changes in protein structure that result in loss of apparent function without apparently affecting binding to downstream partners. For example, a point mutation in the small G-protein p21Ras at residue 17 (result-ing in replacement of serine with asparagine, N17Ras), produces a dominant negative form which can still interact with upstream signalling molecules such as mSOS, but does not transmit the signal to downstream components such as c-Raf. Other point mutations of p21Ras have been shown to lead to con-stitutive activation of the molecule (see below) and these have been very useful in determining the role of this protein. These were initially identified as being disease-causing mutants (e.g. oncogenic forms), or were produced by random mutation followed by analysis in cell-free assays.

The use of chemical inhibitors to block protein function is becoming more common (see Chapters 2 and 3). Initially, there was some scepticism as to the potential for specificity in such an approach. However, it is becoming clear that selective inhibitors can add greatly to the understanding of signalling systems, especially when used in conjunction with the molecular biological approaches discussed in this chapter. The growing involvement of the pharmaceutical industry in the design and generation of specific inhibitors has already led to insights into the molecular mechanism underlying the inhibition of signalling molecules. For example, a family of pyridinyl imidazoles can specifically inhibit the protein kinases SAPK2α and SAPK2β, but not the highly related SAPK1 (JNK), SAPK3, and SAPK4 (9). The basis for this selectivity has been largely ascribed to a single amino acid within subdomain V of the kinase domains, i.e. threonine-106 (SAPK2α). Although the kinase domain is highly conserved between all serine/threonine protein kinases, the equivalent residue in SAPK1/JNK, SAPK3, and SAPK4 is methionine rather than threonine. Replacing the methionine with threonine in any of these three SAPK isoforms produces a protein kinase with equivalent drug sensitivity to SAPK2α and SAPK2β (9). The particular residue located at this position in the ATP-binding pocket therefore appears to be diagnostic for sensitivity to this class of compound. Potentially, a mutated form of any protein kinase could be produced, exhibiting sensitivity to these drugs. This would allow examination of the requirement of a protein kinase in a given signalling system, assuming the mutant has normal wild-type activity in the absence of the drug. As more inhibitor molecules are characterized, it should become possible to design mutant proteins with a distinct spectrum of sensitivity to a

Table 2. Different types of gene knock-out approaches

Type	Target	Reference
Antisense	Protein kinase C-α	12, 45
	c-raf	13, 46
Ribozyme	p21ras	45
	p300	53
	bcr-abl	14
	mdr-1	15
	IRS1	16
	c-myb	52, 55
Gene targetting	IRS1	17

variety of inhibitors, aiding in the search for specific function of proteins of interest *in vivo*.

The activity of numerous proteins (including signalling molecules) is regulated by phosphorylation on serine, threonine, or tyrosine residues. Thus, mutation of the phosphorylation site to a residue that is not a target for protein kinases (e.g. glycine or alanine) produces a mutant that should no longer respond to the signalling pathway. However, such mutants often behave as activators of the pathways when overexpressed, since they may contain some low, but finite, activity. Alternatively, if the phosphorylation of the protein is an absolute requirement for activity [e.g. Tyr-215 in GSK3β, or Thr-198 in cAMP-dependent protein kinase (10, 11)], then these mutants are 'kinase-dead' and can act as dominant negative molecules (*Table 1*).

A distinct approach to inhibiting signalling pathways is to deplete a protein by antisense or gene targetting (*Table 2*) (12–17). These approaches rely on the target molecule not being essential for viability of the cell or organism. The use of antisense technology is discussed in Section 5, while gene targetting methodology is the subject of another volume in this series (18).

2.3 Induction of signalling by expression

The function of a signalling molecule can be introduced into a cell by expression of a constitutively active form of the molecule (*Table 3*). Once again, some knowledge of the molecular mechanisms of function/regulation of the protein can aid in the design of inherently active molecules. Overexpression of wild-type proteins can often produce increased activity, even in the absence of the inducing signal. However, this protein will be regulated in a similar fashion to the endogenous molecule and thus may not be as effective as a mutant that contains activity that cannot be down-regulated. In addition, overexpression of wild-type proteins may cause them to act as partial dominant negative molecules if their activity is low. It is therefore usually more effective to design a mutant protein that contains inherent constitutive activity. Examples of active forms of signalling molecules (*Table 3*) include truncated forms that

Table 3. Different types of constitutively active forms of signalling molecules

Mutation/treatment	Example	Reference
Truncations		
Lacking inhibitory	cAMP-dependent protein kinase	1
domain or subunit	MEKKc	19, 20
	Tyrosine kinase receptors, e.g. EGF/PDGF receptor	21
Point mutations		
Regulated S/T/Y to D/E	PKB T308 and S473 to D	31
Conformational change:	e.g. p21Ras/Rac	27, 29; 30
	p21Ras (V61)-active	
	p21Ras (V12)-active	
	p21Rac (V12)-active	
	p21Ras (V12 S35)-activates Raf only	
	p21Ras (V12 E38)-activates Raf only	
	p21Ras (V12 G37)-Ral-GDS only	
	p21Ras (V12 C40)-activates PI 3-kinase	
Addition of membrane localization motifs		
	CAAX motif, from H-Ras	
	Myristoylation motif (e.g. PKB, p110)	32
	Gag, from retrovirus (e.g. PKB)	22
Addition/expression of upstream activator		
Second messenger	AICA riboside (AMP-activated protein kinase)	23
	cAMP analogue (cAMP-dependent protein kinase)	24
Kinase kinase	MEKK	25
	AMPKK	26

lack inhibitory/regulatory domains or subunits, and single/multiple amino acid mutations that gain activity due to a conformational change in the structure of the protein (e.g. that mimic a post-translational modification) (19–26). Point mutants of p21Ras and Rac have been found that contain the ability to transform cells due to an inherent constitutive activation of the molecule [e.g. p21Ras and Rac, with valine replacing glycine at residue 12 (V12), and p21Ras, with valine replacing glutamate at position 61 (V61)] (27, 28). Interestingly, point mutations of p21Ras have been isolated that contain constitutive activity towards one downstream signalling pathway but not another [e.g. V12/S35–Ras (27), V12/E38–Ras (29), V12/G37–Ras (27) and V12/C40–Ras (30)].

Proteins whose activity is induced by phosphorylation of serine, threonine, or tyrosine may exhibit constitutive activity when the residue responsible for mediating the regulation (following phosphorylation) is mutated to a negatively charged residue such as aspartate or glutamate. The negative charge presumably mimics the charge of the phosphate group. For example, mutation of Thr-308 and Ser-473 of protein kinase B to aspartate residues produces an active form of PKB (see ref. 31 for review).

Certain signalling molecules are activated by translocation from one cellular compartment to another, so constitutively active molecules can be produced by fusing a specific localization signal to the wild-type protein and overexpressing the fusion protein. For example, the protein kinase c-Raf is recruited to the cell membrane following the activation of p21Ras, and an active form of c-Raf can be produced by fusion with a membrane localization motif, such as CAAX (from H-Ras) or the viral protein Gag. This is true of other signalling molecules such as PI 3-kinase (28) or PKB (32). These mutants will exhibit activity in a prolonged fashion throughout their expression, as opposed to endogenous signalling molecules that are usually activated more transiently. Thus, interpretation of the results of this type of approach must be conducted with some caution (Section 3.3). One method to overcome this problem is the use of inducible expression systems, where the active mutant is only produced (and therefore active) when the inducing agent is present.

3. Expression of recombinant DNA in living cells

3.1 Introduction of DNA into cultured cells

The cell membrane is the first barrier to manipulation of signalling pathways when using cultured cells. DNA constructs must be transported through this barrier, and into the nucleus, in order for expression of the molecule to occur. There are several well-established techniques that allow transient transfection of recombinant DNA of cells in tissue culture. These methods generally involve the permeabilization of cell membranes by chemical or electrical means, or the use of viral constructs that can recognize specific receptors on the cell surface, resulting in cellular uptake.

3.1.1 Transient transfection of naked DNA

There are at least three methods available for transient transfection of naked DNA into cultured cells:

(a) *Calcium phosphate precipitation (Protocol 1)*, where precipitation of the DNA using calcium phosphate, and its application to target cells, results in attachment of the DNA to the cell surface and subsequent uptake into the cell. The permeabilization of the cells using either dimethyl sulfoxide, or glycerol shock, results in an increased efficiency of DNA uptake. The degree of success of this method is cell-type dependent, with optimization of the protocol recommended for each cell-type being studied.

(b) *Liposome-mediated transfection (Protocol 2)*, where a mixture of cationic and neutral lipid molecules transport the DNA through the cell membrane.

(c) *Electroporation (Protocol 3)*, where electrical stimulation causes opening of pores in the membrane, allowing diffusion of the DNA into the cell.

Electroporation appears to be less cell-type dependent than other methods, possibly because it involves a passive uptake of the DNA. However, this is not the method of choice when co-transfection experiments are planned, because transfer of more than one DNA construct is not as efficient as with the calcium phosphate precipitation protocol.

Protocol 1. Transient transfection using calcium phosphate precipitation

Equipment and reagents

- exponentially growing cells
- tissue culture medium
- trypsin/EDTA
- phosphate-buffered saline (PBS)
- 2.5 M $CaCl_2$ (183.7g $CaCl_2.H_2O$ + H_2O to 500ml: sterilize through 0.45 μM filter, store in aliquots at −20°C)

- purified DNA[a]
- 2 × Hepes buffer[b] (8.2 g NaCl, 5.95 g Hepes, 0.105 g Na_2HPO_4 + 400 ml H_2O; pH to 7.05 with NaOH, take to 500 ml with H_2O and sterilize through a 0.45 μM filter, store in aliquots at −20°C)

Method

1. Preparation of cells:

    1.1 *Adherent cells:* cells are seeded in 10 cm dishes 24–48 h prior to transfection, cells should be 50–70% confluent at the time of transfection.

    1.2 *Cells in suspension:* cells that grow in suspension are harvested by centrifugation. Some adherent cell lines can also be transfected more efficiently in suspension as follows. Cells are seeded 24 h prior to transfection to give 70–80% confluence at the time of harvest. Cells are washed with PBS and harvested by incubation with trypsin/EDTA for 1–5 min at 37°C. Trypsin is neutralized by addition of excess serum-containing medium and the cells are pelleted by centrifugation at 1000 *g* for 5 min.

2. *DNA precipitation:* the DNA(s) to be transfected (2.5–20 μg per DNA construct, to a maximum of 50 μg per dish) are ethanol precipitated (add 0.1 vol 3 M Na acetate and 2 vol ethanol). This is not an essential step, but it serves to sterilize the DNA. Resuspend the DNA in 450 μl of sterile water and add 50 μl of 2.5 M $CaCl_2$. Add the DNA solution to 500 μl of 2 × Hepes buffer dropwise (aerating the Hepes can increase the transfection efficiency). This mixture is left at room temperature for 20 min.

3. Cell treatment:

    3.1 *Adherent cells:* remove medium, and place 4 ml of complete medium on to each 10cm dish. Pipette 0.5 ml of the DNA solution on to the dish evenly, and mix by swirling. The cells are left at 37°C for 4 h.

Protocol 1. *Continued*

3.2 *Cells in suspension:* the DNA solution (1 ml per 10^6 cells) is added to the cells which have been pelleted as in step 1.2, and the mixture pipetted up and down several times. After 15 min at room temperature, 1 ml of this mixture is added to a 10 cm dish containing 9 ml of complete medium, the dish is swirled and left at 37 °C for 4 h.

4. *DMSO/glycerol shock (optional):* the transfection efficiency of many cell types can be very low ($<0.1\%$), but this can be improved by a 'DMSO/glycerol shock' of the cells 4 h after DNA treatment. Remove the medium and replace with 5 ml of either 20% DMSO or 10% sterile glycerol in complete medium.[c] After 5 min at room temperature the DMSO or glycerol medium is removed, the cells are washed once with 5 ml complete medium, and once with 5 ml of PBS. Finally, add 10 ml of complete medium.

5. *Harvesting:* cells are harvested at the desired time after transfection. The timing is dependent on the cell type and particular DNA constructs transfected (i.e. on the promoter and reporter system employed). Standard transient transfection incubation times are 20–48 h. Cells can be harvested by trypsin/EDTA treatment, as above, or by addition of a detergent-containing lysis buffer, dependent on the analysis required.

[a] DNA for transfection should be purified by CsCl banding or by use of commercially available kits.
[b] Alternative buffer systems can be used [e.g. 2 × BES-buffer: *N,N*-bis(2-hydroxyethyl)-2-aminoethanesulfonic acid (BES, 50 mM), NaCl (280 mM), Na$_2$HPO$_4$ (1.5 mM), pH 6.95]
[c] the DMSO should be added to cold medium, as the temperature will increase following DMSO addition.

Protocol 2. Transient transfection using electroporation

Equipment and reagents

- cells, medium, and DNA, as for *Protocol 1*
- electroporation buffer:[a]
either phosphate-buffered saline (no calcium or magnesium ions)
or Hepes-buffered saline (HBS; 8.2 g NaCl, 5.95 g Hepes, 0.105 g Na$_2$HPO$_4$ + 400ml H$_2$O; pH to 7.05 with NaOH, dilute to 1000 ml with water and sterilize through 0.45 μM filter, store in aliquots at −20°C)

or tissue culture medium (no serum)
or phosphate-buffered sucrose [1mM MgCl$_2$, 272mM sucrose, 7mM K$_2$HPO$_4$, pH 7.4 (with phosphoric acid)]
- electroporation cuvettes (e.g. EquiBio)
- electroporation device (e.g. EquiBio)

Method

1. *Cell culture:* as in *Protocol 1*, cells should be exponentially growing prior to transfection. Cells are harvested by centrifugation (cells grown in suspension) or by addition of trypsin/EDTA as in *Protocol 1*, step 1. For each transfection 10^6–10^7 cells are required.

2. *Electroporation*: the cell pellet is resuspended in 5 ml of ice-cold electroporation buffer and then the cells reharvested by centrifugation. The pellet is resuspended at 10^7–10^8 cells/ml in ice-cold electroporation buffer, and 0.5 ml aliquots transferred to sterile electroporation cuvettes on ice. DNA(s) for transfection are added to each aliquot (10–40 μg of purified DNA). Pipette up and down briefly and incubate on ice for 5 min. Place the cuvette in the electroporator and shock one or more times at the desired voltage and capacitance settings. Standard settings are 1200 V/25 μF, but this is dependent on cell type. Following shock treatment, the cuvette is placed on ice for 10 min.

3. *Replating*: the cells are resuspended in 10 ml of complete medium (wash out the cuvette with an aliquot of this medium). Cells are placed in a 10 cm dish and incubated for 24–48 h prior to harvesting and assay (as in *Protocol 1*, step 5).

[a] The choice of electroporation buffer depends on the cell type being employed. Most cell types will transfect equally well with any of the above buffers, but some (including primary cells) require lower voltage and higher capacitance settings (250 V/ 960 μF) compared with the standard settings (1200 V/25 μF). In these cases, tissue culture medium, phosphate-buffered sucrose or HBS is recommended.

Protocol 3. Transient transfection by lipofection

Equipment and reagents

- cells, medium, and DNA, as for *Protocol 1*
- liposome suspension [e.g. commercially available kits such as Lipofectin (Gibco)]
- polystyrene containers (lipid–DNA complexes bind to polypropylene)
- 2 × serum-free tissue culture medium

Method

1. *Cell culture:* as in *Protocol 1*, cells should be exponentially growing prior to transfection. Cells are seeded in 10 cm dishes 24–48 h prior to transfection and should be 50–80% confluent at the time of transfection.

2. *Liposome–DNA suspension:* DNA(s) should be mixed with the liposome suspension as per the manufacturer's instructions: a ratio of 1:10 (w/v) DNA to liposomes is standard. As with the previous protocols, 2–50 μg of total DNA can be transfected per 10 cm dish. The mixture is vortexed briefly and incubated at room temperature for 10 min.

3. *Transfection:* remove medium from the cells, wash once with 5 ml serum-free medium (or PBS) and add 5 ml of liposome–DNA suspension. Incubate for 3–5 h at 37 °C. Add 5 ml of 2 × medium containing serum and incubate for a further 16–20 h at 37 °C. Remove the medium

Protocol 3. *Continued*

and replace it with 10 ml of medium containing serum. Incubate for 20–48 h.

4. *Harvesting:* cells are harvested and assayed as in *Protocol 1*, step 5.

3.1.2 DNA transfer using viral vectors

Recently, a variety of viral systems have become available, including adenoviruses (*Protocols 4–7*) and retroviruses, that will carry recombinant DNA into cells (33). The DNA can either be incorporated into the viral genome or can be chemically linked to the exterior of the virion. The virus enters the target cell via membrane receptors, permitting expression of the recombinant DNA(s). The success of this technique requires the presence of receptors for the virus but, at least in the case of adenovirus, these receptors appear to be relatively common across a wide spectrum of cell types. Several different but related methods and vectors are available for preparation of recombinant adenoviruses. The most common methodologies are described in *Protocols 4–7* and are reviewed in detail elsewhere (34). More recently, a bacterial system has been developed that may prove very useful in the preparation of recombinant adenovirus (35).

Protocol 4. Preparation of recombinant adenovirus

Equipment and reagents

- purified DNA
- adenovirus genome vector (e.g. pJM17[a])
- shuttle vector (e.g. pACCMV[a])
- appropriate restriction enzymes
- embryonic kidney cells (293 or 911)
- tissue culture medium
- phosphate-buffered saline (PBS)

Methods

1. *Subcloning:* the DNA intended for transfection is subcloned into the pACCMV shuttle vector by conventional cloning. The vector contains a bacterial origin of replication, promoter, and polyadenylation sequences that allow growth and expression of the recombinant plasmid following subcloning. Once subcloned the recombinant vector should be amplified and purified. This can be carried out by standard $CsCl_2$ banding or using commercial kits (e.g. Qiagen or Wizard).

2. *Cell culture:* ten dishes (60 mm) of 293 (or 911) cells are seeded to provide 60% confluency on the day of vector transfection.

3. *Vector transfection:* using calcium phosphate precipitation (*Protocol 1*) 4 μg of pJM17 and 4 μg of recombinant pACCMV are co-precipitated in a total of 5 ml (the Stratagene MBS Mammalian Transfection Kit, cat. no. 200388 can be used for this purpose). Fresh complete medium

(2 ml) is placed on the cells and 0.5 ml of the precipitated DNA is added to each of the 10 dishes. The dishes are swirled and incubated for 3 h at 37°C. The cells are washed twice with PBS and then incubated at 37°C in 3 ml of complete medium.

4. *Identification of virus:* complete medium should be replaced every 3–4 days. It is important to remove only 70–80% of the liquid when replenishing the medium. If recombination occurs, competent virions will produce cell lysis within 7–10 days of transfection. Virus is harvested by pipetting the lysed cells and infected medium into a 5 ml tube and freezing. Two freeze–thaw cycles releases any virus in incompletely lysed cells. The cell debris is removed by centrifugation at 1000 *g* for 5 min. This lysate usually has a titre of around 10^7–10^8 pfu/ml and can be used as a stock to confirm the identity of the recombinant virus (*Protocol 7*), isolate a clone of the recombinant virus (*Protocol 6*), and prepare high titre virus (*Protocol 5*).

a A full description of these plasmids is given by Becker *et al.* (34): alternative plasmids and modified protocols are also available (35). The quality of the pJM17 DNA preparation is very important to the success of the recombination event. *Hind*III digestion of pJM17 should produce a DNA ladder of 8 or 9 bands with no fragments less than 2 kb.

Protocol 5.   Preparation of high titre recombinant adenovirus

Equipment and reagents
- virion stock of interest (10^7–10^8 pfu/ml)
- embryonic kidney cells (293 or 911)
- DMEM (Dulbecco's modification of Eagle's medium)
- FCS (fetal calf serum)
- phosphate-buffered saline (PBS)
- NP40 (Nonidet P-40, 10% v/v)
- CsCl (solid)
- PEG/NaCl (20% (w/v) polyethylene glycol 8000, 2.5 M NaCl)
- ultracentrifuge
- PD10 Sephadex columns (Pharmacia)
- gel filtration buffer (10 mM Tris, pH 7.4, 137 mM NaCl, 5 mM KCl, 1 mM $MgCl_2$; sterilize by passing through a 0.45 μm filter)
- BSA [0.1% (w/v), sterile]

Method

1. *Cell culture:* plate out 20 dishes (150 mm) or 20 × T150 flasks of 293 (or 911) cells in DMEM containing 10% FCS. The cells should reach 70–80% confluence prior to infection.

2. *Infection:* wash the cells once with 15 ml PBS. Dilute the viral stock to 10^6–10^7 pfu/ml in DMEM containing 2% FCS, place 10 ml into each dish/flask, and incubate at 37°C. Cells should lyse within 36–50 h.

3. *Harvesting:* once lysis has occurred, add 0.5 ml of NP40 to each dish/flask, mix well, and combine lysates. Pellet cell debris by centrifugation at 15000 *g* for 10 min at room temperature.

4. *Purification:* add 0.5 vols of PEG/NaCl to the cleared lysate, mix well,

Protocol 5. *Continued*

and incubate on ice for at least 30 min. Centrifuge at 15 000 *g* for 10 min at 4°C, and discard the supernatant. Resuspend the viral pellet, by pipetting up and down in 7 ml of PBS and centrifuge at 5000 *g* for 10 min at room temperature. Add 3.5 g of CsCl to the supernatant making sure the CsCl dissolves completely, and then centrifuge at 40 000 *g* for at least 4 h. Remove the white adenovirus band using a needle and syringe (as with DNA purification techniques) and desalt using a PD-10 column equilibrated in gel filtration buffer according to the manufacturer's instructions. Fractions of 0.5 ml are collected.

5. *Approximate determination of virus titre:* an aliquot of each fraction is diluted 50-fold and the A_{260} measured (as an approximation of titre, 1 OD_{260} = 10^{11} pfu/ml). Peak fractions are made up to 1mg/ml with BSA and stored at –80°C in 50–100 µl aliquots. The virus titre can be functionally determined by the plaque assay (*Protocol 6*).

Protocol 6. Measurement of titre and isolation of individual clones by the plaque assay

Equipment and reagents
- virion stock of interest
- embryonic kidney cells (293 or 911)
- DMEM (Dulbecco's modification of Eagle's medium)
- FCS (foetal calf serum)
- phosphate-buffered saline (PBS)
- agarose (Aldrich, molecular biology grade)
- 2 × DMEM [DMEM powder (Gibco No. 52100) to half the normal volume, plus 4% FCS, 200 units/ml penicillin/streptomycin and 30 mg/l phenol red]
- Giemsa's stain (Sigma GS500)

Method
1. *Cell culture:* for each virus to be titrated prepare 15 × 60 mm dishes of 80% confluent 293 (or 911) cells. Wash with PBS on day of infection.
2. *Infection:* dilute the stock virus solution 10^4, 10^5, 10^6, 10^7, and 10^8-fold, in a total of 6 ml complete medium. Place 2 ml of each dilution on 3 × 60 mm dishes of cells. Incubate for 2 h at 37°C. Aspirate and gently wash twice with PBS.
3. *Agar application:* boil a solution of 1.3% agarose, cool, and reboil (boiling twice kills fungal spores). Mix the melted agarose with an equal volume of 2 × DMEM (pre-warmed to 40°C) (at least 100 ml total) and incubate at 40°C for 10 min (the agarose should not solidify). Aspirate the final PBS wash from the infected cells and gently add 6 ml of the agarose mix to each plate (care should be taken to ensure the agarose does not set prior to completing all 15 dishes). Allow the agarose to solidify at room temperature and incubate at 37°C for 7–10 days.

4. *Plaque counting or clone isolation*:

either (a) *Plaque counting:* carefully remove the agarose and add enough Giemsa's stain to cover the cells. Incubate at room temperature for at least 4 h and then wash several times with large volumes of water and allow to dry. Count the clear plaques, divide by two, and multiply by the appropriate dilution to obtain the plaque-forming units per ml (pfu/ml).

or (b) *Clone isolation:* carefully take a core (using a sterile pipette tip) of agarose from the centre of an isolated plaque (a small region of cell lysis, viewed in the light microscope). The virions are eluted from the core by multiple freeze–thaw cycles in 2 ml of DMEM containing 10% FCS. The 2 ml of medium is used to infect a 60 mm dish of 293 cells (80% confluent). After 2 h the cells are washed with PBS and 2 ml of DMEM containing 10% FCS is added. Cell lysis will occur after 5–7 days. This will be a relatively low titre solution (about 10^7 pfu/ml) but can be amplified as in *Protocol 5*. Viral DNA can be isolated as described in *Protocol 7*, and restriction digests and/or Southern blotting used to identify the success of the cloning.

Protocol 7.   Isolation of viral DNA

Equipment and reagents

- virion stock of interest
- embryonic kidney cells (293 or 911)
- DMEM (Dulbecco's modification of Eagle's medium)
- FCS (foetal calf serum)
- complete medium [DMEM containing 10% (v/v) FCS]
- phosphate-buffered saline (PBS)
- lysis buffer (100μg/ml proteinase K, 0.6% SDS, 10mM EDTA)
- inoculation loop
- sodium acetate (3 M)

- phenol/chloroform (phenol (containing 1g/l of 8-hydroxyquinoline) is equilibrated to pH 8 by addition of an equal vol of Tris base (pH 10.5), stirred for 10 min, and allowed to stand for 30 min at room temperature. The upper, aqueous phase is removed and an equal vol of 50 mM Tris/HCl, pH 8.0, added. The bottom phenol phase (25 ml) is mixed with 24 ml chloroform and 1 ml isoamyl alcohol
- ethanol (95%)
- 5 M NaCl (sterile)

Method

1. *Cell culture:* for each viral DNA preparation plate out 2 × 60 mm dishes of 293 (or 911) cells in complete medium. The cells should reach 70–80% confluence prior to infection.

2. *Infection:* infect both plates with 10–50 μl of viral lysate (10^8–10^9 pfu/ml), in a final volume of 2 ml complete medium. Incubate for 2–3 h at 37°C, wash twice with PBS, and replace with 2 ml of complete medium.

3. The cells will become round and refractile, but will still be adherent to

Protocol 7. *Continued*

the dish 24–48 h later. The medium is removed and the adherent cells washed carefully with PBS. Remove the PBS completely.

4. *Cell lysis:* add 800 μl of lysis buffer/dish and incubate for a further 1 h at 37°C. Add 200 μl/dish of 5 M NaCl while gently swirling the dish, and incubate on ice for 1 h. Centrifuge the lysate in a 1.5 ml microcentrifuge tube at 12 000 *g* for 30 min at 4°C.

5. *Purification of viral DNA:* remove the pellet using an inoculation loop and extract the supernatant once with an equal volume of phenol: chloroform. Centrifuge once more for 5 min at 4°C, and precipitate the DNA from the upper phase using 0.1 vol sodium acetate and 2 vols of 95% ethanol, followed by centrifugation at 4°C (>12 000 *g*) for 5 min. Resuspend in 10–50 μl of sterile water and measure the OD_{260} to determine the concentration of the viral DNA.

Alternatively, the recombinant DNA can be stably integrated into the genome of the cell (*Protocol 8*), producing a subclone of the cell line that expresses the molecule of interest. Expression can either be continuous, or it can be regulated if the recombinant DNA is placed under a regulatable promoter [e.g. the *TET* on/off system (36)]. A variety of selection media are available for this purpose [(37–41); *Table 4*].

Table 4. Commonly used selection media for preparation of stable cell lines: the plasmid required for co-transfection is presented, in addition to the media required for selection of positive clones

Selection plasmid	Selection medium (concentration)	Reference
Adenosine deaminase	Thymidine (10 μg/ml) Hypoxanthine (15 μg/ml) 9-β-D-xylofuranosyl adenine (4 μM) 2'-Deoxycoformycin (0.3μM)	37
Aminoglycoside phosphotransferase (neo, G418)	G418 (0.5 mg/ml in highly buffered solution pH 7.3)	38
Cytosine deaminase	*N*-(phosphonacetyl)-L-aspartate (1 mM) Inosine (1 mg/ml) Cytosine (1 mM)	39
Dihydrofolate reductase	Methotrexate (1–300 μM) (nucleoside-free medium)	40
Thymidine kinase (TK⁻ cells only)	Hypoxanthine (100 μM) Aminopterin (0.4 μM) Thymidine (16 μM) Glycine (3 μM)	41

Protocol 8. Stable transfection of eukaryotic cells

Equipment and reagents
- exponentially growing cells
- complete medium
- appropriate plasmids
- appropriate selection media (*Table 4*)
- cloning cylinders (available from Sigma)
- trypsin/EDTA
- transfection reagents (*Protocols 1, 2* or *3*)

Method

1. *Cell culture:* the initial step involves a transient transfection of the recombinant gene of interest along with an appropriate selection plasmid (*Table 4*). Any of the transient transfection protocols (1–3) can be used to introduce the plasmids into the cells. Incubate the transfected cells for 48 h in complete medium.

2. *Selection of transfected cells:* incubate the cells in complete medium containing selection reagent. Cell death in non-transfected cells should become obvious within 2–3 days. Replace with fresh medium (containing the selection reagent) and again every 4 days for up to 4 weeks. Colonies of resistant cells should be visible within 3–4 weeks.

3. *Cloning and harvesting:* using sterile forceps, place a sterile cloning cylinder over the growing colony. Add 1–2 drops of trypsin/EDTA into the cylinder and leave for 2–5 min. Add a few drops of complete medium to the cylinder and using a sterile Pasteur pipette remove the medium (and cells). Place in 10 ml of complete medium on a 10 cm dish, swirl, and incubate at 37°C until confluent.

3.2 Introduction of recombinant DNA *in vivo*

3.2.1 Introduction

Manipulation of signalling pathways in tissue culture has provided many insights into the mechanisms required for the regulation of a variety of metabolic processes. However, the question of physiological relevance requires a more physiological system, i.e. studies *in vivo*. The techniques to introduce recombinant DNA and manipulate gene expression in whole animals have developed rapidly in recent years. These types of experiments are a major undertaking and are more expensive than manipulation of cells *in vitro*, but they are vital to the proper understanding of the role of a signalling pathway in a physiological background.

3.2.2 General approaches

Two main approaches are taken to manipulate signalling pathways *in vivo*:

- the 'knock-out' of a gene or family of genes;
- the 'transgenic' introduction of a gene (or 'knock-in'), thus enabling over-expression of a wild-type protein, or a mutant with novel properties.

The techniques of inserting or deleting a gene in the host genome in a manner that allows expression or suppression of the protein of interest have been reviewed extensively previously (18). The targetting of the gene insertion to a specific cell type can be achieved relatively simply, by use of a tissue-specific promoter. More recently, adenoviral vectors (discussed above) have been used to infect whole animals and produce transient expression of recombinant protein. Direct injection of the virus into the tissue of interest can lead to a tissue-selective expression, while intravenous injection targets the adenovirus mainly to the liver.

3.3 Potential drawbacks/problems with transfection techniques

Any transient transfection techniques utilized should be monitored for the efficiency of DNA transfer using a reporter system (e.g. co-transfection with an expression plasmid for a fluorescent protein such as green fluorescent protein, or an easily assayable protein such as β-galactosidase, luciferase or chloramphenicol-acetyl transferase). The efficiency will vary from cell type to cell type, and with the transfection method of choice. For instance, in differentiated cells, such as isolated hepatocytes, the efficiency of transfection with non-viral methods is usually less than 1%, while with the adenoviral transfection method it can be as high as 100%. Thus, if a high efficiency of transfection is of particular importance in the interpretation of the results, viral vectors should be considered as the method of choice. A high efficiency of transfection allows measurement of endogenous cellular parameters (e.g. enzyme activities, gene expression) which cannot be readily assessed in single cells.

There is usually little control over the number of copies of the cDNA entering each cell. For most cell types it is likely that there will be transfer of between 10^4 and 10^6 copies of each molecule of DNA. This can sometimes result in artefactual results owing to the presence of a large amount of exogenous DNA in the cell. For example, if the goal of the experiment is the investigation of the activity of a DNA-binding protein that, coincidentally, can interact with the transfected DNA, then there will be a 'sequestering' of the protein by the recombinant plasmid. This in itself can affect the experimental results. Alternatively, expression of constitutively active signalling molecules will result in continuous induction of the signalling pathway, which does not necessarily exactly mimic the effects of transient activation by physiological agents.

The chemicals or reagents used in the transfection protocols can also affect signalling pathways. For example, the use of calcium phosphate precipitation will activate a number of potential signalling molecules such as calcium/calmodulin-dependent protein kinases and phosphatases. The techniques are quite invasive and may thus also induce stress-activated protein kinases.

The success of stable transfection of cells often depends on the site of integration of the recombinant gene. For instance, integration behind a strong promoter will increase the expression of the gene of interest, while, conversely, the presence of a 5'-repressor element will result in low expression. In addition, stably transfected cells can lose the recombinant gene expression following a number of passages; thus it is important to maintain a large stock of initial transformants.

4. Antisense technology

4.1 Introduction

Expression of proteins requires that the transcription/translation machinery gain access to the gene and the mRNA, respectively. Antisense oligonucleotides are usually relatively short pieces of synthetic nucleic acid designed to hybridize with single-stranded cellular target nucleic acid and prevent proper interaction between the transcription/translation machinery and the target nucleic acids (they may also interact with double-stranded nucleic acids, with the formation of triple helical structures). Theoretically, this lends a high degree of specificity when compared with chemical inhibitors, since oligonucleotides specific for individual genes, isoforms, or protein families can be designed. The target sequence can be in the coding region of the gene (intron or exon) preventing elongation of the transcript; in the promoter of the gene (triple helix), interfering with transcription factor binding; or, as is most common, in a stretch of sequence within the target mRNA, resulting in reduced translation and degradation of the message by RNAase H. This high degree of selectivity raises the possibility that this technology may be developed as a new class of pharmaceutical agents.

4.2 Design and application of antisense oligonucleotides

The criteria for designing successful antisense oligonucleotides are becoming clearer:

(a) *Metabolic stability:* nucleotides are notoriously unstable in biological fluids and medium. Use of oligonucleotides with a nuclease-resistant phosphorothiorate backbone is recommended.

(b) *Uptake and compartmentalization:* the rate of cellular uptake of the oligonucleotides will strongly influence the efficacy of the reagent. High intracellular concentrations are required for successful antisense effects. High extracellular concentrations are therefore recommended, since passive diffusion is the standard method for introduction of oligonucleotide into cells. Methods for obtaining co-localization of the target and the antisense oligonucleotide are discussed below.

(c) *Affinity of the interaction:* high-affinity binding is required for successful antisense experiments. This can be achieved by increasing the length of the antisense oligonucleotide.

(d) *Nature of the target sequence:* target sequences at the 5'-end, 3'-end, or the full length of mRNA have all proved successful in the nucleus, whereas 5'-end targetted antisense RNAs may be more efficacious in the cytoplasm. However, the major criterion for successful target selection appears to lie in locating an accessible region of the target sequence (i.e. no secondary structure).

(e) *Ribozymes:* the use of synthetic antisense ribozymes can remove the problems of uptake and reduce the number of molecules required for efficacy. However, RNA–RNA duplex formation *in vivo* is thought to be relatively inefficient.

Phosphothiorate antisense oligonucleotides (in which one of the oxygen atoms of the internucleotide phosphate group is replaced by a sulfur atom) are much more nuclease resistant than phosphodiester antisense oligonucleotides, and are therefore generally now the agents of choice. The other two major hurdles that require to be overcome before antisense oligonucleotides can be effective are: (1) transport of the antisense oligonucleotides into the cells; and (2) proper compartmentalization and annealing with the target mRNA, which may have a high degree of secondary structure (42, 43). Thus, although the approach can be effective, there is still an element of trial and error in obtaining a successful antisense oligonucleotide. It is advisable to examine the target mRNA sequence and design antisense oligonucleotides to areas lacking potential secondary structure. In addition, it is no longer clear that the most successful target sites for antisense oligonucleotides lie at the 5'-end of the mRNA. However, there is a chance that truncated protein may be produced from mRNA targeted at the 3'-end of the sequence.

Passive diffusion remains the main method for applying antisense oligonucleotides to tissue culture cells. This is a simple approach but requires relatively large amounts of antisense oligonucleotide, and different antisense oligonucleotides may have distinct uptake (and stability) profiles. Antisense structures (including ribozymes) as part of larger vector/plasmids can be introduced into cells using transfection techniques (as described above). In this case the antisense sequence is encoded by RNA transcribed from the plasmid. This approach requires the antisense sequence to be transcribed under the control of a very strong promoter, as high levels of expression are needed for most antisense effects. However, this does allow compartmentalization with co-transfected target genes. For instance, the antisense sequence can be fused to the target sequence such that only one RNA species is produced from the gene, thus producing instant co-localization of each transcript. Alternatively, a common localization sequence can be added to the 3'-end of both the target gene and the antisense/ribozyme construct (44).

There are many reports of these types of approaches producing up to 90% reduction in target protein levels. For example, the antisense oligonucleotide ISIS 3521 [GTT CTC GCT GGT GAG TTT CA (45)] reduces PKC-α gene expression by 90%, while ISIS 5132 [TCC CGC CTG TGA CAT GCA TT (46)] can severely reduce c-Raf gene expression. Both antisense oligonucleotides are potent inhibitors of tumour growth and have entered clinical trials (45). However, although many antisense oligonucleotide constructs are in clinical trials as therapeutic agents towards a variety of targets, there is now evidence that many of their effects are not actually due to antisense effects (Section 4.3).

4.3 Potential drawbacks/problems in the antisense approach

A major drawback of antisense technology is the inability to predict whether a given antisense oligonucleotide will be successful or not. In general, trial and error is still required to test the efficacy of these reagents. Prediction of a lack of secondary structure in the target sequence can improve the chances of success, but does not guarantee that there will be a reduction in the level of the target mRNA.

The main method for introduction of antisense oligonucleotides into cells is passive diffusion, and although this is technically straightforward, it requires large quantities of nucleotide, increasing the cost. In addition, although antisense techniques can produce >90% reduction in the level of the target protein, there is always a degree of 'leakiness'. This means that the target protein will continue to be expressed at a reduced level, and this may be sufficient for its normal function, calling into question any experiments where antisense oligonucleotides have no apparent effect.

The turnover rate of the target protein will also determine the rate of reduction of cellular protein levels and thus the success of the approach. For instance, if the half-life of a target protein is more than 4 days, then it becomes difficult to deplete the cell of the protein using antisense methodology, since the antisense oligonucleotides would have to be continuously present in the culture medium for at least 4–7 days.

Possibly more worrying are reported effects of antisense oligonucleotides which may not be related to binding to the target RNA (43, 47–49). As antisense oligonucleotides are large, complex molecules with multiple charges, they can interact with a broad spectrum of cellular proteins, resulting in non-specific effects unrelated to interference with the target RNA (*non-antisense effects*). Such effects include stimulation of B-cell proliferation and inhibition of viral entry into cells (47, 48). These reports question the initial claims of 'high specificity' attributed to antisense techniques, although the non-antisense effects of these reagents may still prove to have therapeutic value (43). They remain of great interest to the pharmaceutical industry in themselves, although their mechanism of action remains unclear.

5. Ribozyme technology

5.1 Introduction

The discovery of RNA molecules that contain an intrinsic ribonuclease activity (*ribozymes*) has led to their development as tools to deplete cells of specific proteins. The specific degradation of a target mRNA species can be achieved, thus reducing the translation of target proteins. These molecules were first discovered as part of the cellular RNA splicing and processing

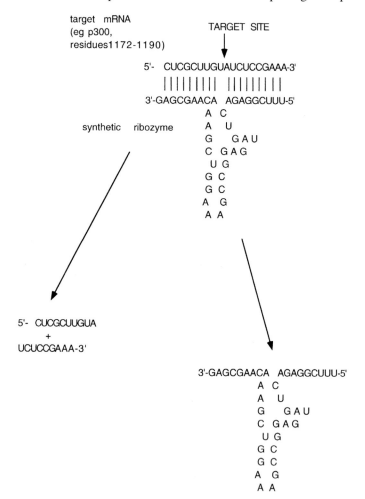

Figure 3. Model of ribozyme action. A synthetic ribozyme is engineered that contains the catalytic core of a naturally occurring ribozyme (e.g. hammerhead ribozyme) flanked by sequences complementary to the target mRNA. The target site for cleavage depends on the ribozyme employed. Hammerhead ribozymes cleave preferentially at GUX sequences, where residue X is A>U>C>G (*in vitro*).

machinery, but several unrelated small RNA species that contain RNAase activity have subsequently been identified. They have been described as 'molecular dissection kits' because of their potential as isoform-specific targetting molecules. A number of naturally occurring ribozymes have been identified (50). The most widely employed in mRNA-targeting experiments to date is the hammerhead ribozyme, owing to its relatively small size and the fact that it has been well characterized. It has a central core motif of 15

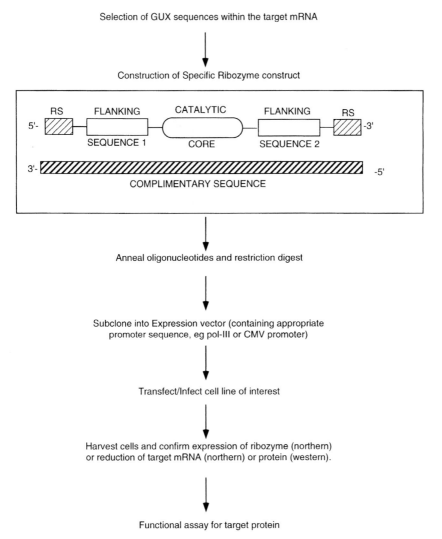

Figure 4. Flow diagram for preparation and application of ribozyme molecules. RS: appropriate restriction site. 'Flanking sequence 1' refers to the sequence immediately 5' to the GUX target sequence (including GU), while 'flanking sequence 2' refers to the sequence immediately 3' to the X residue within the target sequence.

residues that are essential for catalytic activity (51, 52). These residues form complex structures that mediate RNA folding and thus catalysis.

5.2 Theory of ribozyme action

The hammerhead ribozyme is made up of two domains, a catalytic core and a specifier sequence (*Figure 3*). As mentioned above, the catalytic core contains 15 conserved residues, while the specifier sequence consists of two 'arms' of variable length flanking this core (around 12 bases total is optimal *in vitro* (52)). These sequences hybridize to a complementary target sequence in the substrate nucleic acid, allowing catalytic cleavage of the substrate. The products are then free to dissociate from the ribozyme allowing a further cycle of catalysis of a new substrate molecule (*Figure 3*). This ability to cycle through a number of substrates suggests that ribozymes should be more efficient than conventional antisense oligonucleotides in depleting a cell of mRNA substrates, since antisense oligonucleotides are degraded by RNAase H following dimerization with their target. Each class of ribozyme catalyses hydrolysis of specific nucleotide sequences. For instance, the hammerhead ribozyme cleaves the sequences GUC>GUA>GUU *in vitro*. However, *in vivo* this order of preference appears different (GUA being most efficiently cleaved), at least with respect to the sequences examined to date (53). Thus preparation and application of ribozyme methodology is relatively straightforward (*Figure 4*).

An elegant procedure, designed to determine the best target site in a given molecule for ribozyme attack, was recently reported by Kawasaki *et al.* (53). A modified version of this procedure is presented (*Protocol 9*). This technique should be very useful in determining whether the target mRNA is accessible to the ribozyme. A similar strategy can be developed to determine the efficacy of antisense constructs.

Protocol 9. Identification of the most efficient target site for a ribozyme

Equipment and reagents

- cultured cells, growing exponentially
- expression plasmid, containing a promoter and polyadenylation signal
- luciferase cDNA
- appropriate restriction endonucleases
- luciferase assay kit

Method

1. *Fusion gene preparation:* subclone a region of the gene of interest (GI) containing a number of potential ribozyme target sites, 3′ of a promoter and 5′ of a luciferase reporter gene (*luc*) to produce a construct capable of expressing the message for *GI-luc*. Transfection of this construct into cells should produce a high level of luciferase activity.

---

2. *Ribozyme preparation:* a series of ribozymes, each targetted to a specific site in the GI sequence, should be inserted into an expression plasmid, under the control of a promoter at least as strong as that of the *GI-luc* construct.

3. *Co-transfection:* equal amounts of the *GI-luc* construct and each individual ribozyme construct are co-transfected into cells using one of *Protocols 1–3*. After 24, 48, and 72 h, the cells are harvested and assayed for luciferase activity. Luciferase activities in the presence of each ribozyme are compared with control levels in order to determine the most accessible site within the GI sequence and the optimal time of incubation.

4. The most effective ribozyme can then be used in transient transfection studies to reduce endogenous levels of GI. This may require further subcloning into a viral transfection system if the cell line used has a low transfection efficiency.

5.3 Examples of successful ribozyme design

There are a number of successful examples of this type of approach (*Table 5*). For instance, expression of an anti-*Ras* ribozyme by adenoviral-mediated delivery produced neoplastic reversion in oncogenic H-Ras-expressing cells (54), while transfection of an anti-*IRS1* ribozyme affected insulin regulation of GLUT4 translocation (55). However, it is generally accepted that the success rate of ribozymes as tools for reducing specific mRNAs is relatively low. An accurate assessment of the frequency of success with this approach is not available, since unsuccessful attempts to produce such molecules are not generally published.

5.4 Potential drawbacks/problems

The criteria for designing successful molecules are somewhat elusive, but as many more researchers attempt to employ the improving technology, these

Table 5. Examples of successful ribozyme molecules targetted to signalling molecules

Example	Reference
(1) Ribozyme against p300: gagcgaacaAAGCUGGAAAGCCAGGAGUAGUCagaggcuuu-5′ (residue 1181) Inactive ribozyme: gagcgaacaAAGCUGGAAAGCCAGGAGUAAUCagaggcuuu-5′	53
(2) Ribozyme against IRS1: ccggucucaAAGCAGGAGUGCCUGAGUAGUCaucgcagugua-5′ (residue 3051)	55
(3) Ribozyme against bcr-abl: ccuucuucgggaAAGCAGGAGUGCCUGAGUAGUCgucgccg -5′ (residue 8, exon 2)	15

should become clearer. Several problems are associated with the use of this approach:

(a) As with antisense protocols (Section 4), the delivery of the nucleic acids into the cell presents a serious challenge. However, as the ribozyme is RNA rather than DNA, it is feasible to deliver it using an expression system.

(b) A major drawback has been the low transfer efficiency of available transfection techniques, resulting in the inability to determine with certainty the amount of ribozyme expressed in a given cell, or to reduce the target mRNA in a whole population of cells. The availability of high efficiency transfection systems (e.g. adenovirus systems, Section 3.1.2) should aid in overcoming this problem.

(c) Like antisense technology, the ribozyme approach may be 'leaky'. Although the expression levels of a ribozyme required to reduce the target mRNA level should be lower than those required for antisense technology, it is unlikely to be high enough to achieve 100% loss of the mRNA species. Thus a complete 'knock-out' is unlikely.

(d) Just as non-antisense effects are being reported in experiments involving antisense constructs, it is possible that non-ribozyme effects could occur with some nucleotides designed as ribozymes. Such non-specific effects are, however, currently less well documented than for the antisense approach.

References

1. Hardie, D.G. and Hanks, S. (ed.) (1995) *The protein kinase facts book*, Vol. 1. Academic Press, London.
2. Brown, P.H., Alani, R., Preis, L.H., Szabo, E., and Birrer, M. J. (1993) *Oncogene* **8**, 877.
3. Akimoto, K., Takahashi, R., Moriya, S., Nishioka, N., Takayanagi, J., Kimura, K., Fukui, Y., Osada, S.-I., Mizuno, K., Hirai, S.-I., Kazlauskas, A., and Ohno, S. (1996) *EMBO J.* **15**, 788.
4. Brown, P.H., Chen, T.K., and Birrer, M.J. (1994) *Oncogene* **9**, 791.
5. House, C. and Kemp, B.E. (1987) *Science* **238**, 1726.
6. Gschwendt, M., Jahannes, F.-J., Kittstein, W., and Marks, F. (1997) *J. Biol. Chem.* **272**, 20742.
7. Walaas, O., Horn, R.S., and Walaas, S.I. (1997) *FEBS Lett.* **413**, 152.
8. Bushdid, P.B., Brantley, D.M., Yull, F.E., Blaeuer, G.L., Hoffman, L.H., Niswander, L., and Kerr, L.D. (1998) *Nature* **392**, 615.
9. Eyers, P.A., Craxton, M., Morrice, N., Cohen, P., and Goedert, M. (1998) *Curr. Biol.* **5**, 321.
10. Shoji, S., Titani, K., Demaille, J.G., and Fischer, E.H. (1979) *J. Biol. Chem.* **254**, 6211.
11. Hughes, K., Nikolakaki, E., Plyete, S.E., Totty, N.F., and Woodgett, J.R., (1993) *EMBO J.* **12**, 803.

12. Dean, N.M., McKay, R., Condon, T.P., and Bennett, C.F. (1994) *J. Biol. Chem.* **269**, 16416.

13. Monia, B.P., Johnston, J.F., Geiger, T., Muller, M., and Fabbro, D. (1996) *Nat. Med.* **2**, 668.

14. Leopold, L.H., Shore, S.K., Newkirk, T.A., Reddy, R.M.V., and Reddy, E.P. (1995) *Blood* **85**, 2162.

15. Pachuk, C.J., Yoon, K., Moelling, K., and Coney, L.R. (1994) *Nucl. Acids Res.* **22**, 301.

16. Palfner, K., Kneba, M., Hiddemann, W., and Bertram, J. (1995) *Biol. Chem. Hoppe-Seyler* **376**, 289.

17. Bruning, J.C., Winnay, J., Bonner-Weir, S., Taylor, S.I., Accili, D., and Kahn, C.R. (1997) *Cell* **88**, 561.

18. Joyner, A.L. (ed.) (1997) Gene targetting: a practical approach (ed. D. Rickwood and B.D. Hames). Oxford University Press, Oxford.

19. Deak, J.C., Cross, J.V., Lewis, M., Qian, Y., Parrott, L.A., Distelhorst, C.W., and Templeton, D.J. (1998) *Proc. Natl. Acad. Sci. USA* **95**, 5595.

20. Khokhlatchev, A., Xu, S., English, J., Wu, P., Schaefer, E., and Cobb, M. (1997) *J. Biol. Chem.* **272**, 11057.

21. Deuel, T.F. (1987) *Annu. Rev. Cell. Biol.* **3**, 443.

22. Ueki, K., Yamamoto-Honda, R., Kaburagi, Y., Yamauchi, T., Tobe, K., Burgering, B.M.T., Coffer, P.J., Komuro, I., Akanuma, Y., Yazaki, Y., and Kadowaki, T., (1998) *J. Biol. Chem.* **273**, 5315.

23. Merrill, G.F., Kurth, E.J., Hardie, D.G., and Winder, W.W. (1997) *Am. J. Physiol.* **273**, E1107.

24. Sutherland, C., O'Brien, R.M., and Granner, D.K. (1995) *J. Biol. Chem.* **270**, 15501.

25. Merrall, N.W., Plevin, R.J., Stokoe, D., Cohen, P., Nebrada, A.R., and Gould, G.W. (1993) *Biochem. J.* **295**, 351.

26. Hardie, D.G. and Carling D. (1997) *Eur. J. Biochem.* **246**, 259.

27. White, M.A., Nicolette, C., Minden, A., Polverino, A., Vanaelst, L., Karin, M., and Wigler, M.H. (1995) *Cell* **80**, 533.

28. Khwaja, A., Lehmann, K., Marte, B.M., and Downward, J. (1998) *J. Biol. Chem.* **273**, 18793.

29. Krengel, U., Schlichting, L., Schere, A., Schumann, R., Frech, M., John, J., Kabsch, W., Pai, E.F., and Wittinghofer, A. (1990) *Cell* **62**, 539.

30. Joneson, T., White, M.A., Wigler, M. H., and Bar-Sagi, D. (1996) *Science* **271**, 810.

31. Downward, J. (1998) *Science* **279**, 673.

32. Cichy, S.B., Uddin, S., Danilkobvitch, A., Guo, S., Klippel, A., and Unterman, T.G. (1998) *J. Biol. Chem.* **273**, 6482.

33. Yeh, P. and Perricaudet, M. (1997) *FASEB J.* **11**, 615.

34. Becker, T.C., Noel, R.J., Coats, W.S., Gomez-Foix, A.M., Alam, T., Gerard, R.D., and Newgard, C.B. (1994) *Meth. Cell Biol.* **43**, 161.

35. He, T.-C., Zhou, S., Da Costa, L.T., Yu, J., Kinzler, K.W., and Vogelstein, B. (1998) *Proc. Natl. Acad. Sci. USA* **95**, 2509.

36. Gossen, M., Freundlieb, S., Bender, G., Muller, G., Hillen, W., and Bujard, H. (1995) *Science* **268**, 1766.

37. Kaufman, R.J., Murtha, P., Ingolia, D.E., Yeung, C.-Y., and Kellems, R.E. (1986) *Proc. Natl. Acad. Sci. USA* **83**, 3136.

38. Southern, P.J. and Berg, P. (1982) *J. Mol. Appl. Gen.* **1**, 327.
39. Wei, K. and Huber, B.E. (1996) *J. Biol. Chem.* **271**, 3812.
40. Simonsen, C.C. and Levinson, A.D. (1983) *Proc. Natl. Acad. Sci. USA* **80**, 2495.
41. Littlefield, J.W. (1964) *Science* **145**, 709.
42. Epstein, P.M. (1998) *Methods: a companion to 'Methods in enzymology'* **14**, 21.
43. Branch, A.D. (1998) *Trends Biochem. Sci.* **23**, 45.
44. Arndt, G.M. and Rank, G.H. (1997) *Genome* **40**, 785.
45. Dean, N.M. McKay, R., Miralgia, L., Geiger, T., Muller, M., Fabbro, D., and Bennett, C.F. (1996) *Biochem. Soc. Trans.* **24**, 623.
46. Monia, B.P. (1997) *Ciba Found. Symp.* **209**, 107.
47. Krieg, A.M., Yi, A.-K., Matson, S., Waldschmidt, T.J., Bishop, G.A., Teasdale, R., Koretzky, G.A., and Klinman, D.M (1995) *Nature* **374**, 546.
48. Stein, C.A. (1996) *Trends Biotechnol.* **14**, 147.
49. Branch, A.D. (1996) *Hepatology* **24**, 1517.
50. Scott, W.G. and Klug, A. (1996) *Trends Biochem. Sci.* **21**, 220.
51. Haseloff, J. and Gerlach, W.L. (1988) *Nature* **334**, 585.
52. Jarvis, T.C., Wincott, F.E., Alby, L.J., McSwiggen, J.A., Biegelman, L., Gustofsen, J., DiRenzo, A., Levy, K., Arthur, M., Karpiesky, A., Gonzalez, C., Woolf, T.M., Usman, N., and Stinchcomb, D.T. (1996) *J. Biol. Chem.* **271**, 29107.
53. Kawasaki, H., Ohkawa, J., Tanishige, N., Yoshinari, K., Murata, T., Yokoyama, K.K., and Taira, K. (1996) *Nucl. Acids Res.* **24**, 3010.
54. Feng, M., Cabrera, G., Deshane, J., Scanlon, K.J., and Curiel, D.T. (1995) *Cancer Res.* **55**, 2024.
55. Quon, M.J., Butte, A.J., Zarnowski, M.J., Sesti, G., Cushman, S.W., and Taylor, S.I. (1994) *J. Biol. Chem.* **269**, 27920.

5

Phosphopeptide mapping and phosphoamino acid analysis on cellulose thin-layer plates

PETER VAN DER GEER, KUNXIN LUO, BARTHOLOMEW M. SEFTON, and TONY HUNTER

1. Introduction

Phosphopeptide mapping enables the investigator to study the phosphorylation of a protein of interest in detail. The technique can be used to estimate the number of sites of phosphorylation in a protein, to determine the stoichiometry of phosphorylation at particular sites, and to deduce the identity of protein kinases responsible for the phosphorylation of specific sites. Additionally, comparative phosphopeptide mapping is an invaluable tool for the determination of the identity, or lack of identity, of phosphoproteins obtainable only in trace amounts.

In two-dimensional phosphopeptide mapping the protein of interest is labelled with [^{32}P]orthophosphate and digested to completion with site-specific proteases or chemicals. The resulting peptides are separated in two dimensions on thin-layer cellulose (TLC) plates, by electrophoresis in the first dimension and chromatography in the second dimension. The radioactive phosphopeptides are then visualized by exposure to X-ray film.

This technique has the advantage that it is extremely sensitive, requiring only several hundred disintegrations per minute of labelled protein (the amount of radioactivity needed to obtain readable phosphopeptide maps depending, obviously, on the total number of phosphorylation sites present in the protein). In addition, individual phosphopeptides can be isolated from the inert cellulose coating of the plates, and used for phosphoamino acid determination, amino-terminal sequencing, and secondary digestion with additional proteases.

2. Protein labelling *in vivo* and *in vitro*

Phosphopeptide mapping and phosphoamino acid analysis depend on visualization of phosphopeptides or phosphoamino acids by autoradiography.

Because of its high specific activity, the use of [^{32}P]orthophosphate makes it possible to analyse extremely small amounts of protein. Proteins can be labelled *in vivo* by incubating cells with [^{32}P]orthophosphate, or by phosphorylation of the isolated protein with a protein kinase in the presence of [γ-^{32}P]ATP.

For a more detailed discussion of ^{32}P-labelling techniques with intact cells see Chapter 1. Cells can be labelled by incubating in medium containing up to 2 mCi/ml [^{32}P]orthophosphate in phosphate-free medium (phosphate-free serum, to add to the phosphate-free medium, can be obtained by dialysing regular serum against Tris-buffered saline). Usually 4 h of ^{32}P-labelling in phosphate-free medium will be sufficient, but if steady-state labelling is required, cells can be labelled for 24 h or more. However, for these extended periods of labelling, the use of medium totally deficient in phosphate may be deleterious to the cells. The appropriate length of the labelling and the concentration of unlabelled phosphate in the medium depends on the abundance of the protein of interest, and the sensitivity of the cells to radiation damage (and cell cycle checkpoint arrest) and phosphate depletion. After labelling, the cells are rinsed with cold Tris- or phosphate-buffered saline, to remove serum and residual label, and the cells are lysed in lysis buffer at 4°C. If tyrosine phosphorylation sites are being analysed, we include 100 μM sodium orthovanadate, a protein–tyrosine phosphatase inhibitor, in the lysis buffer. The protein of interest can be purified by conventional purification techniques or by immunoprecipitation and SDS–PAGE.

It is important to bear in mind that large amounts of [^{32}P]phosphate are used during labelling of cells, and that proper shielding is essential. To protect the investigator, at all times during the labelling procedures we use shields that are a sandwich of 1 inch thick Plexiglas with 1/4 inch lead sheeting on the outside. Incubation of cells with [^{32}P]orthophosphate is done in incubators reserved for such labelling procedures. Tissue culture dishes containing radioactive medium are housed in Plexiglas boxes during the labelling, and while being transported from the incubator to a properly shielded area that is set up specifically for handling and lysing ^{32}P-labelled cells.

If the effect of growth factors or protein kinase activators on protein phosphorylation is being studied, cells are starved in low serum (0.1–1.0%) or serum-free medium. The cells are subsequently labelled with [^{32}P]orthophosphate in low serum or in serum-free medium, and protein kinase activators or peptide growth factors are added to the labelling medium for the final minutes of labelling.

Proteins can also be labelled *in vitro* by incubation with a protein kinase and [γ-^{32}P]ATP. Both purified proteins or immunoprecipitated proteins can be used as substrates. If the substrate protein is isolated by immunoprecipitation, it may be necessary to wash the immunoprecipitates once or twice with Tris-buffered saline, to remove detergent or other reagents that are present in the lysis and washing buffers, and that may inhibit the protein kinase that is used for phosphorylation. We use up to 100 μCi [γ-^{32}P]ATP in 25 μl reactions. Kinase buffers usually contain only buffer and 5–10 mM divalent cation. In general, protein–serine/threonine kinases prefer Mg^{2+}, whereas protein–tyrosine kinases prefer Mn^{2+}. The phosphorylation reaction is usually terminated by adding 2 × SDS–PAGE sample buffer containing up to 10 mM EDTA (the EDTA chelates divalent cations: boiling proteins with [γ-^{32}P]ATP, SDS, and free divalent cations can result in chemical phosphorylation). The samples are boiled and analysed by SDS–PAGE. The radioactive ATP runs with, or in front of, the dye front, and will end up in the lower buffer chamber if the dye front is allowed to run off. In such a case the buffer should be disposed of as radioactive waste. Since protein kinase preparations usually contain contaminating substrates, it is important to run a kinase reaction without the substrate as a control, to ensure that the phosphorylated polypeptide is not derived from the kinase preparation.

3. Digestion of ^{32}P-labelled proteins

3.1 Recovery of phosphoproteins from polyacrylamide gels

Prior to phosphopeptide mapping, complex mixtures of proteins must usually be resolved by SDS–PAGE. Two strategies can be used for recovery of the protein of interest from the gel. The protein can be eluted from the gel, precipitated with trichloroacetic acid (TCA) or acetone, and oxidized (*Protocol 1*) before digestion (*Protocol 3*) (see, for example, ref. 1). Alternatively, the fractionated proteins can be transferred electrophoretically to a nitrocellulose, nylon, or Immobilon membrane and subjected to digestion while bound to the membrane (*Protocols 4* and *5*) (see, for example, ref. 2). The digestion of membrane-bound proteins using these latter protocols is rapid, and the procedure minimizes losses. The membrane digestion procedures, however, do have several potential drawbacks. First, they are not suitable for proteins that transfer inefficiently to membranes. Secondly, the recovery of hydrophobic peptides may be poor, although the hydrophilicity provided by the phosphate moieties on phosphopeptides may render them particularly suitable for analysis by this procedure. The usefulness of this technique should be evaluated in each case.

3.2 Elution of proteins from polyacrylamide gels

After electrophoresis the gel is dried (*Protocol 1*). Staining and fixing is not necessary: some investigators feel that fixation reduces the recovery of proteins from the gel, but good recoveries can still be obtained after fixation. The ^{32}P-labelled protein of interest is located by autoradiography, or by using a phosphorimager. Protein bands are excised from the gel, and ground up in ammonium bicarbonate buffer containing SDS in a 1.5 ml microcentrifuge tube. The protein is eluted twice for at least 90 min each (*Protocol 1*); for convenience, one of the elution steps can be carried out overnight. The combined eluates (volume ~1.3 ml) are then cleared of particulate matter and residual gel bits by centrifugation, and the protein is concentrated by TCA precipitation in the presence of a carrier, usually boiled RNAase, followed by an ethanol wash. When used as a carrier, RNAase eliminates residual ^{32}P-labelled RNA contaminating the protein. Immunoglobulins and bovine serum albumin (BSA) can also be used as carriers for protein precipitation. Some investigators prefer precipitation with 4 vols cold acetone at –20°C. The protein is recovered by precipitation from the gel eluate, rather than by lyophilization, because of the SDS in the elution buffer. The SDS is removed during the precipitation step. It is not uncommon that the protein pellet, obtained after the first centrifugation step following TCA precipitation, is hardly visible. The pellet obtained after the ethanol wash should be visible as a white speck. Precipitated protein can now be digested or hydrolysed. The recovery of radioactivity during protein isolation, and during subsequent steps, should be monitored by Cerenkov counting.

Protocol 1. Elution of proteins from SDS–polyacrylamide gels

Equipment and reagents
- apparatus and reagents for SDS–PAGE
- microcentrifuge (~7000 g)
- ammonium bicarbonate buffer (50 mM, pH 7.3–7.6, freshly prepared)
- disposable tissue grinder pestle (Kontes)
- SDS (20%, w/v)
- carrier protein (RNAase A, bovine serum albumin, or immunoglobulin)
- trichloroacetic acid (TCA, 100% w/v)

Method

1. Resolve ^{32}P-labelled samples by SDS–PAGE. Dry the gel and mark the dried gel around the edges with ^{35}S-labelled India ink (10 μCi/ml) or fluorescent ink. Use thin ink marks so that the alignment in step 2 can be achieved accurately.

2. Expose the gel to X-ray film (or a phosphorimager plate). Line up the markers around the gel with their images on the film (or a transparent copy of the phosphorimage), staple or tape the gel and film together, and place this sandwich on a light box.

3. Identify and cut out the protein bands from individual lanes using a single edge razor blade. Try not to shave off any pieces from the gel.

4. Peel the paper backing from the gel and remove residual paper by scraping with a razor blade.

5. Place the pieces of gel in a 1.5 ml screw-cap microcentrifuge tube, rehydrate in 500 μl ammonium bicarbonate buffer for 5 min at room temperature. A 0.5 cm × 0.7 cm piece from a 1 mm thick gel can be easily ground up in 500 μl buffer.

6. Grind the gel pieces using a disposable Kontes tissue grinder pestle until they can be passed through the tip of a 200 μl adjustable pipette. Add 500 μl more ammonium bicarbonate, 10 μl 2-mercaptoethanol, and 5 μl 20% SDS, boil for 2–3 min, and incubate for at least 90 min at room temperature (or overnight if convenient), or at 37 °C on a shaker.

7. Centrifuge for 2–3 min at room temperature in a microcentrifuge and transfer the supernatant, using a disposable transfer pipette, to a new 1.5 ml microcentrifuge tube. A low speed swinging bucket centrifuge (~2000 g) can also be used, and this may be advantageous because it packs the ground gel down with a flat surface, making it easier to transfer the supernatant without disturbing the pellet. Measure the volume of eluate and add enough 50 mM ammonium bicarbonate to the pellet in the original tube so that the final combined eluate volume will be ~1300 μl. Add SDS and 2-mercaptoethanol so that the final concentrations are 0.1%, vortex thoroughly, and incubate again for at least 90 min.

8. Centrifuge for 2–3 min in a microcentrifuge at room temperature, remove the supernatant, and combine the two eluates. Centrifuge the combined eluates for 5–10 min in a microcentrifuge, and transfer the supernatant to a new microcentrifuge tube, making sure you leave all the gel fragments behind. Monitor by Cerenkov counting to ensure that the majority of the ^{32}P radioactivity has been eluted.

9. Cool the combined eluates on ice, add 20 μg carrier protein, mix well, add 250 μl ice-cold 100% TCA, mix well, and incubate for 1 h on ice. Centrifuge for 10 min in a microcentrifuge at 4 °C, decant the supernatant, centrifuge again for 1 min, and remove the last traces of TCA with a disposable microtube transfer pipette. Add 500 μl 96% ethanol, vortex, and centrifuge for 5 min at 4 °C, decant the bulk of the supernatant, remove the residual liquid with a disposable micro-pipette and air-dry (do not lyophilize). Use Cerenkov counting to make sure that the majority of the ^{32}P radioactivity has been recovered in the precipitate.

3.3 Oxidation with performic acid

Methionine (Met) and cysteine (Cys) residues can exist in several oxidation states, and this markedly affects the chromatographic properties of peptides. Samples that are used for peptide mapping are therefore generally oxidized, using performic acid, to prevent separation of oxidation-state isomers of peptides in the chromatography dimension. However, if it is known, or if there is reason to suspect, that the peptides containing phosphorylated residues do not contain Met or Cys, then it is not necessary to carry out the oxidation step. If TCA-precipitated protein is to be used for peptide mapping,

Peter van der Geer et al.

the precipitate is dissolved directly in ice-cold performic acid and incubated for 60 min on ice (*Protocol 2*). Never let the sample warm up, since incubation with performic acid at higher temperatures may result in unwanted side reactions, such as the oxidation of histidine and tryptophan. The performic acid is subsequently diluted with deionized water and evaporated in a centrifugal vacuum concentrator (e.g. SpeedVac concentrator, Savant). After evaporation the sample should appear as a small white cotton-like ball.

Protocol 2. Oxidation by incubation with performic acid

Equipment and reagents
- formic acid (98%)
- hydrogen peroxide (30%, w/v)

Method
1. Mix 9 parts formic acid with 1 part 30% hydrogen peroxide. Incubate for 60 min at room temperature to allow performic acid to form.
2. Place the tube on ice and let the performic acid cool down.
3. Add 50 μl ice-cold performic acid to the ethanol-washed TCA precipitate or dried peptide mix (see *Protocol 5*) in a microcentrifuge tube. Vortex well to dissolve, and incubate for 60 min on ice.
4. Add 400 μl deionized water, mix, and freeze on dry ice.
5. Evaporate the performic acid under vacuum in a centrifugal vacuum concentrator.

3.4 Proteolytic and chemical digestion of phosphoproteins in solution

To obtain a characteristic fingerprint, the phosphoprotein of interest is digested with sequence-specific proteases, or by incubation with chemical compounds that cleave proteins at certain residues or sequences. In our laboratories, trypsin and chymotrypsin are used most often. Trypsin is pretreated with *N*-tosyl-L-phenylalanine chloromethyl ketone (TPCK) to inactivate contaminating chymotrypsin activity. Trypsin cleaves proteins following arginine or lysine residues. Chymotrypsin cleaves following phenylalanine, tyrosine, or tryptophan residues, but not following phosphotyrosine. Both enzymes can be used to obtain complete digests of TCA-precipitated proteins, and show high levels of specificity. Be aware that partial digestion products can be generated with trypsin when multiple adjacent arginine or lysine residues are present. In addition, trypsin inefficiently cleaves peptide bonds between arginine/lysine and aspartic acid/glutamic acid, and does not cleave peptide bonds between arginine/lysine and proline. Other enzymes are available that may not produce complete digests, but will generate reproducible fingerprints (*Table 1*).

Table 1. Conditions for enzymatic and chemical digestion of phosphoproteins

Reagent	Reaction conditions	Reference
Carboxypeptidase B	50 mM NH_4HCO_3 pH 8.0, 37 °C	10
α-Chymotrypsin	50 mM NH_4HCO_3 pH 8.0, 37 °C	11
Endoproteinase Asp-N	50 mM NH_4HCO_3 pH 7.6, 37 °C	12
Endoproteinase Glu-C (V8)	50 mM NH_4HCO_3 pH 7.6, 37 °C	13
Proline-specific endopeptidase	50 mM NH_4HCO_3 pH 7.6, 37 °C	14
Thermolysin	50 mM NH_4HCO_3 pH 8.0, 1 mM $CaCl_2$, 55 °C	15
Trypsin	50 mM NH_4HCO_3 pH 8.0, 37 °C	11
Cyanogen bromide	50 mg/ml CNBr in 70% formic acid, 90 min at 21 °C	16
Hydroxylamine	2 M guanidine, 2 M NH_2OH, 0.2 M K_2CO_3 pH 9.0, 4 h at 45 °C	17
Formic acid	70% formic acid for 24–48 h at 37 °C	18

These enzymes are extensively used for secondary digestion, or for digestion of isolated phosphopeptides.

Trypsin and chymotrypsin are often initially dissolved in 0.1 mM HCl, which prevents self-digestion, but we dissolve all proteolytic enzymes in deionized water at a concentration of 1 mg/ml. Stocks are frozen and stored in liquid nitrogen in small aliquots (100–200 μl). For proteolytic digestion with trypsin or chymotrypsin, the TCA-precipitated and oxidized samples are suspended and digested in ammonium bicarbonate at pH 8.0–8.3 (*Protocol 3*). Other enzymes have different pH optima and require different reaction conditions (*Table 1*). As an alternative to steps 1–3 in *Protocol 3*, larger numbers of shorter incubations with smaller amounts of protease may be used (four incubations of 30–60 min at 37 °C with 5 μl protease stock) and will yield reproducible results. Enzyme stocks should be refrozen, since self-digestion will lead to loss of activity. After protease digestion, the ammonium bicarbonate, which will interfere with electrophoresis, is removed by repeated lyophilization steps. The procedure for digestion with chymotrypsin is essentially the same as for trypsin (*Protocol 3*).

Protocol 3. Tryptic digestion of TCA-precipitated phosphoproteins

Equipment and reagents

- ammonium bicarbonate buffer (50 mM, pH 8.0–8.3)
- microcentrifuge (~7000 g)
- trypsin (TPCK-treated, available from several commercial suppliers)

Method

1. Dissolve the protein pellet in 50 μl ammonium bicarbonate buffer. Add 10 μg trypsin (10 μl of a 1 mg/ml stock).

Protocol 3. *Continued*

2. Incubate for 3–4 h at 37 °C (or overnight if more convenient).

3. Add a second 10 μg of trypsin and incubate for another 3–4 h at 37 °C.

4. Add 400 μl deionized water and lyophilize in a centrifugal vacuum concentrator.

5. Dissolve the pellet in 400 μl deionized water and lyophilize. This step should be repeated if a significant amount of white salt residue (ammonium bicarbonate) remains.

6. Dissolve the pellet in 400 μl pH 1.9 buffer or pH 4.72 buffer (for electrophoresis at pH 1.9 and pH 4.72, respectively) or deionized water (for electrophoresis at pH 8.9). Centrifuge for 5 min in a micro-centrifuge, transfer the supernatant, and lyophilize. It is very important that there is no particulate matter in this final supernatant. Estimate the amount of ^{32}P radioactivity in the final digest by Cerenkov counting.

7. Dissolve the pellet in at least 5 μl electrophoresis buffer (for electro-phoresis at pH 1.9 or 4.72) or deionized water (for electrophoresis at pH 8.9), centrifuge briefly, and spot the sample on the TLC plate. Ideally, at least 1000 cpm should be loaded, but if there are fewer cpm the whole sample should be loaded.

4. Proteolytic and chemical digestion of immobilized proteins

The previous section introduced the conventional method of sample prepara-tion for peptide mapping, in which the samples are eluted from a preparative SDS–PAGE gel and then precipitated with TCA in the presence of carrier protein. The disadvantage of this procedure is that it can be laborious and may result in losses of radioactive material. We have developed an alternative method, in which proteins are transferred to a membrane (nitrocellulose, Immobilon-P, or nylon) after separation by SDS–PAGE. Peptides generated by proteolytic or chemical digestion of these immobilized proteins will elute from the membrane during digestion, and are suitable for both one- and two-dimensional peptide analyses.

4.1 Tryptic peptide mapping of immobilized proteins

After separation of proteins by SDS–PAGE, proteins are transferred electro-phoretically to nitrocellulose, nylon, or Immobilon-P membranes (*Protocol 4*). Other protocols for electrophoretic transfer of proteins from polyacrylamide gels to membranes are available. Proteolysis can be carried out with all three types of membranes, but we usually use nitrocellulose. If both phospho-peptide mapping and phosphoamino acid analysis will be done with the same sample, Immobilon-P membrane should be used (nitrocellulose and nylon dissolve in the 6 N HCl used for acid hydrolysis). After transfer, the piece of membrane is wrapped in Saran wrap (cling film) to keep it moist during autoradiography. Drying of the membrane may lead to inefficient elution of material from the membrane.

Protocol 4. Transfer of proteins from polyacrylamide gels to membranes

Equipment and materials

- transfer buffer [dissolve 10.5 g Tris base and 50.4 g glycine in 2.8 litres deionized water and 0.7 litres methanol, degas using a vacuum line for 5–10 min, add 17.5 ml 20% (w/v) SDS]
- membrane (nitrocellulose, nylon, or Immobilon-P)
- blotting apparatus

Method

1. Soak the gel for 5–10 min in transfer buffer.

2. Assemble the transfer sandwich: ScotchBrite pad, three layers of Whatman 3MM paper, the gel, the membrane, three layers of 3MM paper, and another Scotch Brite pad. Transfer gel-fractionated proteins electrophoretically to the membrane, under conditions that are best for the protein of interest. We usually transfer at 60 volts for 1 h for a 60 kDa protein using a Bio-Rad TransBlot apparatus. A semi-dry blotting apparatus can also be used.

3. Disassemble the apparatus, rinse the blot with water, wrap it in Saran wrap (cling film), apply alignment markers (see *Protocol 1*), and expose to X-ray film (or a phosphorimager plate). Identify and cut out the protein of interest.

Prior to proteolytic digestion, the piece of membrane is treated with PVP-360 (Sigma, St. Louis, MO) to block non-specific binding of the enzyme later during digestion (*Protocol 5*). The type of polyvinylpyrrolidone (PVP) to use in this step may not be important. Subsequently, the protein is digested (*Protocol 5*), membrane pieces of 3 mm × 8 mm can be easily incubated in 200 μl buffer. Peptides are eluted from the membrane during digestion and are recovered by lyophilization. Oxidation with performic acid must be done after the digestion of immobilized proteins and elution of the phosphopep-tides from the membrane (oxidation may not always be necessary, see Section 3.3). No peptides are eluted from Immobilon-P by trypsin or chymotrypsin if the oxidation is carried out before proteolytic digestion: strong acid may stabilize the binding of the protein to the Immobilon-P membrane. Both nylon and nitrocellulose membranes dissolve in the strong acid.

Chymotryptic peptide mapping of proteins immobilized to membranes can also be carried out using this procedure.

Protocol 5.   Proteolytic digestion of immobilized proteins

Equipment and materials
- PVP solution (0.5% PVP-360 in 100 mM acetic acid)

Method

1. Soak the pieces of membrane, containing the protein of interest (*Protocol 4*), in PVP solution for 30 min at 37°C in a 1.5 ml microcentrifuge tube.

2. Aspirate the liquid. Wash the membrane with water extensively (5 × 1 ml) and then once or twice with freshly made 50 mM ammonium bicarbonate.

3. Incubate the piece of membrane with 10 μg TPCK–trypsin for 2 h in 200 μl 50 mM ammonium bicarbonate (pH 8.0) (see *Protocol 3*) at 37°C. The membrane should be completely submerged.

4. Vortex, add another 10 μg TPCK-trypsin, and incubate for an additional 2 h at 37°C.

5. Add 300 μl water to the sample, and centrifuge in a microcentrifuge for 5 min.

6. Transfer the liquid to a new microcentrifuge tube. At least 90% of the ^{32}P radioactivity should be present in the liquid.

7. Lyophilize in a centrifugal vacuum concentrator. It will take about 3–4 h to dry.

8. If necessary, the dried peptides can now be oxidized in 50 μl performic acid as described in *Protocol 2*.

9. At the end of oxidation, add 400 μl of water, freeze, and lyophilize in a centrifugal vacuum concentrator, and repeat once. Dissolve the sample in at least 5 μl electrophoresis buffer (*Protocol 3*, step 7) and spot the sample on a TLC plate.

4.2 Cyanogen bromide (CNBr) cleavage of immobilized proteins

CNBr cleavage is often a useful method to study sites of phosphorylation. CNBr reacts specifically with methionine residues in proteins and leads to the cleavage of the peptide bond at the C-terminal side of methionine under acidic conditions (pH <3.0), generating a homoserine lactone and thereby eliminating the sulfur atom. Because of the low abundance of methionine in proteins, CNBr fragments are often large and can be analysed by SDS–PAGE (2).

CNBr cleavage can be done either with proteins eluted from a preparative gel (*Protocol 1*) or with immobilized proteins. In the latter case, proteins must be transferred electrophoretically to a nitrocellulose membrane after separation by SDS–PAGE (*Protocol 4*). Neither Immobilon-P nor nylon membrane can be used in this procedure. Nylon dissolves in formic acid, and no radioactivity can be recovered after CNBr digestion of proteins bound to Immobilon-P. Nitrocellulose membrane should be wrapped in Saran wrap (cling film) to keep it moist during autoradiography.

CNBr cleavage can then be carried out by incubating a piece of membrane containing the protein of interest with CNBr in 70% formic acid (*Protocol 6*); note that 70% formic also cleaves Asp–Pro bonds (*Tables 1* and *4*). Increased reaction time will not increase the yield. The volume of the reaction can be varied depending on the size of the piece of nitrocellulose. Enough solution should be used to cover the piece of membrane completely. A larger volume usually increases the recovery of the peptides released from the membrane during the reaction. We generally use a volume of 200 μl for a 3 mm × 8 mm piece of nitrocellulose. CNBr is extremely toxic and should be handled only in a fume hood.

During digestion, peptides are released into solution, and can be recovered after the formic acid and CNBr are removed by two rounds of lyophilization. Depending on the protein, recoveries of 85–95% can be obtained. It is important to remove the formic acid completely, since residual acid will interfere with subsequent SDS–PAGE analysis. For one-dimensional analysis by SDS–PAGE, peptides are dissolved in standard SDS–PAGE sample buffer. If the bromophenol blue in the sample buffer turns yellow, it indicates the presence of residual formic acid in the sample. Small amounts of concentrated Tris–base (pH 9) can be added to raise the pH. Low molecular mass fragments can be resolved by electrophoresis on a 24% SDS gel. A tricine cathode buffer will increase the resolution of low molecular mass peptides.

Protocol 6. CNBr cleavage of immobilized proteins

Equipment and materials

- CNBr/formic acid (50 mg/ml in 70% formic acid; make a 300 mg/ml stock solution by adding 70% formic acid directly to the container in which the CNBr is supplied; this solution can be stored in aliquots stably at –70°C, and then diluted with 70% formic acid before use)
- Tricine cathode buffer [0.1 M *N*-tris-(hydroxymethyl)methylglycine (tricine), 0.1% SDS, 0.1 M Tris–base, pH 8.25]
- SDS–PAGE sample buffer (755 mg Tris–base, 10 ml glycerol, dissolve in 35 ml H_2O, adjust to pH 6.8 with 6 M HCl, add 2 g SDS, 5 ml 2-mercaptoethanol, 1 mg bromophenol blue, dilute to 100 ml with H_2O)

Protocol 6. *Continued*

Method

1. Incubate the piece of nitrocellulose membrane, containing the protein of interest (*Protocol 4*), with CNBr/formic acid for 1–1.5 h at room temperature in a capped 1.5 ml microcentrifuge tube.

2. Centrifuge the samples in a microcentrifuge at ~7000 g for 5 min.

3. Transfer the liquid to a new microcentrifuge tube.

4. Lyophilize the sample in a centrifugal vacuum concentrator. It will take about 30 min to dry a sample contained in 200 μl 70% formic acid.

5. Redissolve the residue in 30–40 μl deionized water and lyophilize; repeat once.

6. Dissolve the CNBr fragments in SDS–PAGE sample buffer. Analyse the peptides by electrophoresis on a 24% acrylamide, 0.054% bisacrylamide gel with the tricine cathode buffer.

5. Separation of phosphopeptides in two dimensions on TLC plates

Peptide maps can be obtained by reversed phase high-performance liquid chromatography, described in *Chapter 6* of this volume, or by two-dimensional separation on TLC plates. If digestion generates very large peptides, analysis can be carried out by SDS–PAGE. Here we will describe the analysis of phosphopeptides on TLC plates by electrophoresis in the first dimension, followed by chromatography in the second dimension.

5.1 Choice of electrophoresis buffer

The digestion products are separated in the first dimension by electrophoresis. The mobility in the electrophoresis dimension is determined by the ratio of the charge and the mass of a peptide and can be described by the formula $m_r = keM^{-2/3}$ in which m_r is the relative mobility, k is a constant, e is the electrical charge, and M is the mass of the molecule (3). Three different buffers are commonly used for separation in the first dimension: pH 1.9 buffer, pH 4.72 buffer, and pH 8.9 buffer (*Table 2*). These buffers are composed of volatile solvents that can be evaporated completely by air-drying the plates after electrophoresis. This allows chromatography or electrophoresis in a different buffer system in the second dimension. In general, phosphopeptides containing basic residues will be resolved better at lower pH. Basic amino acid side chains are protonated and thus charged at lower pH, resulting in a greater mobility towards the cathode. Conversely, peptides containing acidic residues will be charged at higher pH, resulting in higher

Table 2. Composition of electrophoresis buffers[a]

Buffer	Amount	Component
pH 1.9 buffer	50 ml	Formic acid (88% w/v)
	156 ml	Glacial acetic acid
	1794 ml	Deionized water
pH 3.5 buffer	100 ml	Glacial acetic acid
	10 ml	Pyridine[b]
	1890 ml	Deionized water
pH 4.72 buffer	100 ml	n-Butanol
	50 ml	Pyridine[b]
	50 ml	Glacial acetic acid
	1800 ml	Deionized water
pH 6.5 buffer	8 ml	Glacial acetic acid
	200 ml	Pyridine[b]
	1792 ml	Deionized water
pH 8.9 buffer	20 g	Ammonium carbonate
	2000 ml	Deionized water

[a] For 2 litres, made up using reagent-grade solvents and deionized water, and stored in glass bottles with air-tight lids.
[b] Pyridine is unstable and should be not be used if obviously yellow in colour. Store pyridine under nitrogen, by flushing the bottle with nitrogen and capping tightly.

mobility towards the anode. Optimal conditions for separation by electrophoresis, however, are usually determined empirically. Most phosphopeptides contain both acidic and basic groups and in order to find which system gives the best separation, each of the buffers should be tried. After choosing a particular buffer system to analyse the protein of interest, the separation in the electrophoresis dimension can be improved by changing the electrophoresis time. This may require a change in the position of the origin to prevent peptides from moving off the TLC plate. If no one particular buffer system appears to be superior, it may be best to use pH 1.9 buffer. Most peptides appear to be soluble in pH 1.9 buffer and streaked maps are obtained less often with this buffer.

Protocol 7. Loading TLC plates for two-dimensional separation of phosphopeptides

Equipment and materials

- electrophoresis buffers[a] (see *Table 2*)
- glass-backed TLC plates (20 cm × 20 cm, 100 μm cellulose, e.g. from E. M. Science; check before use to make sure that they do not have gross irregularities in the cellulose coating, discard imperfect plates; plastic-backed plates can also be used)
- green marker dye (5 mg/ml ε-dinitrophenyl-lysine (yellow) and 1 mg/ml xylene cyanol FF (blue) in 50% pH 4.72 buffer in deionized water)

Protocol 7. *Continued*

Method

1. Dissolve samples in at least 5 μl pH 1.9 buffer, 4.72 buffer, or deionized water, using a vortex mixer.

2. Mark the sample and dye origins on the cellulose side of a TLC plate with a small cross using an extra soft blunt-ended pencil, making sure not to perturb the cellulose layer [or mark on the reverse (glass) side with a permanent marker] by placing the plate over a marking template on a light box (*Figure 1*). Samples are usually electro-phoresed in the direction in which the plates were poured (i.e. the sides where the cellulose extends to the end of the plate are in contact with the buffer wicks during electrophoresis), and the plate should be oriented accordingly for marking.

3. Spot the sample on to the origin using a fixed volume micropipette fitted with a disposable capillary tip. To keep the sample spot small (which improves resolution), apply 0.2–0.5 μl drops, and dry between applications using an air-line fitted with a filter to trap aerosols and particulate matter, and a 1 ml syringe or a pasteur pipette to focus the flow. Avoid touching the plate with the air nozzle or the pipette tip, since gouges on the cellulose may affect electrophoresis or the chromatography.

4. Spot 0.5 μl of green marker on the dye origin at the top of the plate (*Figure 1*). This marker dye is green, but separates into its blue and yellow components during electrophoresis. The blue dye gives a visual indication of the progress of electrophoresis, while the yellow compound defines the position to which neutral peptides should migrate, except at pH 1.9 where ε-dinitrophenyl-lysine is positively charged.

[a] Check the pH of the buffer before use; if it varies by >0.2 units from that specified, *do not* adjust the pH but make up fresh buffer.

5.2 Electrophoresis

There are many commercial flat-bed thin-layer electrophoresis systems available. In our laboratories we use the Hunter Thin Layer Electrophoresis apparatus HTLE 7000, which is available from CBS Scientific, Inc. The HTLE 7000 has several important features that may be missing in other systems (e.g. the restraining lid or cooling plate). During electrophoresis the apparatus is closed using a clamping system. An airbag, which is connected to an air-line delivering 10 pounds per square inch, is inflated, thus removing excess buffer from the plate as well as preventing buffer from siphoning up from the buffer tanks on to the TLC plate. Excess buffer on the plate may result in fuzzy

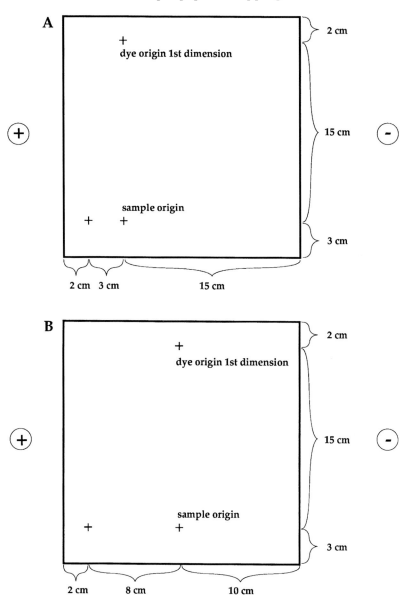

Figure1. Location of sample and dye origins for two-dimensional separation of phospho-peptides on 20 cm × 20 cm TLC plates. Panel A is for electrophoresis at pH 1.9 or 4.72, and panel B for electrophoresis at pH 8.9. Marker dye for the second dimension can be spotted on the left- (as shown here) or right-hand margin of the plate once the plate has been dried after electrophoresis.

maps. In addition, this system features a cooling system that prevents over-heating of the plate during the run. The method for carrying out electrophoresis is described in *Protocol 8*.

Protocol 8. First dimension thin-layer electrophoresis

Equipment and materials
- electrophoresis buffers[a, b] (see *Table 2*)
- a power pack capable of delivering at least 100 mA at 1.5 kV
- electrophoresis apparatus (the protocol below is for the HTLE 7000 from CBS Scientific Inc.: this requires running water at ~16°C, and a constant pressure air-line with a regulator valve; other types of apparatus may lack the cooling plate and/or airbag)

Method

1. Fill both buffer tanks with ~600 ml of the appropriate electrophoresis buffer.

2. Cut two 35 cm × 25 cm sheets of thin (0.1 mm) polyethylene sheeting. These sheets can be re-used several times, but should be discarded after running samples with large amounts of radioactivity (>25 000 cpm).

3. Set up the apparatus as shown in *Figure 2*, but without the thin layer plate. The polyethylene sheets should stick out ~4 cm at both sides and ~1 cm at the front and back. Place one sheet on the Teflon sheet over the cooling plate and tuck the ends down between the cooling plate and the buffer tanks.

4. Cut two sheets (20 cm × 28 cm) of Whatman 3MM paper and fold lengthwise to give double thickness sheets (20 cm × 14 cm) for the electrophoresis wicks. Wet the wicks in electrophoresis buffer and place them into the slots of the buffer tank next to the cooling plate, with the folded edge up. Fold the ends of the wicks over the polyethylene sheet on the cooling plate.

5. Place the second polyethylene sheet over the cooling plate so that it covers the cooling plate, the wicks, and part of the buffer tank.

6. Place the second Teflon sheet and the neoprene pad on top, close the apparatus, and secure the lid with the two pins.

7. Turn the air pressure up to 10 pounds per square inch, which inflates the airbag and squeezes out excess buffer from the wicks.

8. When ready to start, shut off the air pressure, and open the apparatus. Remove the pad, the Teflon sheet, and the polyethylene sheeting, and wipe the sheeting dry with tissue paper.

9. Make a blotter from two layers of Whatman 3MM paper (*Figure 3*), stitched together with a zigzag stitch around the edges. The holes

(1.5 cm) that surround the origins are cut with a sharp cork borer. These blotters can be re-used many times, but it is best to keep a separate blotter for each buffer. Soak the blotter briefly in electrophoresis buffer at the chosen pH, and remove excess liquid by blotting briefly on another piece of 3MM paper.

10. Wet the TLC plate by placing the wetted blotter over it, with the sample and marker origins in the centres of the two holes, respectively. Press the blotter on to the plate around the sample and marker origins with your fingertips to ensure uniform concentration of the sample as the buffer front moves towards the origins.

11. Examine the plate: it should be dull grey with no shiny puddles of buffer. Allow excess buffer to evaporate or blot very carefully with tissue paper.

12. Place the plate on the apparatus, fold the wicks over the plate so they cover ~1 cm of the plate at each end, and carefully reassemble the apparatus as in *Figure 2*. Avoid lateral movement of the polyethylene sheeting when it is in contact with the TLC plate.

13. Secure the lid with the pins, inflate the airbag to 10 pounds per square inch, and then turn on the cooling water flow.

14. Turn on the power. With the HTLE 7000 system, electrophoresis is typically for 20–27 min at 1.0 kV.

15. Air-dry the plates for ~20 min using a fan (do not oven-dry).

[a] Check the pH of the buffer before use: if it varies by >0.2 units from that specified, do not adjust the pH but make up fresh buffer.
[b] Pyridine is unstable and should not be used if yellow in colour; flush buffer bottles containing pyridine with nitrogen after use and cap tightly.

5.3 Separation in the second dimension by chromatography

Separation in the second dimension is achieved by ascending chromatography. The hydrophobicity of the different peptides determines their ability to partition between two phases. The chromatography buffer that migrates slowly to the top of the plate is the mobile phase; the cellulose on the plate forms the stationary phase. The migration of a particular peptide determines the ratio of time spent in the mobile and stationary phases. A hydrophobic peptide that spends a greater part of the time during the separation in the mobile phase will migrate relatively fast towards the top of the plate. Conversely, a hydrophilic peptide that spends a greater part of the time bound to the stationary phase will move only slowly towards the top of the plate.

We use three different buffer systems (*Table 3*). 'Regular chromatography buffer' is most often used for fingerprinting [35]S- or ^{125}I-labelled proteins. 'Phosphochromo buffer' is a more hydrophilic mixture and is the buffer of

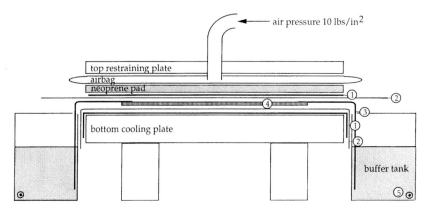

①   teflon insulators

②   polyethylene protector sheets

③   Whatman 3MM wick (double layer)

④   TLC plate

⑤   electrode

Figure 2. Layout of the HTLE 7000 flat-bed apparatus for electrophoresis.

Table 3. Composition of chromatography buffers[a]

Buffer	Amount (ml)	Component
Regular chromatography	785 ml	*n*-Butanol
	607 ml	Pyridine[b]
	122 ml	Glacial acetic acid
	486 ml	Deionized water
Phosphochromatography	750 ml	*n*-Butanol
	500 ml	Pyridine[b]
	150 ml	Glacial acetic acid
	600 ml	Deionized water
Isobutyric acid buffer	1250 ml	Isobutyric acid
	38 ml	*n*-Butanol
	96 ml	Pyridine[b]
	58 ml	Glacial acetic acid
	558 ml	Deionized water

[a] For 2 litres, made up using reagent-grade solvents and deionized water, and stored in glass bottles with air-tight lids.
[b] Pyridine is unstable and should be not be used if obviously yellow in colour. Store pyridine under nitrogen, by flushing the bottle with nitrogen and capping tightly.

---

choice for analysing phosphopeptides. 'Isobutyric acid buffer' is sometimes used to resolve extremely hydrophilic phosphopeptides, but is foul smelling. The best buffer system for a particular separation should be determined by preliminary trials.

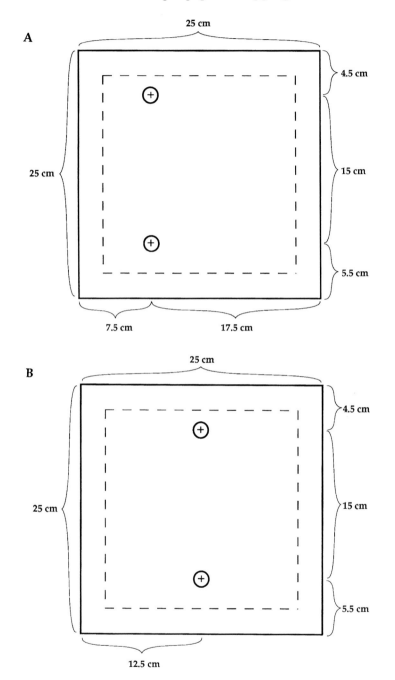

Figure 3. Dimensions of blotters for wetting TLC plates before electrophoresis at pH 1.9 and 4.72 (A) or pH 8.9 (B).

Protocol 9. Second dimension: thin-layer chromatography

Equipment and materials

- chromatography buffers (see *Table 3*: they should be made up using reagent-grade solvents and deionized water, and stored in amber-coloured glass bottles with air-tight lids)
- chromatography tanks[a] (sizes holding from two to eight 20 cm × 20 cm plates are

available, e.g. from CBS Scientific Inc; the LCT-100 holds eight plates: they should have tight-fitting lids sealed with vacuum grease; smaller tanks should be lined with Whatman 3MM paper to ensure equilibration of the buffer and vapour phases)

Method

1. Spot a drop (~0.5 μl) green marker dye (*Protocol 7*) in the left- or right hand-margin of the plate (see *Figure 1*) at the same level as the sample origin. The dye components may be used to monitor the progress of chromatography and to calculate relative mobilities of peptides.

2. Place the dried plates in an almost upright position in the tanks and replace the lid (do not disturb or open a tank while chromatography is in progress).

3. Allow the buffer to run to within 1 cm of the top of the plate. This usually takes 8–12 h, but the time depends on the batch of plates, the buffer system, the quality of reagents used to make up the buffer, and the ambient temperature.

4. Remove all the plates (irrespective of the position of the buffer front on individual plates) and allow to dry in a fume hood or 65°C oven. An oven should not be used if peptides are to be extracted from these plates for further analysis.

5. Mark the dried plates with radioactive or fluorescent ink around the edge.

6. Expose the plates to presensitized X-ray film with an intensifier screen for autoradiography, or to a phosphorimager screen.

---

[a] Chromatography tanks can be used for several months with the same buffer, but be aware that the buffer composition changes if the lid is left off for too long or is not tightly sealed.

---

6. Phosphoamino acid analysis

Phosphoamino acid analysis is accomplished by partial hydrolysis of the purified ^{32}P-labelled protein. Incubation of proteins or peptides in concentrated acid or base results in hydrolysis of peptide bonds and release of peptides and, eventually, individual amino acids. Phosphodiester bonds, however, are also unstable under these reaction conditions and dephosphorylation of the phosphoamino acids and release of free [^{32}P]phosphate also occur. The time of hydrolysis is therefore critical.

The protein of interest can be purified by gel electrophoresis and isolated from the gel by elution, followed by TCA precipitation as described above for peptide mapping (*Protocol 1*). Alternatively, proteins can be transferred from the gel to Immobilon-P (*Protocol 4*) and hydrolysed on the membrane (4). Nitrocellulose or nylon membranes are not stable in concentrated acid and are, therefore, not suitable for this purpose. Hydrolysis of Immobilon-bound protein is recommended except in cases where the protein of interest transfers poorly.

6.1 Acid hydrolysis of purified phosphoproteins

Before hydrolysis, resolve complex protein mixtures by SDS–PAGE and transfer to Immobilon-P (*Protocol 4*). To remove residual salt and detergent the membrane should be washed several times with a large volume of water after transfer and the optional staining with India ink. The membrane should be wrapped in Saran wrap (cling film) to keep it moist while exposing it to X-ray film. Proteins on membranes are hydrolysed as shown in *Protocol 10*. Alternatively, TCA-precipitated phosphoproteins can be subjected to acid hydrolysis. Dissolve the TCA-precipitated protein, after the ethanol wash, in 50 μl hydrochloric acid and hydrolyse for 1 h at 110 °C (*Protocol 10*, step 2). Remove the hydrochloric acid by evaporation (*Protocol 10*, step 4) and the samples are ready for analysis.

^{32}P-labelled phosphoamino acids can also be released by hydrolysis in strong base. This method is useful when looking for small amounts of phosphotyrosine (P.Tyr), since P.Tyr is considerably more stable in strong base than it is in strong acid (5). This technique is usually carried out by incubation of TCA-precipitated or membrane-bound protein with 5 M KOH at 110°C. The drawback is that the KOH needs to be removed by anion-exchange chromatography to allow electrophoresis on TLC plates.

Protocol 10. Acid hydrolysis of membrane-bound phosphoproteins

Equipment and materials

- 5.7 or 6 N HCl
- 110 °C oven

Method

1. Place the strip of membrane in a screw-cap, 1.5 ml microcentrifuge tube and rinse the membrane several times with deionized water.

2. Add 200 μl 5.7 (constant boiling) or 6 N HCl, secure the cap, and incubate for 1 h at 110°C.

Protocol 10. *Continued*

3. Centrifuge for 5 min at ~7000 g in a microcentrifuge at room temperature, and transfer the supernatant to a new microcentrifuge tube. There may be a small residue at the bottom of the tube; avoid transferring this.

4. Evaporate HCl in a centrifugal vacuum concentrator.

6.2 Separation of phosphoamino acids by electrophoresis in two dimensions on TLC plates

Phosphoamino acids can be separated from each other and from possible contaminating nucleotides by electrophoresis in two dimensions (*Protocol 11*). Phosphothreonine (P.Thr) and P.Tyr co-migrate during electrophoresis at pH 1.9, but separate well from phosphoserine (P.Ser). P.Thr separates well from P.Tyr during electrophoresis at pH 3.5. Phosphate, phosphopeptides, and the individual phosphoamino acids from relatively pure samples can be resolved adequately by electrophoresis in one dimension at pH 3.5. This method may be preferred when working with *in vitro* phosphorylated proteins or purified peptides, where phosphate, phosphopeptides, and phosphoamino acids are the only labelled compounds present in the hydrolysate. The principal advantage is that as many as 16 samples can be analysed on a single plate. When working with *in vivo* ^{32}P-labelled material, 3'-UMP and ribose 3'-phosphate, generated from ^{32}P-labelled RNA, are major contaminants, and co-migrate with P.Tyr during electrophoresis at pH 3.5. For further information see Cooper *et al.* (6) and Duclos *et al.* (5). Phosphoamino acids can also be resolved by HPLC using anion-exchange columns (7).

Protocol 11. Two-dimensional electrophoresis of phosphoamino acids

Equipment and materials

- pH 1.9 and pH 3.5 buffers (*Table 3*)
- phosphoamino acid standards (1 mg/ml P.Ser, 1 mg/ml P.Thr, 1 mg/ml P.Tyr dissolved in water; this mixture is stable for years at 4°C)
- radioactive ink (Indian ink containing 10 μCi/ml ^{35}S)
- ninhydrin solution (0.25% ninhydrin in acetone)

Method

1. Dissolve the protein hydrolysate samples using a vortex mixer in at least 5 μl pH 1.9 buffer containing ~0.5 μl unlabelled phosphoamino acid standards (i.e. use 15 parts pH 1.9 buffer plus 1 part phosphoamino acid standard mix). Centrifuge for 30–60 sec in a microcentrifuge at room temperature to remove particulate material.

2. Fours samples can be run on one TLC plate (*Figure 4*). Mark each origin with a cross using a blunt, extra soft pencil on the cellulose, or a marker pen on the reverse side (see *Protocol 7*, step 2).

3. Spot a drop (~0.5 μl) of green marker (*Protocol 7*) at the marker origin at the top of the plate, and load the samples at the other four origins as in *Protocol 7*. The whole sample can be spotted, but ~50–100 cpm is sufficient for reproducible phosphoamino acid analysis, bearing in mind that after partial acid hydrolysis only ~20% of the radioactivity will be present as phosphoamino acids, the remainder being mainly inorganic phosphate plus some incompletely hydrolysed phospho-peptides.

4. Wet the plate with pH 1.9 buffer, using a blotter with five holes (*Figure 5A*). These are made from two layers of Whatman 3MM paper stitched together as described in *Protocol 8*, step 9. Soak the blotter briefly in pH 1.9 buffer, remove excess buffer by blotting it briefly on another sheet of Whatman 3MM paper, and place on the TLC plate, with the holes over the sample and dye origins.

5. Press around the circumference of each hole with your fingertips to aid in the even flow of buffer towards the origins, thereby con-centrating each sample on its origin. The plate should be uniformly wet with no puddles of excess buffer.

6. Carry out electrophoresis (see *Protocol 8*) for 20 min at 1.5 kV.[a] The phosphoamino acids move towards the anode (*Figure 4*).

7. Remove the plate from the electrophoresis apparatus and dry the plate using a fan for ~30 min.

8. The dried plate is wetted with pH 3.5 buffer as in steps 4 and 5 but using three single layer strips of Whatman 3MM paper designed to wet the areas between the samples (*Figure 5B*). When the edges of these strips are pressed down, the buffer will flow towards and concentrate the sample separated in the first dimension. Sometimes a sharp brown line will be seen where the buffer fronts meet, but this does not affect subsequent electrophoresis.

9. Carry out electrophoresis for 16 min at 1.3 kV[a] as before but in a direction at 90° with respect to the first dimension. The phospho-amino acids move towards the anode (*Figure 4*).

10. Dry the plate using a fan or in a 65°C oven for several minutes.

11. Spray the plate with ninhydrin solution and heat in a 65°C oven for 5–10 min. The internal phosphoamino acid standards will appear as purple spots, which should be no more than 5 mm in diameter if the procedure has been done properly.

12. Mark several spots around the edge of the plate with radioactive ink

Protocol 11. *Continued*

or fluorescent ink and expose to pre-sensitized X-ray film at –70°C with an intensifier screen for autoradiography, or to a phosphor-imager plate.

13. Align the autoradiogram or phosphorimage with the ink marks on the plate, and determine whether there are radioactive spots coincident with the stained phosphoamino acid marker spots. Radioactive phosphoamino acids released by hydrolysis will co-migrate precisely with the stained markers (N.B. upon acid hydrolysis some P.Ser-containing proteins generate a ^{32}P-labelled product that runs very close to phosphotyrosine).

[a] Electrophoresis conditions are for the HTLE 7000 electrophoresis system and are chosen to ensure that the migration of inorganic phosphate is confined to the quadrant of the plate reserved for each sample (*Figure 4*); conditions may need to be optimized for other systems.

7. Further analysis of phosphopeptide maps

Phosphopeptide mapping enables the investigator to estimate the number of phosphorylation sites present in a particular protein, assuming that all phosphorylation sites are separated from each other by proteolytic cleavage sites. By comparing phosphopeptide maps of the same protein isolated from cells grown under different conditions one can find out whether growth conditions affect the phosphorylation status of the protein that is being studied. Sometimes it may be possible to correlate a difference in phosphorylation status with a change in stability or enzymatic activity. Phosphopeptide maps can also be used to investigate whether phosphorylation of a particular protein with a purified protein kinase involves sites that are also phosphorylated *in vivo*. If one wants to identify a specific phosphate-acceptor site, further analysis is usually required.

7.1 Isolation of individual phosphopeptides from TLC plates

Individual phosphopeptides can be isolated from TLC plates for further analysis. Pipette tips containing porous polyethylene disks (Omnifit, Atlantic Beach, NY) are used to scrape the cellulose off the plates (*Protocol 12*), and phosphopeptides are eluted using pH 1.9 buffer (*Table 2*). The recovery of hydrophobic peptides may be improved by the use of 0.1% trifluoroacetic acid/20% acetonitrile for elution. The small amount of cellulose that makes its way through the porous disk is cleared from the eluate by centrifugation in a microcentrifuge and the supernatants are lyophilized. The elution tips can be

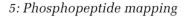

A

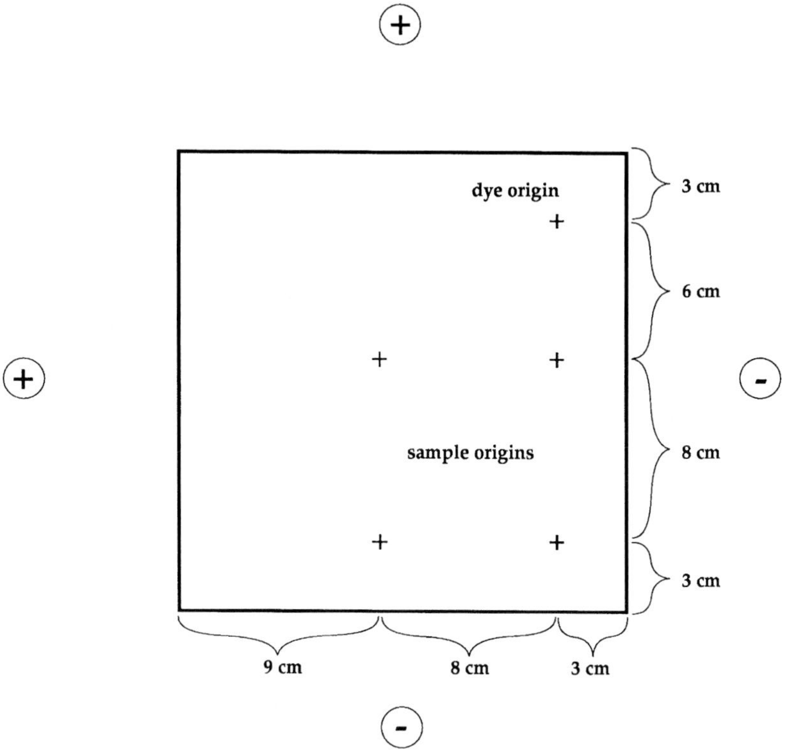

B

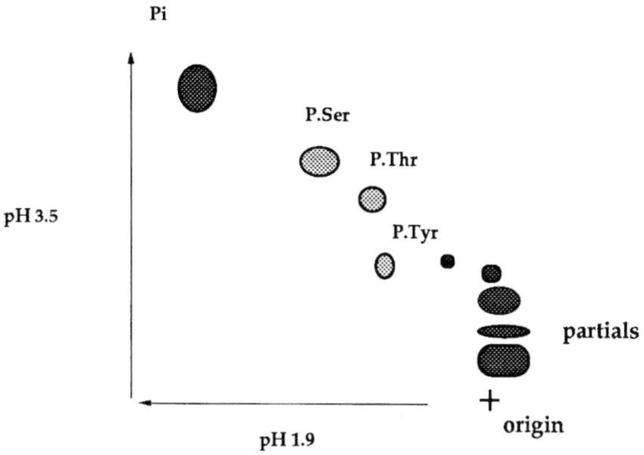

Figure 4. (A) Locations of sample and dye origins for two-dimensional electrophoresis of phosphoamino acids on 20 cm × 20 cm TLC plates. Four samples can be spotted per plate. (B) Position of the individual amino acids after two-dimensional electrophoresis. After partial acid hydrolysis of ^{32}P-labelled phosphoproteins (*Protocol 10*) about 20% of the radioactivity migrates as phosphoserine (P.Ser), phosphothreonine (P.Thr), or phosphotyrosine (P.Tyr), the remainder being present as inorganic phosphate (Pi) or peptide products derived by partial digestion (partials).

121

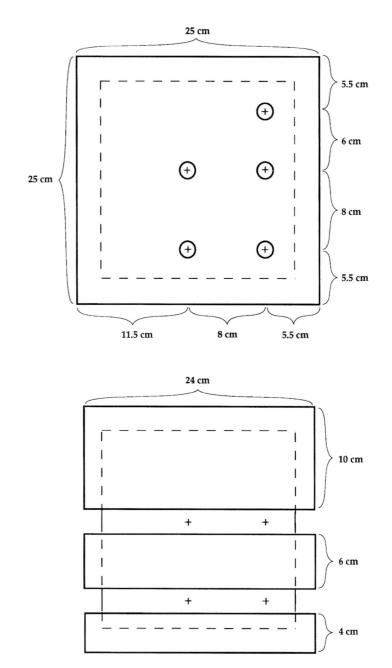

Figure 5. Dimensions of blotters for wetting TLC plates before separation of phospho-amino acids by electrophoresis in the first dimension at pH 1.9 (Panel A) and in the second dimension at pH 3.5 (Panel B).

re-used several times. To clean them, attach a vacuum line connected to a liquid trap to the small end of the tip and flush then sequentially with deionized water and 96% ethanol. Dry the tips under vacuum.

Protocol 12. Isolation of individual phosphopeptides from TLC plates

Equipment and materials

• pH 1.9 buffer (see *Table 2*)

Method

1. Align the TLC plate with the film using the the radioactive or fluorescent ink marks on the plate. Hold the sandwich over a light-box and localize phosphopeptides of interest. Mark the outline of the peptide with a soft pencil on the cellulose or on the back of the plate with a marker pen.

2. Slice the first 3–4 mm off the pointed end of a blue disposable pipette tip for a 1 ml adjustable pipette and push a 6.5 mm diameter porous polyethylene disk into the large end with a glass rod until it is firmly in place.

3. Attach the large end of an elution tip to a vacuum line, turn on the vacuum, and, after making sure that the disk remains in place, scrape the cellulose containing an individual phosphopeptide off the plate with the cut end of the pipette tip. The dislodged cellulose will be sucked into the tip and caught against the disk.

4. Remove the tip from the vacuum line, and place the tip, large end down, in a microcentrifuge tube. Add 100 μl pH 1.9 buffer (or 0.1% trifluoroacetic acid/20% acetonitrile) to the small end of the tip, and let it stand at room temperature for 5 min. Centrifuge for 5–10 sec at room temperature in a microcentrifuge at ~500 g. Be careful the tip does not hit the lid. Repeat this process twice using pH 1.9 buffer and twice using deionized water (or four times with 100 μl 0.1% trifluoroacetic acid/20% acetonitrile), collecting each eluate sequentially in the same tube.

5. Centrifuge the combined eluates for 5 min at ~7000 g in a microfuge at room temperature and transfer the supernatant to a new tube. Lyophilize in a centrifugal vacuum concentrator.

6. Count the radioactivity in the cellulose and in the eluate by Cerenkov counting. If elution was inefficient, re-extract with 0.1% trifluoroacetic acid/20% acetonitrile.

7.2 Experimental manipulations of individual phosphopeptides

To determine the phosphoamino acid content of purified phosphopeptides simply follow *Protocols 10* and *11*. To characterize sites of phosphorylation, the susceptibility of phosphopeptides to digestion with different proteases or chemicals can be tested. Incubate individual phosphopeptides, isolated from thin layer plates as described in *Protocol 12*, with proteases or chemicals (for the specificity and conditions of use of different reagents see *Tables 1* and *4*), and analyse by running the digested phosphopeptide, the undigested phospho-peptide and a mix of these two in one or two dimensions on TLC plates. A shift in the mobility is indicative of the presence of a cleavage site. For analysis of mixtures of digested and undigested peptides, it may be important to inactivate the protease by boiling the digest for 2 min in 1% (v/v) 2-mercaptoethanol. Active proteases can digest the undigested phosphopeptide in a mixture while being spotted onto the TLC plate.

It is sometimes possible to identify the position of a phosphorylated residue within a phosphopeptide by subjecting a purified phosphopeptide to manual Edman degradation. This can be done using the conventional Edman reagent or a newly developed volatile reagent (8). The reaction products of each cycle can be analysed by electrophoresis at a suitable pH in one dimension on TLC plates (9). It is best to analyse the reaction products of all of the cycles on a

Table 4. Specificity of different cleavage reagents

Reagent	Specificity	Reference
Carboxypeptidase B	X–K–COO⁻; X–R–COO⁻	10
α-Chymotrypsin	F–X; W–X; Y–X	11
Endoproteinase Asp-N	X–D; X–C$_{SO3H}$	12
Endoproteinase Glu-C (V8)	E–X	13
Proline specific endopeptidase	P–X	14
Thermolysin	X–L; X–I; X–V	15
TPCK–Trypsin	K–X; R–X	11
Cyanogen bromide	M–X	16
Hydroxylamine	N–G	17
Formic acid	D–P	18

single plate. Run free [^{32}P]orthophosphate (50–200 cpm) and a sample taken from the starting material as markers. The cycle where the label is released from the peptide identifies the position of the phosphoamino acid within the peptide. If the phosphorylated residue is P.Ser or P.Thr, the label comes off as free [^{32}P]orthophosphate; if the phosphorylated residue is P.Tyr, the label comes off as PTH-P.Tyr (9). Reliable analysis requires at least 50–100 cpm of purified phosphopeptide per cycle. It is difficult to run more than five cycles. It is also possible to do the same procedure with peptides covalently linked to an inert membrane such as Sequelon (Millipore, Bedford, MA), which may be useful since more cleavage cycles can generally be achieved.

In order to identify a particular phosphorylation site the following approach may be useful. Based on the primary sequence derived from DNA sequencing, make a list of all the possible phosphopeptides that will be generated by digestion. Make a table listing the properties of the individual candidate peptides (see, for example, ref. 1). List the position and nature of the phosphate acceptor site within the candidate peptides and the theoretical susceptibilities to digestion with secondary proteases or chemicals (*Table 4*), for comparison with experimentally determined phosphoamino acid content and susceptibilities to secondary digestion. In addition, note the presence of Cys and Met within candidate peptides. The presence of Met or Cys residues can potentially be determined by double labelling procedures followed by peptide mapping. Such maps are first exposed for ^{32}P with an intensifier screen and aluminium foil shielding the low energy radiation emitted by the ^{35}S. After decay of the ^{32}P and treatment with a fluorographic enhancer, the maps can be exposed for ^{35}S. The two exposures can be lined up with the help of radioactive marks on the TLC plates. This method, however, will only yield reliable results if the phosphopeptide that is studied is phosphorylated to a high stoichiometry (9). Predicted mobilities under the chosen conditions of electrophoresis and chromatography for all the possible phosphorylated peptides can be calculated (9) (this can be done via the Internet at http://www.genestream.org) and then plotted. In combination with additional analyses, such plots can be useful in helping to determine the identities of phosphopeptides. Using the procedures described above it should be possible to eliminate most of the candidate peptides. This method has been successfully used for identification of phosphorylation sites in several different proteins.

Ultimately it is desirable to confirm the identification of a phosphorylation site by co-migration of the phosphopeptide in question with a synthetic peptide of known sequence that was phosphorylated *in vitro*. It is possible that a synthetic peptide is not a good substrate for the phosphorylation by a protein kinase. The synthesized peptide may be too small and lacking part of the protein kinase recognition sequence. Synthesis of a longer peptide followed by phosphorylation and trimming with the appropriate protease may circumvent this problem. As an alternative to this approach it is possible to chemically synthesize phosphopeptides directly and test for co-migration with a ^{32}P-labelled peptide, detecting the phosphorylated standard by staining with ninhydrin (*Protocol 11*). The identity of a phosphorylation site can also be confirmed by mutating the putative acceptor site and analysis of the phosphorylation of the mutant protein after transfection and expression in an appropriate cell line.

References

1. van der Geer, P. and Hunter, T. (1990). *Mol. Cell. Biol.*, **10**, 2991.
2. Luo, K., Hurley, T. R., and Sefton, B. M. (1990). *Oncogene*, **5**, 921.
3. Offord, R. E. (1966). *Nature*, **211**, 591.
4. Kamps, M. A. and Sefton, B. M. (1989). *Anal. Biochem.*, **176**, 22.
5. Duclos, B., Marcandier, S., and Cozzone, A. J. (1991). In *Methods in enzymology* (ed. T. Hunter and B. M. Sefton), Vol. 201, p. 10. Academic Press, New York.
6. Cooper, J. A., Sefton, B. M., and Hunter, T. (1983). In *Methods in enzymology* (ed. J. D. Corbin and J. G. Hardman), Vol. 99, p. 387. Academic Press, New York.
7. Ringer, D. P. (1991). In *Methods in enzymology* (ed. T. Hunter and B. M. Sefton), Vol. 201, p. 3. Academic Press, New York.
8. Fischer, W. H., Hoeger, C. A., Meisenhelder, J., Hunter, T., and Craig, A. G. (1997). *Protein Chem.*, **16**, 329.
9. Boyle, W. J., van der Geer, P., and Hunter, T. (1991). In *Methods in enzymology* (ed. T. Hunter and B. M. Sefton), Vol. 201, p. 110. Academic Press, New York.
10. Folk, J. E. (1971). In *The Enzymes* (ed. P. D. Boyer), Vol. 3, p. 57. Academic Press, New York.
11. Smyth, D. G. (1967). In *Methods in Enzymology* (ed. C. H. W. Hirs), Vol. 11, p. 214. Academic Press, New York.
12. Drapeau, G. R. (1980). *J. Biol. Chem.*, **255**, 839.
13. Drapeau, G. R. (1977). In *Methods in enzymology* (ed. C. H. W. Hirs and S. N. Timasheff), Vol. 47, p. 189. Academic Press, New York.
14. Yoshimoto, T., Walter, R., and Tseru, D. (1980). *J. Biol. Chem.*, **255**, 4786.
15. Heinrikson, R. L. (1977). In *Methods in enzymology* (ed. C. H. W. Hirs and S. N. Timasheff), Vol. 47, p. 175. Academic Press, New York.
16. Stark, G. R. (1977). In *Methods in enzymology* (ed. C. H. W. Hirs and S. N. Timasheff), Vol. 47, p. 129. Academic Press, New York.
17. Bornstein, P. and Balian, G. (1977). In *Methods in enzymology* (ed. C. H. W. Hirs and S. N. Timasheff), Vol. 47, p. 132. Academic Press, New York.
18. Landon, M. (1977). In *Methods in enzymology* (ed. C. H. W. Hirs and S. N. Timasheff), Vol. 47, p. 145. Academic Press, New York.

6

Phosphorylation site analysis by mass spectrometry

KEN I. MITCHELHILL and BRUCE E. KEMP

1. Introduction

The determination of phosphorylation sites on proteins is a vital part of their characterization, and an important step in understanding the structural basis of how phosphorylation modulates their functions. With the recent explosive growth in the use of mass spectrometry in the structural analysis of proteins and peptides, this approach has also become the method of choice for the elucidation of post-translational modifications, including phosphorylation. Many developments have taken place since early studies using fast-atom bombardment mass spectrometry (FABMS) in 1985 (1). Matrix-assisted laser desorption/ionization time-of-flight (MALDITOF), electrospray ionization (ESI) triple quadrupole mass spectrometry and its low flow rate version, nano-ESI mass spectrometry, are currently the most widely used forms of instrumentation in these studies. They are both sensitive and highly precise, and this chapter will be confined to these instrumentation types, with which we have practical experience. Hybrid instruments e.g. Q-TOF (quadrupole time-of-flight) are now available, or in advanced stages of development, from several manufacturers, as well as both electrospray and MALDITOF instruments with Fourier transform detection at the frontiers of precision. Other mass spectrometric analytical methodologies, including ion-trap methods, have been applied to phosphorylation site analysis. It is now essential for all laboratories involved in protein phosphorylation studies to have access to mass spectrometry if they are to contribute to understanding this pre-eminent regulatory mechanism.

2. Choice of instrumentation and methodology

Mass spectrometry equipment is expensive and technically demanding and is often under the control of mass spectrometry service laboratories, so the type of instrumentation available to individual investigators may not be specifically focused on the needs of protein phosphorylation projects. We have both

MALDI–MS and ESI–MS available in our laboratory, and have performed phosphorylation site studies on both types of instrument. Our preference is for a combination of the specificity of Liquid Chromatography/Mass Spectrometry for phosphopeptide detection, the exquisite sensitivity of linear MALDI analysis, and the structural information available from nano-ESI tandem MS. Other mass spectrometric approaches to phosphorylation site analysis have been reviewed elsewhere (2).

A systematic survey of laboratories, which employ different types of instruments and methodologies in phosphopeptide analysis, was undertaken by the Association of Biomolecular Resource Facilities (ABRF) Mass Spectrometry Research Committee in 1997. This evaluation involved the independent analysis of a protein digest spiked with two unknown phosphopeptides (LFTGHPETpLEK and ASpEDLK) by 28 laboratories. The second of these phosphopeptides is small and hydrophilic, as can often occur in phosphoprotein analysis. This study revealed that the most successful approach to the identification of phosphopeptides was the use of negative-ion, high orifice fragmentation by ESI–MS and the method of choice in the structural analysis of the identified peptides is MS–MS, typically on a nano-ESI instrument. It was noteworthy that no laboratory using MALDITOF instrumentation alone successfully characterized both phosphopeptides. The study concluded that 'identification and characterization of phosphorylation sites in proteins using mass spectrometric approaches can be a challenging task and still cannot be considered a routine analysis for most laboratories'. The conclusions of the survey may be viewed on the ABRF WWW site at: http://www.abrf.org/ABRF/ResearchCommittees/msrcreports/abrf97ms.html.

The approach chosen for phosphorylation site analysis will, to a great extent, depend on the availability of instrumentation, the nature and abundance of the sample, and whether the phosphoproteins are isolated as native peptides or radiolabelled with ^{32}P, either *in vitro*, or more typically from cultured cells. Mass spectrometry is increasingly permitting native phosphoprotein characterization, but the majority of investigations at this time use ^{32}P-labelled phosphoproteins.

Two analytical techniques that are complementary to mass spectrometric analysis of radiolabelled phosphoproteins are two-dimensional phosphopeptide mapping on cellulose thin-layer plates (3, 4) (see *Chapter 5*) and solid-phase Edman sequencing with detection of phosphate release, covered in detail in the previous edition of this book (5). Alternative protocols for the latter technique are given in Section 8 of this chapter.

3. Phosphoprotein purification

The isolation of phosphoproteins for mass spectrometric analysis is similar to those methods used in the past to obtain salt- and detergent-free protein samples for Edman sequencing. The advantage of utilizing ^{32}P-labelled protein is that it facilitates both analysis and non-destructive recovery monitoring using Cerenkov counting. The phosphorylated form of a protein may behave differently from the native protein during purification, resulting in low recovery of phosphoproteins, for example:

- phosphorylation may inactivate the protein making it difficult to monitor in a functional assay
- phosphorylation may block antibody binding if the site is located within the epitope, or otherwise modifies the epitope, making it difficult to monitor in an immunoassay
- the phosphoprotein may be in low abundance; or may specifically bind to other proteins

Whichever purification protocols are used, care needs to be taken that the phosphorylated and dephosphorylated proteins are recovered in similar yield. Where affinity chromatography and immunoprecipitation are used, it is important to be aware that phosphoproteins may behave differently from their dephospho forms, and that differences in behaviour may arise from phosphorylation at different sites. During the purification procedure, it is desirable to pool fractions either side of the peak protein fraction, in order to include as much protein as possible and not exclude phospho forms of the protein that typically migrate at the leading or trailing edges of the dephosphoprotein peak. Techniques favoured for early stages of purification include reversed phase and gel permeation chromatography, and SDS–PAGE (6).

4. Determination of the phosphorylation state of the protein

The direct determination of the mass of a phosphoprotein will provide information about how many phosphates are esterified, and the complexity of the sample. For larger proteins, electrospray MS is required to provide sufficient resolution to observe mass differences corresponding to a single phosphate group, i.e. 80 units. MALDI–MS can provide sufficient resolution to allow such analysis on peptides and lower molecular weight proteins. The amount of protein required, and the data quality, may vary depending on the sample heterogeneity and the propensity of the protein to ionize in the mass spectrometer. The presence of lower molecular weight contaminating proteins frequently suppresses the ionization of larger proteins. Nevertheless, it is always worthwhile trying to obtain a whole protein mass spectrum prior to phosphorylation site analysis, as this may reveal unexpected features, such as errors in the published sequence or other post-translational modifications. It is important to apply the protein sample to the electrospray MS in a salt-free form. Typically, chloroform/methanol precipitation (described in *Protocol 1*) is useful if the sample is contaminated with detergent.

Protocol 1. Concentration and desalting of sample for electrospray mass spectrometry

Equipment and reagents
- The choice of tubes in which to carry out sample preparation prior to mass spectrometry is very important. Tube manufacturers may use mould-wetting agents based on polyethylene, and these compounds can swamp the mass spectrum. We find the screw-capped 1.5 ml tubes from

QSP (Quality Scientific Plastics, California, USA) are excellent for this and other applications, but there are many other manufacturers of suitable tubes on the market, and local mass spectrometry laboratories will give advice on choice of tubes before sample preparation.

Method
1. Place 100 µl sample in a screw-capped tube on ice. Place a container of the highest purity water, HPLC grade methanol, and AR grade chloroform on ice and allow to cool.
2. Add 400 µl ice-cold methanol to the tube, cap, and vortex. Add 100 µl ice-cold chloroform, cap, and vortex. Add 300 µl ice-cold water, cap, and vortex.
3. Centrifuge for 1 minute in a bench-top microcentrifuge so that phases separate; the protein may be visible as an insoluble layer at the phase interface.
4. Remove the majority of the upper phase with a Pasteur pipette without disturbing the interface.
5. Add 500 µl ice-cold methanol to the tube, cap, and vortex, and centrifuge for 1 minute in a bench-top microcentrifuge.
6. Remove the supernatant and dissolve the pellet in 10 µl 30% acetic acid. Add 10 µl acetonitrile to the tube, and the sample is ready for infusion into the electrospray mass spectrometer.

The operation conditions of electrospray mass spectrometers are varied, and the most appropriate conditions are established by optimizing the experimental conditions with a calibration protein such as horse skeletal muscle myoglobin, a widely used standard in biomolecular mass spectrometry.

Figure 1 shows an example of the raw and deconvoluted mass spectra for a three subunit protein, the AMP-activated protein kinase (AMPK) isolated from rat liver (7). The final preparation was 30 µg protein contained in 100 µl 30% ethylene glycol, 2 M NaCl, 50 mM Tris–HCl pH 7.5 desalted and concentrated according to *Protocol 1*. We obtained interpretable spectra for the two smaller subunits showing the expected mass of the N-terminally acetylated AMPK-γ subunit and the three phosphoforms of the N-terminally myristylated AMPK-β subunit. The larger catalytic AMPK-α subunit was not detected, which is indicative of substantial mass heterogeneity, or the quenching of its ionization by the smaller subunits.

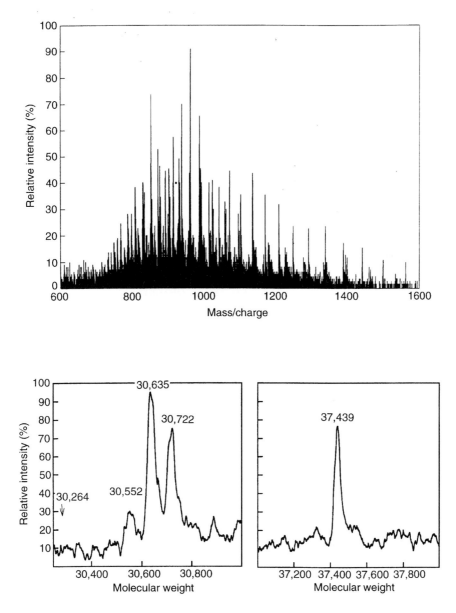

Figure 1. Raw ESI–MS data for native AMP-activated protein kinase heterotrimer (α, β, γ) prepared from rat liver and desalted and concentrated as in *Protocol 1* (upper panel). The lower two panels are reconstructed mass spectra derived from the above raw data for the β-subunit (lower left) and γ-subunit (lower right). The arrow in the left panel at 30 264 represents the molecular weight calculated from the nucleotide sequence of the β-subunit, the species observed represent the myristoylated mono-, di-, and tri-phosphorylated forms of the protein.

Caution. Since β-mercaptoethanol adducts of proteins increase their mass by 76 units it is important, if this reagent has been used during the protein preparation, not to confuse these adducts with phosphorylation. Prior reduction of the protein with DTT, with or without cysteine alkylation, will eliminate these adducts.

5. Protein digestion

5.1 Generation of phosphopeptides

Determination of sites of phosphorylation of a protein by mass spectrometry requires the generation of peptides from the phosphoprotein. This is achieved either in solution or from an in-gel digest. For tandem mass spectrometric analysis, tryptic fragmentation remains the ideal, and should be the first choice because each peptide contains a C-terminal basic residue (see discussion of alternative endoproteases below). Prior disulfide reduction and cysteine alkylation is essential to ensure against loss of disulfide-bonded peptides.

Protocol 2. Phosphoprotein reduction, alkylation, and digestion in solution

Equipment and reagents

- digestion buffer (10 × concentrate: 2 M Tris–HCl, 20 mM EDTA, pH 8.5: dilute 1:10 before use)
- DTT (dithiothreitol, 1 M, freshly prepared)
- iodoacetamide (1M in water, freshly prepared)

- indicator paper strips
- modified sequencing grade trypsin (Promega, reconstituted and stored frozen as a 0.5 mg/ml stock in 50 mM acetic acid to suppress autocatalysis)

Method

1. (a) If the protein is already in a volume less than 80 μl, adjust pH to 8.5 by the addition of 10 μl digestion buffer concentrate, and add 1 μl DTT. Cap and heat at 50°C for 30 min.

   (b) If the protein is in a volume greater than 80 μl, adjust to 10% trichloroacetic acid (TCA) and centrifuge in a bench-top microcentrifuge for 2 min. Discard the supernatant and dissolve the pellet in 90 μl digestion buffer. Check the pH on indicator paper by spotting an aliquot on the indicator strip with an automatic pipette (this can consume as little as 50 nl). If the pH is around 8.5, add 1 μl DTT, cap, and heat at 50°C for 30 min. If the pH is below 8, simply add a small aliquot of 10 × digestion buffer before proceeding with the reduction step.

2. Remove from the heat and allow to cool for a few min. Add 5 μl iodoacetamide. Allow the reaction to proceed in the dark (wrap the tube in aluminium foil) for 15 min at room temperature.

---

3. Quench the reaction by the addition of 4 μl 1M DTT.

4. Precipitate the sample with chloroform/methanol (see *Protocol 1*).

5. Redissolve the pellet in 10 μl 6 M guanidine hydrochloride (GuHCl), add 10 μl digestion buffer concentrate, add 80 μl water, and, finally, add 1 μg modified sequencing grade trypsin (Promega).

6. Incubate at 37 °C overnight.

7. Prior to HPLC, the solution is acidified with TFA to a final concentration of 2%.

If the protein of interest is insufficiently pure to proceed with *Protocol 2*, then in-gel digestion is the method of choice. There are, however, problems with achieving sufficiently high concentrations of some proteins on gels, especially in the presence of other proteins in the mixture that may be present in large excess. Resolubilization of large TCA or chloroform/methanol pellets in SDS sample buffer may fail. The following protocol can be used for the concentration of samples prior to SDS–PAGE.

Protocol 3. Concentration of samples for SDS–PAGE

Equipment and reagents

- low SDS concentration buffer (100 mM Tris–HCl pH 6.8, 10% glycerol, 10 mM DTT, 2 mM EGTA, 0.01% SDS: this low SDS concentration is important to avoid concentrating micelles)
- centrifugal concentrator (Centricon, Amicon) with an appropriate molecular weight cut-off to retain the protein of interest
- angle rotor, high-speed centrifuge

Method

1. Concentrate the required amount of protein using a Centricon according to the manufacturer's instructions. The residual volume will be less than 50 μl with a Beckman JA25 rotor (45 degree angle).

2. Desalt the sample with at least three vols (2 ml each unit) of low SDS concentration buffer, ensuring the volume is reduced to the minimum between each spin.

3. Add 10 μl 10% SDS in water containing a little bromphenol blue and vortex several times.

4. Leave capped and standing upright in a rack at 4 °C overnight.

5. Vortex vigorously and recover the concentrate according to the manufacturer's instructions.

6. Add 1 μl 1 M DTT to each tube, boil, and subject to SDS–PAGE as usual.

In order to generate peptides from a Coomassie Blue-stained gel slice containing protein separated by SDS–PAGE, it is necessary to reduce and alkylate the protein, destain the gel slice, digest the protein, and extract the peptides as described in *Protocol 4.*

Protocol 4. In-gel reduction, alkylation, tryptic digest, and peptide extraction

Equipment and reagents

- gel reduction buffer (10 mM DTT, 0.2 M Tris pH 8.5, 2 mM EDTA)
- gel destain solution (50 mM ammonium bicarbonate, 50% acetonitrile)
- stock trypsin (see *Protocol 2*)
- gel digestion buffer (0.2 M Tris pH 8.5, 2 mM EDTA or 50 mM ammonium bicarbonate/10% acetonitrile)

Method

1. SDS gels of any acrylamide concentration are generally poured late one afternoon and then stored in the cold overnight before running the next day (as an alternative, the pre-cast, commercially available gels from both Novex and BioRad have proved satisfactory for this technique). Pouring the gels in advance ensures more complete polymerization of the gel and reduces the possibility of N-terminal blocking or other involvement in the polymerization process.

2. The peptide recoveries from in-gel digests are adversely affected by the amount of gel. Therefore, minigels are ideal and large gels inappropriate. Recoveries using this technique exhibit a threshold of the order of 1–10 pmol per 1 mm $\times$ 7 mm $\times$ 0.75 mm gel slice, with poorer recoveries of higher molecular weight proteins. It is therefore imperative to get the protein as concentrated as possible. If the band runs sharper on a gradient gel, this is preferred. Use thin gels where practical, with as much protein as possible per lane. It is far better to overload a couple of lanes than to distribute the preparation over many lanes. See *Protocol 3* for a sample concentration guide. **Caution.** After the gel is run, it should only be handled with gloves, and only come into contact with apparatus that has been extensively washed in detergent and rinsed in methanol.

3. The gel should be lightly stained in freshly prepared Coomassie Blue solution and destained using standard protocols. There have been modifications to the staining/destaining process suggested in the literature to maximize peptide yields, but our experience is that these are not necessary. The gel can be photographed or scanned at this stage.

4. The proteins are reduced and alkylated *in situ*, either before or after excision. Wash the gel or gel slice with several changes of distilled

---

water. Incubate the gel, preferably at elevated temperature (45°C) with shaking, in gel reduction buffer for 1 hour. Add stock iodoacetamide, or 4-vinylpyridine, to 50 mM and incubate with shaking at room temperature in the dark (covered in foil, preferably in a fume cupboard) for 1 hour. Quench the alkylating reagent by making the solution 1% with respect to mercaptoethanol, discard the solution, and wash with several changes of water. The gel is placed on a clean glass plate and the bands are excised as precisely as possible with a clean razor blade: leaving excess gel around the stained bands will lead to losses. The gel bands are destained in destaining solution. This works faster at elevated temperatures but can also be done in scintillation vials (1 band per vial) on the bench, overnight at room temperature.

5. The destained bands (1 band per 0.5 ml microcentrifuge tube) are dried in a centrifugal freeze-drier (Savant) and the dried polyacrylamide gel slices can be stored until required at room temperature.

6. The dry polyacrylamide slices are rehydrated individually with trypsin, at around 0.5–1 μg per gel slice, in 20 μl of gel digestion buffer. The bands are allowed to completely hydrate with protease before being pooled into a single tube. Additional digestion buffer (usually no more than 200 μl) is added to the tube to cover the gel slices, and is incubated at 37 °C overnight.

7. The excess buffer containing the peptides is transferred to a second screw-capped tube. The peptides are eluted from the gel pieces with three sequential 30 min washes of 200 μl 2% TFA, 30% acetonitrile/0.1% TFA, and 60% acetonitrile/0.1% TFA using a sonicating water-bath. At the completion of each elution step, the supernatant is re-moved and pooled with the previous extracts.

8. The combined extracts are partially dried in a centrifugal concentrator (Savant) to remove the acetonitrile. Avoid completely drying the ex-tract, as phosphopeptide recovery can be compromised by complete drying.

5.2 Choice of proteases

Trypsin is the ideal protease, especially when contemplating tandem mass spectrometry experiments. There are times, however, when trypsin does not generate a suitable peptide for analysis; they are either too small and too hydrophilic to be amenable for subsequent handling, or too long to extract efficiently from the gel slice. In those cases, alternative endoproteases should be sought. Those suitable for in-gel digest are more limited, with endoprotein-ase Arg-C and Glu-C being ineffective in our experience. Endoproteinase lysine-C from Achrombacter (Wako) (other Lys-C sources are unsatisfactory) in 50 mM Tris pH 9.5, 10% acetonitrile, and Asp-N protease in the same buffer as for the trypsin digestions are good alternatives to trypsin for in-gel digestions.

5.3 Control digestions

Since there are multiple steps involved in a protein digestion it is recommended that investigators, especially those less experienced with the technique, include a positive control in their experiments. Human serum albumin is highly recommended: a half μl of your own serum run on a gel will give an intensely stained band suitable for use as a control. Albumin, with its 35 cysteines involved in 17 disulfide bonds, makes a good control for the reduction/alkylation technique, protease activity, and peptide extraction efficiency. The extracted digest also provides an effective control for the following HPLC purification.

5.4 Monitoring of recoveries

If a radiolabelled protein is available, the efficiency of each step can be monitored by Cerenkov counting of fractions and gel slices. If the number of counts in the peptide extract is less than 95% of the total radioactivity, the possibility of phosphopeptide loss in the gel slice should be considered. In the event, this occurrence may be turned to advantage by redigesting the gel slice with an alternative protease such as Asp-N protease.

5.5 Phosphopeptide maps

As noted previously, it is advantageous to follow the phosphopeptide characterization with the two-dimensional (2D) high voltage electrophoresis/TLC mapping strategy (see Chapter 5). At each stage, a small aliquot of the digest can be mapped and compared with the original map. Much information can be deduced from the position of spots on these maps, adding confidence to the assignment. Importantly, it allows the investigator to account for all the spots having regard for the small hydrophilic peptides that may not bind to the HPLC column (discussed below) or those that are so long and hydrophobic that they fail to elute. One particularly useful aspect of the 2D maps is that closely related phosphopeptides will migrate as pairs, with the more highly charged peptides in a group also exhibiting higher mobility in the TLC dimension, thus giving chains of spots running at 45 degrees (or thereabouts) from the origin. In our experience, especially where basic residues are close to the phosphorylation site, incomplete digestion series may result because the phosphate influences the protease digestion. For example, in the digestion of phospho-kemptide (LRRASpLG), trypsin preferentially cleaves between the adjacent arginines, whereas the dephosphopeptide cleavage occurs after the second arginine.

Caution. If a phosphopeptide map of a digest is planned, a volatile NH_4HCO_3 buffer, or desalting of the digest, is required.

6. Phosphopeptide purification

A variety of phosphopeptide purification methods have been reported, but we prefer a single reversed phase step. Reversed phase HPLC concentrates and desalts the digest and provides a high resolution separation that is compatible with on-line mass spectrometry, 2D phosphopeptide maps, and Edman sequencing. It is superior to alternatives such as ion-exchange or immobilized metal affinity chromatography, since additional desalting steps are inevitably required in these approaches and, when the sample amount is limiting, any additional chromatographic step may dramatically reduce recovery. In the case of a small protein it may be possible to go directly from the digest to mass spectrometry on the unseparated mixture, but we prefer the additional information available from the HPLC chromatogram. When the sample is extremely limited (e.g. using silver-stained proteins), chromatography of the digest will be prohibitive in terms of recovery, and it may be necessary to employ sophisticated alternatives (8, 9) beyond the scope of this chapter.

6.1 Reversed phase HPLC

Selection of the HPLC system is often a question of availability, but it is essential to select a system capable of performing good chromatography below 50 μl per minute, and with appropriate fraction collection. Our preference is for the Pharmacia SMART system, mainly because of its integrated intelligent fraction collector, but there are other suitable intruments on the market. It is thought that the presence of metal (mostly stainless steel) in the flow path of HPLC systems (including columns and especially column frits) is a primary cause of significant phosphopeptide loss. A system, often labelled as 'biocom-patible', with the flow path consisting of mostly polymers such as PEEK and Teflon, appears to be the best choice for phosphopeptide purification.

HPLC columns should be selected with this 'metal-free' objective in mind. Columns in glass-lined stainless steel with polymeric frits, manufactured by SGE, are excellent for this application. The 1 mm × 250 mm columns packed with a 300 Å, 5 μM, C18 packing serve as our primary separation columns. However, the availability of a higher carbon load and smaller particle size packing is very useful for small hydrophilic peptides that don't stick strongly to the first column. Small diameter columns are used to improve recovery, the higher the concentration of analyte (or, conversely, the smaller the sample volume) the lower the non-specific losses. A well-seasoned column is pre-ferred to a new column when undertaking high-sensitivity purification.

An excellent diagnostic test of the whole system is to run the Sigma HPLC Peptide Standards (H2016). A suitable system should be expected to provide baseline resolution of every component of the standards, the most diagnostic being the separation of the two most hydrophobic components. The dipeptide in that mixture may not bind if lower carbon load columns are employed.

Protocol 5. Reversed phase chromatography of phosphoprotein digest

Equipment and reagents

- HPLC fraction collector
- ESI mass spectrometer

- Buffer A (0.1% TFA in water)
- Buffer B (0.085% TFA, 80% acetonitrile)

Method

1. With an injection loop larger than 100 µl in the HPLC, dilute the almost dried sample to 100 µl with 6 M guanidine hydrochloride and inject on to the column. If part of the stream is directed to the mass spectrometer, it is important at this stage to temporarily divert the column to waste: the guanidine hydrochloride flow through must not be sprayed into the mass spectrometer interface.

2. The 1 mm column is run at 50 µl/min and the flow of buffer A is allowed to continue until the absorbance at 214 nm has stabilized to the expected baseline level. If the protein has been radiolabelled, the injection effluent is collected for counting to check that short hydrophilic phosphopeptides have bound.

3. The chromatogram is developed at 50 µl/min from buffer A to 100% buffer B over 60 min with collection of 1 min fractions.

4. (a) If the stream is to be directed to the mass spectrometer, a post-column, pre-detector split is required (we prefer the use of the Valco low dead volume PEEK Y-connector, cat. no. MY1XCPK). Exit tubing lengths are adjusted to achieve approximately 10% of flow to the electrospray source, with the remainder being collected in the fraction collector after UV detection. Electrospray settings are optimized by infusion of a standard phosphopeptide (use a P.Ser peptide which appears to be the least sensitive) under conditions of orifice fragmentation (we set the orifice voltage to −350 V) monitoring, in negative single-ion mode, for the production of a −79 *m/z* fragment. For further details regarding the mechanism of this fragmentation see refs 10 and 11 and Section 7.2.1. This data will identify peaks that contain phosphopeptides.

   (b) If the phosphopeptides are radiolabelled, fractions are monitored by Cerenkov counting. Individual peaks of radioactivity can be analysed by 2D phosphopeptide maps as described above.

7. Phosphopeptide characterization

The fractions containing phosphopeptides should be evident by a radioactivity trace versus fraction number plot, or, if part of the stream has been directed to the mass spectrometer, a total ion current trace of the production of −79 m/z fragments. The following sections cover the various approaches that can be used to characterize the phosphopeptide.

7.1 MALDITOF mass spectrometry

7.1.1 Peptide mass determination

Our experience is limited to the use of a linear Voyager DE (PerSeptive Biosystems) MALDITOF instrument in delayed extraction mode. Alternative approaches utilizing MALDI reflector technology are discussed in Section 7.1.4. For all our phosphopeptide mass determinations, the matrix material used is α-cyano-4-hydroxycinnamic acid (α-cyano) at 33 mM in acetonitrile/methanol (Hewlett Packard G2054–85010) mixed 1:1 with sample directly on the sample stage and allowed to dry on the bench at room temperature. When on-stage modifications or enzyme digestions are carried out in alkaline buffers, the matrix is made 2% with respect to TFA. It is important to ensure that the instrument is calibrated appropriately. We prefer the use of a multipoint calibration based on a tryptic digest of myoglobin in a location on the sample stage as close as practical to the sample (near-external calibration) to give the highest possible level of mass accuracy without the requirement for internal standards. It is possible to automate the standard and sample data acquisition and analysis in a single process on the Voyager DE instrument. Rather than just acquiring mass data on the peak fraction containing phosphopeptides, we scan adjacent fractions across the peak, to observe the appearance and disappearance of masses accompanying the appearance and disappearance of the phosphopeptide(s).

Peaks for phosphopeptides may be very small relative to dephosphopeptides, so the spectrum needs careful inspection. The temptation, when interpreting data from these spectra, is to look for pairs of peptides differing in mass by 80 units, corresponding to phosphate. It is important to undertake a rigorous assignment of observed masses compared with predicted proteolytic fragments of the protein. This will highlight those unaccounted for, and those corresponding to phosphopeptides. In digestion of large proteins of over 100 kDa, there may be numerous peptides with predicted masses close to the phosphopeptide mass. Phospho- and dephospho-peptides frequently, but not always, chromatograph in the same or adjacent fractions. The dephospho form of a peptide may not elute in the same or a nearby fraction to the phosphopeptide, due to differential proteolytic processing around a phosphorylation site, as described above, or to the difference in reversed phase retention time between the two forms of the peptide.

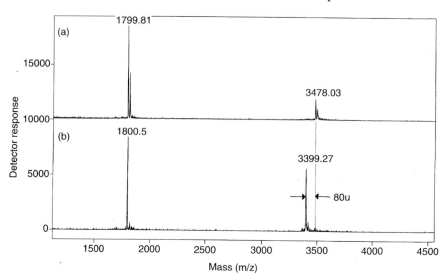

Figure 2. On-MALDI target phosphatase treatment as in *Protocol 6* highlights the phosphorylated peptide at *m/z* 3478. Panel A shows a fraction from reversed phase chromatography and panel B after the same fraction is treated with phosphatase.

The peptide mass assignment analysis is greatly simplified by a good peptide mass calculator program (ProMac program from Perkin–Elmer–Sciex). It is important to take into account the cysteine modification performed (acrylamide modification of cysteines can occur during SDS–PAGE), the possibility of other post-translational modifications (especially N-terminal modifications), and the possibility of methionine oxidation during the digestion (resulting in +16 units, or multiples). Anomalous proteolytic processing, especially the presence of chymotryptic peptides in a tryptic digest, as well as errors derived from DNA sequencing, also need to be borne in mind.

7.1.2 Confirmation of the identity of a putative phosphopeptide

This method involves dephosphorylation of the phosphopeptide using alkaline phosphatase. An example of the analysis is shown in *Figure 2*.

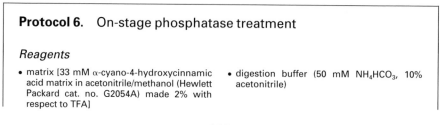

Protocol 6. On-stage phosphatase treatment

Reagents

- matrix [33 mM α-cyano-4-hydroxycinnamic acid matrix in acetonitrile/methanol (Hewlett Packard cat. no. G2054A) made 2% with respect to TFA]
- digestion buffer (50 mM NH$_4$HCO$_3$, 10% acetonitrile)

Method

1. Desalt alkaline phosphatase (Boehringer) into digestion buffer on a Pharmacia SMART fast desalting column, or any other gel-permeation chromatography-based desalting system, such as a Pharmacia PD-10, according to the manufacturer's instructions.[a]

2. Apply 0.5 μl sample of HPLC fraction containing the phosphopeptide (in 0.1% TFA, varying acetonitrile concentrations) on the sample stage followed by 0.5 μl desalted alkaline phosphatase. As a control, add 0.5 μl digestion buffer to 0.5 μl sample.

3. Allow the reaction to proceed at room temperature in a humidified atmosphere (cover the sample stage and a piece of damp paper hand towel with a plastic box) for a period of time.[b]

4. Stop the reaction by addition of 0.5 μl matrix to each spot and allow to dry at room temperature on the bench.

5. Analyse by MALDITOF according to the manufacturer's instructions.

[a] 0.5 ml Boehringer Mannheim Alkaline Phosphatase (cat. no. 713 023, 1U/μl), desalted on a Pharmacia PD-10, results in an enzyme preparation that completely dephosphorylates phosphopeptide substrates at picomole per microlitre concentrations in 10 min at room temperature.
[b] Optimal conditions of enzyme concentration and incubation time need to be determined for each peptide.

7.1.3 Determination of phosphorylation site

With the availability of a linear MALDITOF mass analyser, it may be possible to determine the site of phosphorylation using a variety of enzymatic approaches. For example, if the potential phosphorylation sites are separated by an amino acid that is a candidate for an endoproteinase subdigestion, then the proteolytic fragmentation pattern may be diagnostic of a specific phosphorylation site. An example of this approach is demonstrated in *Figure 3*, where two putative phosphorylation sites in a peptide generated in a lysine-C digest were separated by arginine, and the identity of the phosphorylation site was confirmed by trypsin subdigestion. Alternatively, exoproteinases such as carboxypeptidase Y, carboxypeptidase B, or aminopeptidase M can be used to generate sequence ladders, providing confirmatory sequence information, and aiding in the assignment of phosphorylation sites. In our experience, these exoproteinases stop at phosphorylated residues. Examples of phosphorylation site analysis by aminopeptidase M digestion are shown in *Figure 4*.

Protocol 7.   On-stage endoproteinase subdigestion with trypsin

Reagents

• stock trypsin (see *Protocol 2*)   • digestion buffer (see *Protocol 6*)

Protocol 7. *Continued*

Method

1. Dilute stock trypsin 1:10 into digestion buffer.

2. Follow steps 2–5 of *Protocol 6* using digestion buffer in the control digest, and vary trypsin concentrations and/or digestion times to visualize the desired reaction intermediates.

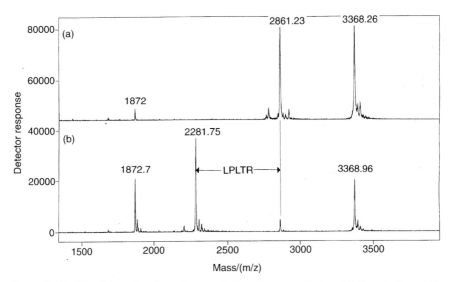

Figure 3. On-MALDI target endoprotease subdigestion as in *Protocol 7*. Trypsin is used to subdigest the Lys-C peptides shown in panel A to confirm the position of phosphorylated serine on the peptide LPLTRSpQNNFVAILDLPEGEHQYK (M^r calc = 2861.4) by generating the fragment SpQNNFVAILDLPEGEHQYK (M^r calc = 2281.1)

Protocol 8. On-stage carboxypeptidase ladder sequencing

Reagents

- stock carboxypeptidase solutions [1 mg/ml in water: we have successfully used carboxypeptidase Y from Pierce (cat. no. 20212), Boehringer Mannheim (cat. no. 238 139) and carboxypeptidase B from Boehringer Mannheim (cat. no. 103 233)]

- digestion buffer (see *Protocol 6*)

Method

1. Dilute 1:10 with digestion buffer just prior to use.

2. Follow steps 2–5 of *Protocol 6* using digestion buffer in the control digest, varying carboxypeptidase concentrations, and/or digestion times to visualize the desired reaction intermediates.

Note. At these carboxypeptidase concentrations, useful ladders can be generated in as little as 10 min incubation at room temperature. There can be gaps in the ladders generated by fast cleavages between rate-limiting cleavage sites. In our experience, termination of the sequential digestion is generally diagnostic of a phosphorylated amino acid one residue N-terminal to the last peptide bond cleaved. A kit comprising all the required reagents for this procedure is now available from PerSeptive Biosystems under the name 'Sequazyme C-Peptide Sequencing Kit'.

Protocol 9. On-stage aminopeptidase M ladder sequencing

Reagents

- aminopeptidase M (lyophilized, Sigma cat. no. 9776, reconstituted at 1 mg/ml in 50 mM ammonium bicarbonate/10% aceto-nitrile just prior to use)
- digestion buffer (see *Protocol 6*)

Method

1. Follow steps 2–5 of *Protocol 6* using digestion buffer in the control digest, and vary aminopeptidase concentrations and/or digestion times to visualize the desired reaction intermediates.

Note. The rate of this reaction can be quite slow and very variable between peptides. Reaction rates can be improved by digestion in the same buffers in a small capped test tube at elevated temperature: for example, the digestion shown in *Figure 4* was incubated at 37 °C over the weekend.

7.1.4 Other MALDI techniques

There are several analytical approaches for phosphopeptide characterization using a MALDITOF mass spectrometer with a reflector. Identification of phosphopeptides can be made in simple mixtures in the positive ion MALDI reflector spectrum by the presence of $[MH–H_3PO_4]^+$ and $[MH–HPO_3]^+$ fragment ions formed by metastable decomposition. Post-source decay can be used to provide sequence information on the identified phosphopeptides. Application of these methods is, as previously discussed, beyond the scope of this chapter but is well described in the literature (12, 13).

7.2 Electrospray mass spectrometry

Our preferred approach to phosphoprotein characterization involves the combination of four separate types of mass spectrometric analysis:

(a) electrospray mass spectrometry of the whole protein to measure phosphorylation stoichiometry (described in detail in Section 4)

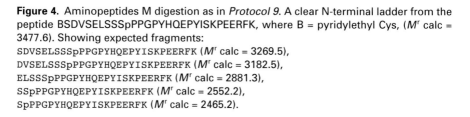

Figure 4. Aminopeptides M digestion as in *Protocol 9*. A clear N-terminal ladder from the peptide BSDVSELSSSpPPGPYHQEPYISKPEERFK, where B = pyridylethyl Cys, (M^r calc = 3477.6). Showing expected fragments:

SDVSELSSSpPPGPYHQEPYISKPEERFK (M^r calc = 3269.5),
DVSELSSSpPPGPYHQEPYISKPEERFK (M^r calc = 3182.5),
ELSSSpPPGPYHQEPYISKPEERFK (M^r calc = 2881.3),
SSpPPGPYHQEPYISKPEERFK (M^r calc = 2552.2),
SpPPGPYHQEPYISKPEERFK (M^r calc = 2465.2).

(b) HPLC of the protein digest with phosphopeptide detection
(c) linear MALDI and nano-ESI mass analysis of the phosphopeptide-containing fractions
(d) tandem nano-ESI–MS for structural analysis to provide sequence signatures to identify the peptide and determine the site of phosphorylation

These techniques were developed by Steve Carr and his colleagues at Smith Kline Beecham (10, 12, 13). We find that, after chromatography and detection as outlined in the next section, thorough examination of the MALDI and positive ESI spectra, followed by tandem nano-ESI–MS characterization of peptides in the fractions of interest, are sufficient to solve most phosphorylation problems.

7.2.1 Phosphopeptide detection using LCMS with high orifice voltage fragmentation

We have found the most sensitive approach to the detection of phosphopeptides in a digest is to subject the mixture to reversed phase HPLC in conventional TFA-containing buffers, and direct a portion of the stream to the ESI source of the mass spectrometer using a flow-splitting, post-column pre-detector. The mass spectrometer is operated in negative-ion mode and is set to monitor two ions, one at m/z –79 (PO_3^-) and the other at m/z -63 (PO_2^-), with the orifice voltage set sufficiently high (we use –350 V) to cause fragmentation of the peptides as they enter the analyser high vacuum chamber. This allows for detection of phosphopeptides in the chromatogram based on their elution time, and leaves 90% of the eluate in the fraction collector for subsequent analysis. Protocols for the operation of the mass spectrometer in this mode are available elsewhere (10). Typically, a split ratio of 10% flow to the electrospray source and 90% to the fraction collector is used, the ratio being set by lengths of fused silica tubing with appropriate internal diameters. A Valco low dead volume PEEK Y-connector (cat. no. MY1XCPK) is suitable as the splitting device. The technique is very sensitive allowing, in control experiments, the detection of a few picomoles of a synthetic phosphopeptide spiked in a digest of hundreds of picomoles of human serum albumin. An example of this type of analysis is shown for an in-gel digest of the (catalytic) α-subunit of the AMP-activated protein kinase (AMPK) in *Figure 5*.

7.2.2 Structural analysis of phosphopeptides

Following identification of fraction(s) containing phosphopeptides, either by the method described above in Section 7.2.1, or by counting fractions from a digest of a radiolabelled protein, the next step is to identify the phosphopeptide(s) in the fraction. Thorough examination of the MALDI spectra has been discussed in detail above in Section 7.1.1. To identify the phosphopeptide and obtain sequence data the same fractions may be analysed by nano-ESI–MS followed by tandem MS. There are mass spectrometric methods for the specific identification of phosphopeptides using parent-ion scanning in the negative-ion mode (10, 14) with which we have little experience.

The use of the nano-ESI source designed by Matthias Mann and colleagues (9) (available commercially from Protana A/S, Odense, Denmark) is currently the most sensitive and specific approach to obtaining structural information on low abundance peptides by mass spectrometry. A portion of the fraction for analysis (1/3) is dried in a centrifugal freeze-drier (Savant) and reconstituted in a small volume (say 5 μl) of 30:70 formic acid/water. After vortexing the sample for at least 10 sec, it is diluted with an equal volume of 20:80 water/acetonitrile and 2 μl of this solution is loaded in the nanospray tip and tandem MS data are acquired with a calibrated instrument. Deconvolution of tandem MS data is not necessarily straightforward and experience with this type of data interpretation is required. With a correctly established spray from the capillary tip, it is possible to acquire data for extended periods of time with flow rates as low as 30 nl/min. This allows for each peptide in the sample mixture to be identified from its tandem MS fragmentation pattern and, if the position of the phosphorylated amino acid is favourable, to determine the exact site of phosphorylation unequivocally.

Figure 6 shows the tandem MS data from the peak marked B in *Figure 5*.

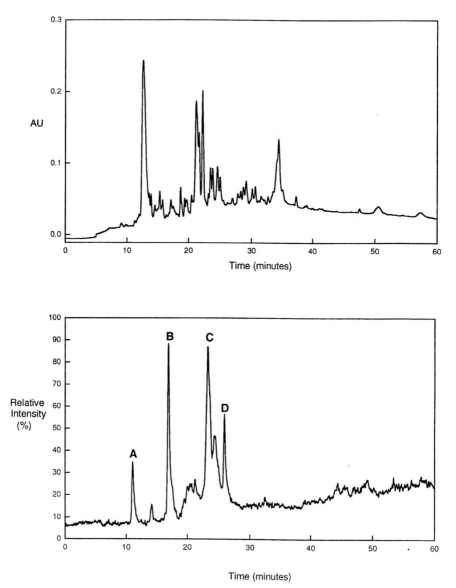

Figure 5. Reversed phase HPLC of an in-gel tryptic digest of the α-(catalytic) subunit of the rat liver AMP-activated protein kinase, diverted in part to the ESI–MS operating in the high orifice voltage fragmentation mode as described in Section 7.2.1. The upper panel is the UV absorbance trace at 214 nm and the lower panel is the total ion current trace for the production of ions –79 (PO_3^-) and –63 (PO_2^-) *m/z* showing four distinct phosphopeptides (reproduced with permission by the authors).

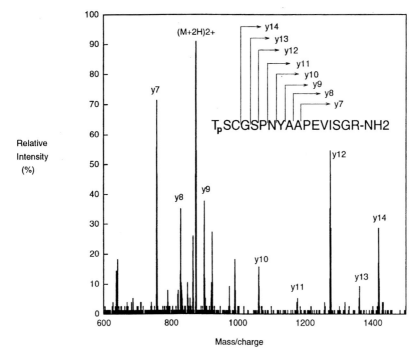

Figure 6. Nano-ESI–MS/MS data from peak B in *Figure 5*. The doubly charge ion ($m/z = 923$) of peptide TpSBGSPNYAAPEVISGR, where B = carbamidomethyl Cys, (M^r calc = 1845.8) is selected in the first quadrupole and fragmented in a field of collision gas (argon) with fragments detected in the third quadrupole. The data have been collected and summed over 10 minutes. The mass spectrum contains a strong y series as follows: (y-ion, M^r calc, M^r obs, respectively: 7, 757.42, 757.6; 8, 828.46, 828.6; 9, 899.50, 899.8; 10, 1062.56, 1063.0; 11, 1176.60, 1178.0; 12, 1273.65, 1273.6; 13, 1360.69, 1361.2; 14, 1417.71, 1417.6 (reproduced with permission by the authors).

This is an example where the phosphopeptide was easy to identify from the data but the site of phosphorylation was not. It was straightforward to eliminate the possibility of phosphorylation at the two C-terminal Ser sites in the peptide, but not possible to discriminate between phosphorylation on the most N-terminal Ser and Thr. Previously, it was shown (15) that the AMP-activated kinase kinase phosphorylated the Thr-172 of the AMPK *in vitro* and this mass spectrometry result demonstrates that the Thr-172-containing peptide is stoichiometrically phosphorylated *in vivo*.

When trying to deconvolute mass spectral data that are difficult to reconcile with the published sequence, it should be considered that the sequence of the protein itself may be incorrect in the database. If this is suspected, then it is useful to prepare an alignment of known sequences to illuminate possible unlikely substitutions. If there is significant error in the sequence, or the protein sequence is unknown, it may be possible to derive considerable protein sequence *de novo* from tandem MS data, but this requires a high level of expertise in data interpretation.

8. Phosphate release sequencing on a Hewlett Packard protein sequencer

Some phosphopeptides are very difficult to characterize by the methods described above, either because of their very low abundance or their complex phosphorylation status. The latter case includes those phosphopeptides that are isobaric (i.e. the same mass), and cases where one or another (but not both) hydroxy-amino acids in the same parent phosphopeptide are alternatively present. These sites are typically adjacent amino acids, and data interpretation of tandem MS data may be complex, and the interpretation equivocal. If a radiolabelled [^{32}P]phosphopeptide can be prepared *in vitro* or *in vivo*, then phosphate release sequencing may be especially effective.

The methodology of phosphate release sequencing has been reviewed elsewhere and in the previous edition of this book (5). This description relates to the application of a specific, commercially available covalent attachment medium and the Hewlett Packard G1000 series of protein sequencers. The radiolabelled phosphopeptide, which must be purified to radiohomogeneity based on the two-dimensional thin layer phosphopeptide maps as described above, is covalently coupled to an aryl-amine membrane and placed in the protein sequencer. The line that transfers the PTC-amino acids to the flask is diverted to the fraction collector. The entire yield of each cycle is counted by liquid scintillation counting and, since it is possible to label proteins with ATP having a specific activity of 10 000 cpm/pmol or higher, it is possible to sequence 200 cpm, or 20 fmol or less, of phosphopeptide.

Protocol 10. Covalent attachment of peptides to Sequelon-AA

Equipment and reagents
- most of the necessary materials and reagents are supplied with the Sequelon-AA membranes available as the 'Sequelon-AA Reagent Kit' from PerSeptive Biosystems
- forceps
- heating block

Method
1. Take a Mylar sheet and place it on a preheated heating block at 55°C.
2. Place the Sequelon-AA membrane on the Mylar sheet with clean forceps.
3. A 20 μl sample is dried on to the membrane.
4. Weigh out 1 mg water-soluble carbodiimide, add 100 μl attachment buffer to the carbodiimide, and vortex to dissolve.

---

5. Remove the Mylar sheet from the heating block.

6. Add 5 μl of the carbodiimide solution to the membrane and allow the membrane to dry for 20 min.

7. Place the membrane in a glass scintillation vial and wash and allow to soak for 5 min in approximately 5 ml of the following solvents, aspirating wash solution between changes: methanol, water, methanol, water, TFA/acetonitrile/water (10:50:40), water, methanol.

8. Slice the membrane into six thin strips and, using a clean forceps, place in an empty upper sequencer column half.

9. The Hewlett Packard G1000A protein sequencer is prepared by replacing the S2A bottle with a bottle containing 90% methanol, 0.015% phosphoric acid. The column half containing the membrane strips is coupled to a second column half containing reversed phase material. (Note: this is *not* the regular male column half used in conventional Edman sequencing which contains ion-exchange material.) The phosphate release sequencer cycles are described in *Table 1*. Extracted cycles were diverted via valve number RV6 (line 61), collected directly into fraction collector tubes, and counted in a liquid scintillation counter. The fraction collector interval was set to 44 min and started manually when about 10 min of the first cycle had elapsed. A simple converter cycle containing a single line coding for a 5 minute wait was written to satisfy the sequencer programming requirements and the sequencer HPLC was left in the 'power off' mode to avoid software conflicts. It is possible to modify the same instrument to deliver the cleavage contents to the reaction flask, carry out the PTC to PTH conversion reaction, and deliver the PTH-amino acid to the HPLC, keeping the loop overfill to count on the liquid scintillation counter, and obtaining amino acid sequence information. In our laboratory, the amounts of phosphopeptide available are generally well below the sensitivity of modern Edman sequencers (less that 500 fmol) and the configuration described above allows the number of counts released at the relevant cycle to be maximized.

Note. Coupling yields vary markedly according to peptide sequence, yields as low as 30% coupled are not uncommon for small tryptic peptides. A few micrograms of an unlabelled synthetic peptide carrier can be added as a precaution against non-specific losses.

9. Concluding remarks

Mass spectrometry is now the method of choice for phosphorylation site identification. It offers specificity and sensitivity beyond that of alternative techniques, it avoids the complications of interpretation of phosphate release data in the face of aberrant proteolytic cleavages, and allows phosphopeptides

Table 1. Column cycle for phosphate release sequencing on a Hewlett Packard G1000A

Sequencer step	Time (sec)	Temperature (°C)
Couple: Meter R2A	52	55
Couple: Deliver R2A DOWN (closed)	200	55
Couple: Dry column DOWN	600	55
Couple: Meter R1	52	55
Couple: Deliver R1 DOWN (closed)	200	55
Couple: Dry column DOWN	600	55
Couple: Meter R1	52	55
Couple: Deliver R1 DOWN (closed)	200	55
Couple: Dry column DOWN	1200	55
Couple: Meter R2	110	55
Couple: Deliver R2 DOWN	200	55
Couple: Dry column DOWN	600	55
Couple: Flush with L1	150	55
Couple: React	3500	55
Couple: Dry column UP	1200	60
Wash: Purge solvent line		50
Wash: Meter S2A	190	50
Wash: Deliver and dry DOWN PVDF (closed)	300	50
Wash: Meter S2A	190	50
Wash: Deliver and dry DOWN PVDF (closed)	300	50
Wash: Meter S2A	190	50
Wash: Deliver and dry DOWN PVDF (closed)	300	50
Wash: Meter S2A	190	50
Wash: Deliver and dry DOWN PVDF (closed)	300	50
Wash: Dry column DOWN	1500	60
Cleave: Purge cleavage line		50
Cleave: Meter R3	40	50
Cleave: Deliver R3 DOWN	300	50
Cleave: React	3500	50
Extract: Meter S2A	300	50
Extract: Deliver solvent UP (closed)	150	50
Cleave: Dry column UP to converter	50	50
Extract: Meter S2A	300	50
Extract: Deliver solvent UP (closed)	150	50
Cleave: Dry column UP to converter	50	50
Extract: Meter S2A	300	50
Extract: Deliver solvent UP (closed)	150	50
Cleave: Dry column UP to converter	1200	50
Extract: Dry column DOWN to waste	2100	55

from previously unsequenced proteins to be characterized. Despite this technology being in its relative infancy, the rapid pace of developments in mass spectrometry, such as MALDI and ESI ion-trap, ESI-TOF and the promise of TOF-TOF means that specific methods for phosphorylation site analysis will evolve rapidly. The choice of methodology is essentially dictated by the type of instrumentation that is available, and the approach we choose combines the specificity of orifice-generated phosphopeptide fragments to detect phosphopeptides in a mixture, followed by linear MALDI analysis and tandem MS experiments on a nano-ESI triple quadrupole instrument. As the genome databases continue to grow at an exponential rate, mass spectrometry is destined to dominate ^{32}P radiolabelling as the major technique in identifying phosphorylation sites in pursuit of the understanding of the regulatory architecture of the cell.

Acknowledgements

We would like to thank Belinda Michell for her constructive proof-reading of the manuscript and Stephen Tindall (Argo BioAnalytica, Inc., Morris Plains, NJ, USA) for providing his PVDF sequencer method on which the phosphate release cycles are based.

References

1. Fenselau, C., Heller, D.N., Miller, M.S., and White, H.B.D. (1986) *Anal. Biochem.* **150**, 309.
2. Resing, K.A. and Ahn, N.G. (1997) In *Methods in enzymology* (ed. William G. Dunphy), Vol. 283, p. 29. Academic Press, New York.
3. van der Geer, P. and Hunter, T. (1994) *Electrophoresis* **15**, 544.
4. van der Geer, P., Sefton, B.M., and Hunter, T. (1993) In *Protein phosphorylation: a practical approach* (ed. D.G Hardie), 1st edn, Vol. 1, p. 31. IRL Press, Oxford.
5. Hardie, D.G., Campbell, D.G., Caudwell, F.B., and Haystead, T.A. (1993) In *Protein phosphorylation: a practical approach* (ed. D.G Hardie), 1st edn, Vol. 1, p. 61. IRL Press, Oxford.
6. Cohen, S.L. and Chait, B.T. (1997) *Anal. Biochem.* **247**, 257.
7. Mitchelhill, K.I., Michell, B.J., House, C.M., Stapleton, D., Dyck, J., Gamble, J., Ullrich, C., Witters, L.A., and Kemp, B.E. (1997) *J. Biol. Chem.* **272**, 24475.
8. Shevchenko, A., Wilm, M., Vorm, O., and Mann, M. (1996) *Anal. Chem.* **68**, 850.
9. Wilm, M., Shevchenko, A., Houthaeve, T., Breit, S., Schweigerer, L., Fotsis, T., and Mann, M. (1996) *Nature* **379**, 466.
10. Huddleston, M.J., Annan, R.S., Bean, M.F., and Carr, S.A. (1993) *J. Am. Soc. Mass Spectrom.* **4**, 710.
11. Carr, S.A., Huddleston, M.J., and Annan, R.S. (1996) *Anal. Biochem.* **239**, 180.
12. Annan, R.S. and Carr, S.A. (1996) *Anal. Chem.* **68**, 3413.
13. Annan, R.S. and Carr, S.A. (1997) *J. Protein Chem.* **16**, 391.
14. Wilm, M., Neubauer, G., and Mann, M. (1996) *Anal. Chem.* **68**, 527.
15. Hawley, S.A., Davison, M., Woods, A., Davies, S.P., Beri, R.K., Carling, D., and Hardie, D.G. (1996) *J. Biol. Chem.* **271**, 27879.

7

Assay and purification of protein (serine/threonine) phosphatases

CAROL MACKINTOSH and GREG MOORHEAD

1. Introduction

1.1. Classification of catalytic subunits by amino acid sequences

The eukaryotic protein phosphatases comprise several families of enzymes that catalyse the dephosphorylation of intracellular phosphoproteins, thereby controlling diverse cellular functions in all eukaryotic cells. The activities of these enzymes are highly regulated in response to extracellular stimuli, and in many cases regulation is mediated by interactions of the catalytic subunits with diverse targetting and regulatory subunits and domains. The protein phosphatases responsible for dephosphorylation of serine and threonine (and histidine) residues in the cytoplasmic and nuclear compartments of eukaryotic cells are derived from two distinct gene families, PPP and PPM, that have been classified both by phylogenetic analysis of their sequences, and by their three-dimensional structures (reviewed in refs 1 and 2).

Table 1 groups the catalytic subunits according to how they are related in amino acid sequence. Closely related members of most of the PPP catalytic subunit subfamilies have been identified in mammals, insects, plants, yeasts, and protozoa, making the PPP family among the most highly conserved of all eukaryotic proteins. The known membership of these families is still growing through the use of molecular genetics and the detection and purification of the proteins from diverse cells and species. Individual protein phosphatases were named according to their original discoveries, either from their biochemical activities [for example, the PP1G_M-complex comprises PP1 catalytic subunit (PP1C) together with its glycogen-binding G_M-subunit from skeletal muscle], as phenotypic mutants [for example, *abi1* mutant *Arabidopsis* plants are insensitive to the hormone abscisic acid, and *Arabidopsis rcn1* plants are defective in auxin transport (3)], or from cDNA and PCR cloning (the so-called novel phosphatases, including PP5, PPQ, PPV, PPX, PPY, PPZ). Many of the protein phosphatases are represented by several isoforms; for example,

Table 1. The PPP and PPM families of protein (serine/threonine) phosphatases [adapted from Barford (1996), ref. 2])

The PPP family (prototypic member in brackets)	Regulatory subunits and domains
PPP1 (PP1)	G$_M$, G$_L$, M110 + M 21, NIPP-1, RIPP-1, R110, p53BP12, L5, sds22, inhibitor-1, DARPP-32, inhibitor-2
PPP2A (PP2A)	A-subunit (PR65), *Arabidopsis* RCN1
	B-subunits (PR55, PR72, PR61), eRF1, PTPA, SET,
	polyoma middle and small T antigens, SV40 small T antigen
PPP5 (PP5)	TPR domain on same polypeptide as catalytic subunit
PPP2B (PP2B)	B-subunit, calmodulin, AKAP-79

Each catalytic subunit is represented by a number of isoforms. For example, there are three human and at least eight *Arabidopsis* PP1 isoforms. In addition, there are novel protein phosphatases of the PPP family. For example, PPY and PPZ belong to the PP1 subfamily, and PP4 to the PPP2A.

The PPM family
PP2C
Arabidopsis ABI1, ABI2, and KAPP
Pyruvate dehydrogenase phosphatase
Bacillus subtilis SpoIIE phosphatase

there are eight PP1 isoforms (termed TOPP1s) in *Arabidopsis*, and three in humans.

1.2 Regulatory subunits

The PPP family catalytic subunits are bound to regulatory or targetting subunits that are very important determinants of their substrate specificities, regulatory properties, and subcellular locations (reviewed in refs 1, 2, and 4).

1.2.1 PP1

PP1 in mammalian cells is largely particulate, because the catalytic subunit is targetted by a variety of regulatory subunits that bind to endoplasmic and sarcoplasmic reticulum, cytoskeleton, myofibrils, glycogen particles, ribosomes, and nuclear structures. Several regulatory subunits of mammalian and yeast PP1s have been purified and/or identified by molecular genetics (named in *Table 1* and reviewed in refs 1 and 2). The known PP1 regulatory subunits are very diverse, but many contain a short motif—(R/K)(V/I)xF in the mammalian proteins—that binds directly to the PP1 catalytic subunit. It is likely that many more regulatory subunits will be identified. For example, PP1 in plant extracts exists in high molecular mass forms that have not been purified and characterized.

There is also an inactive form of PP1, in which the catalytic subunit is complexed with an inhibitory protein, inhibitor-2 (4, 5). Recent evidence suggests that inhibitor-2 may be a chaperone that binds to and aids folding of nascent PP1 catalytic subunits (6, 7).

1.2.2 P2A

The native forms of PP2A that have been characterized to date comprise catalytic (C) subunits bound to an A-subunit (the AC core), together with one of a variety of B-subunits. The 12 or so known B-subunits fall into five gene families that share no obvious sequence similarity. Several studies indicate roles for the B-subunits in determining the substrate specificity of PP2A. For example, association of the C-subunit with the A- and B-subunits increases activity towards some substrates and suppresses activity towards others in *in vitro* assays, and mutations of two different B-subunits affect distinct cellular processes in yeast. PP2AC and PP4, also interacts with Tap 42 and α4, components of the Tor signalling pathway in yeast and mammals respectively.

1.2.3 PP2B (also known as calcineurin)

The catalytic subunit of PP2B contains a long carboxy-terminal extension, which inhibits the catalytic activity. Interaction of the carboxy-terminal extension with a Ca^{2+}-binding regulatory B-subunit relieves the inhibition. This explains why the enzyme is dependent on Ca^{2+} for activity.

1.2.4 PP2C

PP2C is predominantly a soluble enzyme in all tissues examined to date. *KAPP*, a plant PP2C, binds to certain phosphorylated receptor proteins in *Arabidopsis*.

1.3 Inhibitors of protein (serine/threonine) phosphatases

Inhibitors of protein (serine/threonine) phosphatases include endogenous proteins that regulate protein phosphatases in eukaryotic cells, such as inhibitor-1, DARPP 32, and inhibitor-2, which specifically inhibit PP1. The PPP enzymes are also the targets of a diverse group of secondary metabolites produced by bacteria, fungi, dinoflagellates, and insects, whose natural roles may be in defence against competitors or predators. Okadaic acid and related compounds, microcystins, tautomycin, nodularin, cantharidin, endothall, calyculin A, and fostriecin inhibit the PPP enzymes by binding to the same site on the enzymes, but with different relative potencies (8, 9); PP2B is not potently inhibited by these compounds. PP2B has been found to be the target for inhibition by the immunosuppressant drugs cyclosporin and FK506, but only after the drugs have bound to their cognate binding proteins, cyclophilin and FK-binding protein (FKBP), respectively (10). Chapter 3 provides a more detailed discussion of these compounds, and describes their use as pharmacological probes of protein phosphatase function in cells. In this chapter, protocols are presented in which okadaic acid and microcystin are exploited for enzyme identification and for affinity purification of protein phosphatases.

Okadaic acid is a diarrhetic shellfish toxin, and microcystins are potent liver toxins produced by freshwater blue-green algae. Therefore, the ability to detect and quantify these compounds in environmental samples by their inhibition in protein phosphatase assays (*Protocols 3* and *4* and Section 2.4.2) should be of considerable interest to water-associated industries and health authorities.

1.4 Biochemical properties of the protein (serine/threonine) phosphatases

While the specificities of the protein phosphatases are greatly influenced by regulatory subunits *in vivo*, the protein (serine/threonine) phosphatases have broad and overlapping substrate specificities in *in vitro* assays. Individual types of protein phosphatase can, however, be distinguished by testing their substrate specificities, cation dependencies, and sensitivities to inhibitors, as indicated in *Table 2*.

From *Table 2* it can be seen that the active forms of each enzyme can be identified and quantified in dilute extracts of any type of eukaryotic cell in a number of ways, for example:

(a) PP1 can be measured as that proportion of the divalent cation-independent phosphorylase phosphatase activity that is blocked by either 0.2 μM inhibitor-1 or 0.2 μM inhibitor-2, but not by 1 nM okadaic acid.

(b) PP2A is the divalent cation-independent phosphorylase phosphatase, or casein phosphatase activity, that is unaffected by inhibitors-1 and -2, but is completely inhibited by 1 nM okadaic acid (these properties also describe PP4, which has a much more minor activity than PP2A in those extracts that have been examined to date).

(c) PP5 is the divalent cation-independent, microcystin- and okadaic acid-inhibited, and arachidonic acid-activated casein phosphatase activity. In practice, the activity of PP5 may be too low to detect in a crude cell-free extract, but it has been identified after fractionating extracts by anion-exchange chromatography.

(d) PP2B activity is detected as a Ca^{2+}- or Mn^{2+}-dependent, calmodulin-stimulated, and trifluoperazine-inhibited activity measured using ^{32}P-labelled phosphorylase kinase, inhibitor-1, or a ^{32}P-labelled peptide as substrate.

(e) PP2C is a Mg^{2+}-dependent casein phosphatase activity, measured in the presence of 1 μM okadaic acid or 40 nM microcystin-LR.

Classifying any newly identified protein (serine/threonine) phosphatase activity according to these criteria is an extremely useful first step towards its identification. However, unambiguous identification would have to include further structural and functional information. Is the enzyme a free catalytic subunit or is it complexed to other proteins? What is its subcellular local-ization? In addition, the biochemical properties of only a few of the novel protein phosphatases (*Table 1*) have been determined, and it is not clear how far their properties may overlap with those of the better-characterized protein

Table 2. Biochemical classification of the protein (serine/threonine) phosphatases

	PP1	PP2A	PP2B	PP2C
Preference for the α- or β-subunits of phosphorylase kinase	β	α	α	α
Inhibition by I1 and I2	Yes	No	No	No
Absolute requirement for divalent cations	No	No	Yes (Ca^{2+})	Yes (Mg^{2+})
Inhibition by trifluoperazine	No	No	Yes	No
Inhibition by okadaic acid	Yes[a] ($IC_{50} = 10$ nM)	Yes[a] ($IC_{50} = 0.1$ nM)	Yes (weak)	No
Inhibition by microcystin-LR	Yes[a] ($IC_{50} = 0.1$ nM)	Yes[a] ($IC_{50} = 0.1$ nM)	Yes (weak)	No
Phosphorylase phosphatase activity	High	High	Very low	Very low
Activity towards casein phosphorylated by protein kinase A	Very low	High	Very low	High

[a]Concentrations required for 50% inhibition depend on protein phosphatase concentration: quoted values are for 0.2 mU/ml phosphorylase phosphatase activity.

phosphatases listed in *Table 2*. For these reasons, we use the terms 'PP1-like' or 'PP2A-like' to describe an activity that has the properties of PP1 or PP2A according to *Table 2*, but where the enzyme has not been identified by its amino acid sequence.

1.5 Other detection methods

Further aids to enzyme identification that do not depend on measuring activity include the use of commercially available antibodies that recognize catalytic and regulatory subunits. The presence of microcystin-binding, protein phosphatase catalytic subunits can also be demonstrated by labelling with ^{125}I-microcystin and SDS–PAGE (*Protocol 6*). Regulatory subunits can sometimes be identified by running samples on SDS–PAGE, transferring to nitrocellulose blots, and probing the blot for binding to the relevant chemically labelled catalytic subunit (*Protocol 9*).

2. Assay of protein (serine/threonine) phosphatase activities

2.1 Methods available

The two methods used most commonly to measure protein phosphatase activity are:

(a) a single-time point assay of [^{32}P]phosphate released from ^{32}P-labelled substrate (either a phosphorylated protein or a synthetic peptide)

(b) a continuous or single time-point spectrophotometric assay of hydrolysis of a non-proteinaceous compound such as *p*-nitrophenylphosphate (*p*NPP)

A third possibility is to measure a phosphorylation state-dependent change in the function of a phosphorylated protein substrate. This latter method requires that a correlation between degree of phosphorylation and variation in some property of the substrate has been established previously. Examples of such variable functional properties are:

(a) changes in K_m, K_i, or V_{max} if the substrate is an enzyme

(b) changes in binding affinity of the substrate or some other ligand.

2.2 Assays using ^{32}P-labelled phosphorylated protein substrates

2.2.1 Theory of assays

Sufficient ^{32}P-labelled substrate for the required number of phosphatase assays is prepared by phosphorylation of protein by a protein kinase:

$$\text{protein} + [\gamma\text{-}^{32}\text{P}]\text{ATP–Mg} \rightarrow [^{32}\text{P}]\text{phosphoprotein} + \text{ADP–Mg}$$

The [^{32}P]phosphoprotein is freed from residual [γ-^{32}P]ATP–Mg and ADP–Mg by gel filtration on Sephadex G-50, by dialysis, or by a series of concentration and dilution steps in a Centricon concentrator (Amicon). Protein phosphatase activity can then be determined in a fixed-time assay by measuring the release of acid-soluble ^{32}P-radioactivity from ^{32}P-labelled substrate:

$$[^{32}P]\text{phosphoprotein} + nH_2O \rightarrow \text{protein} + n[^{32}P]\text{phosphate}$$

The reaction is stopped by adding trichloroacetic acid (TCA) to inactivate the protein phosphatase and precipitate the unused, ^{32}P-labelled protein. The acid-soluble radioactivity is then determined by scintillation counting. A reaction blank is included from which protein phosphatase is omitted. The following general comments may be helpful:

- It is frequently not possible to use protein substrates at high enough concentrations to attain V_{max} conditions. The assays therefore use a fixed, subsaturating substrate concentration.

- These assays may use physiologically relevant protein substrates such as inhibitor-1 for PP2B, or phosphorylase *a* or phosphorylase kinase for PP1$_G$, or they may use model substrates such as histones or casein.

- The commonly used mammalian protein substrates (phosphorylase *a*, phosphorylase kinase, casein, etc.) are equally useful for assay of protein phosphatases from other eukaryotes such as plants or yeast, whose type 1 and 2 protein phosphatase activities are remarkably similar to those of mammalian cells.

- Some substrates, such as histone H1 and small peptides, are not precipitated by 5% (w/v) TCA. With such substrates, use a modified assay in which acid ammonium molybdate is added and the products extracted into butan-1-ol/heptane, which extracts the complexes formed between molybdate and inorganic phosphate, but not phosphopeptides (*Protocol 5*). This method should also be followed whenever new extracts or new substrates are employed for the first time, because there may be proteolytic enzymes which could be mistaken for protein phosphatase activity if they are capable of releasing small TCA-soluble phosphopeptides from the phosphoprotein substrate.

2.2.2 Assay of PP1 and PP2A using ^{32}P-labelled phosphorylase *a* as substrate

The popular use of skeletal muscle glycogen-metabolizing enzymes (phosphorylase *a* and phosphorylase kinase) as protein phosphatase substrates stems from the early and continued prominence of glycogen metabolism as a model system for studying regulation of cellular metabolism by reversible protein phosphorylation. Nevertheless, phosphorylase *a* has many attributes which make it well suited as a prototype substrate for assaying PP1 and PP2A derived from any organism, cell type, or subcellular compartment:

- Phosphorylase (called phosphorylase *a* in its phosphorylated form) is a good substrate for both PP1 and PP2A, but not for PP2B, PP2C, nor any of the mammalian alkaline or acid phosphatases that have been tested.

- Phosphorylase *b* (the dephosphorylated form) is phosphorylated by phosphorylase kinase at a *single* serine residue (serine-14). This means that phosphorylase *a* with a phosphorylation stoichiometry of one mole phosphate per mole of subunit can be reliably prepared, making quantitation of PP1 and/or PP2A activity reproducible from one batch of substrate to the next.

- The large quantities that can be isolated, and the high solubility or phosphorylase *a*, mean that a near-saturating concentration can be used in assays. Gram quantities of pure phosphorylase *b* can be prepared from 1 kg of rabbit skeletal muscle by a purification procedure requiring only one chromatographic step (11, 12); alternatively, phosphorylase *b* can be purchased from several sources (for example, Life Technologies, Boehringer Mannheim).

- The preparation of phosphorylase kinase (the enzyme which specifically phosphorylates phosphorylase *b* to give phosphorylase *a*) takes only two days and 25 mg can be obtained from 1 kg of rabbit skeletal muscle (12), sufficient for a long-term programme of protein phosphatase experiments. Otherwise, phosphorylase kinase can be purchased from several sources (for example, Sigma, Boehringer Mannheim).

Protocol 1 describes a procedure for making up [γ-^{32}P]ATP and determining its specific radioactivity. *Protocol 2* describes a method for producing sufficient phosphorylase *a* for approximately 60 vials of stock substrate, enough for 3000 assays. The quantities can be scaled down if required. The preparation can be used for up to 6 weeks provided that it is not contaminated with protein phosphatase activity, and provided that the initial specific radioactivity of [γ-^{32}P]ATP is at least 10^6 cpm/nmol (the half-life of ^{32}P is only 14.3 days). Compare a new batch of substrate against the previous one, to check that it behaves identically in the phosphatase assay.

Protocol 1.   Preparation of [γ-^{32}P]ATP

Materials

- radioactive [γ-^{32}P]ATP [Amersham supplies this as a 1 mCi (10 mCi/ml) pack in dry ice, with a negligible chemical content]
- non-radioactive ATP (100 mmol/1, pH adjusted to 7.0 with NaOH: a 1/1000 dilution should have an A_{260} of 1.54)
- Plexiglas shields and remote handling devices, to protect operator from β-radiation

---

Method

1. Allow the [γ-^{32}P]ATP to thaw at room temperature (about 30 min).

2. Transfer [γ-^{32}P]ATP to a 1.5 ml screw-cap microcentrifuge tube, shielded in a Plexiglas box or lead pot.

3. Wash out the original vial from the radioisotope supplier with part of a 0.89 ml aliquot of water and transfer washings to the 1 ml container.

4. Add the remainder of the water plus 10 μl of 100 mM ATP to the container and mix by pipetting up and down gently. Vortex mixing can lead to contamination of the working area and is not recommended.

5. Make a 1/100 dilution in water, measure A_{260} to check concentration, and count 2 × 10 μl aliquots in a scintillation counter to determine specific radioactivity.[a]

6. Divide into aliquots as required and store at –20°C or –70°C.

[a] Note time and date, and refer to a ^{32}P decay chart to work out the specific radioactivity of [γ-^{32}P]ATP after a period of storage; alternatively, keep the 2 × 10 μl aliquots and count them to determine specific radioactivity each time an assay is performed.

Protocol 2. Preparation of ^{32}P-labelled phosphorylase *a*

Materials

- buffer A [50 mM Na 2-glycerophosphate pH 7.5, 10% (v/v) glycerol, 0.1 mM EGTA, 0.1% (v/v) 2-mercaptoethanol]
- phosphorylase kinase (20 mg/ml in buffer A)[a]
- phosphorylase *b* (100 mg/ml in buffer A)[a]
- buffer B (125 mM Na 2-glycerophosphate, pH 8.6)
- buffer C [50 mM Tris–HCl, pH 7.0 (20°C), 0.1% (v/v) 2-mercaptoethanol]

- buffer D [10 mmol Tris–HCl, pH 7.0 (20°C), 0.1% (v/v) 2-mercaptoethanol]
- buffer E [50 mM Tris–HC1, pH 7.0 (20°C)]
- [γ-^{32}P]ATP (*Protocol 1*)
- 90% saturated ammonium sulfate (475 g/l, pH adjusted to 7.0 with NH$_4$OH)
- NaF/EDTA (500 mmol/NaF, 100 mM EDTA, pH adjusted to 7.0 with NaOH)

Method

1. Phosphorylase and phosphorylase kinase preparations are some-times contaminated with traces of protein phosphatases that reduce the useful lifetime of the substrate and cause elevated assay blanks. If protein phosphatase contamination is suspected, pre-incubate phos-phorylase and/or phosphorylase kinase overnight on ice in buffer A containing 50 mM NaF plus 5 mM sodium pyrophosphate, then re-equilibrate into buffer A by gel filtration, or repeated concentration/dilution in a Centricon concentrator (Amicon).

2. Mix together the following components in a 50–100 ml centrifuge tube, adding the ATP last:

phosphorylase kinase	80 μl
Mg acetate (100 mM)	160 μl
CaCl$_2$ (100 mM)	10 μl

Protocol 2. *Continued*

phosphorylase *b*	1 ml
buffer B	4.25 ml
[γ-^{32}P]ATP	2.5 ml

3. Incubate for 1 hour at 30°C.
4. Add an equal vol of ice-cold 90% saturated ammonium sulfate and 0.5 ml NaF/EDTA.
5. Leave on ice for 30 min to precipitate protein.
6. Centrifuge at 12 000 *g* for 10 min at 4°C.
7. Resuspend the pellet gently in buffer C (not more than 5 ml per 100 mg of phosphorylase) using a pipette.
8. Add an equal volume of ice-cold 90% saturated ammonium sulfate and repeat steps 4–6 (this removes excess ATP).
9. Dialyse[b] the suspension for 24 h at 4°C against 2 × 2 litre of buffer D to remove residual [γ-^{32}P]ATP and NaF. Monitor the dialysis buffer for radioactivity at each change. During dialysis the phosphorylase *a* will crystallize, forming a cream-coloured suspension.
10. Carefully empty the contents of the dialysis bag into a 50–100 ml centrifuge tube. Leave on ice for 30 min to ensure complete crystallization. Centrifuge at 12 000 *g* for 10 min at 4°C and resuspend the crystals *gently* using a Dounce homogenizer in ice-cold buffer E (no more than 5 ml for each 100 mg of phosphorylase).
11. Take a sample and dilute 100-fold in buffer E to determine the specific radioactivity. Swirl the concentrated solution before sampling to ensure complete resuspension and warm the diluted sample at 30°C to dissolve the crystals.
12. Determine A_{280} of the 1/100 dilution to estimate protein content (A_{280} is 1.31 for a 1 mg/ml solution) and count triplicate 10 μl aliquots to determine the radioactivity. The specific radioactivity should correspond to 1 mole of phosphate per mole of 97.4 kDa subunit.
13. Pipette aliquots into 1.5 ml microcentrifuge tubes and store at 4°C. DO NOT FREEZE AS THIS DENATURES PHOSPHORYLASE. The size of each aliquot varies depending on the exact protein concentration of the preparation, and is adjusted so that the concentration of phosphorylase *a* will be 3 mg/ml after the addition of 0.5 ml of buffer C and caffeine (see *Protocol 3*).

[a] Life Technologies Inc sells a kit which contains sufficient phosphorylase, phosphorylase kinase and other reagents to provide for two preparations of ^{32}P-labelled phosphorylase *a* and for a total of 300 protein phosphatase assays ([γ-^{32}P]ATP must be purchased separately). Instead of dialysis, their protocol uses successive concentration and dilution in a Centricon 30 concentrator (Amicon, provided in the kit) to remove residual [γ-^{32}P]ATP.
[b] For smaller scale preparations it may be more convenient to remove residual [γ-^{32}P]ATP and NaF by gel filtration on Sephadex G-50 (see *Protocol 4*).

Protocol 3. Assay of phosphorylase phosphatase (PP1 and/or PP2A) activity

Materials

- caffeine (75 mM, pH 7.0; store in the dark at room temperature: caffeine is degraded by light and will crystallize if stored at 4°C)
- ^{32}P-labelled phosphorylase *a* (*Protocol 2*)
- buffer A [50 mM Tris–HCl pH 7.5 (20°C), 0.1 mM EGTA, 0.1% (v/v) 2-mercaptoethanol, 1 mg/ml BSA]
- buffer B [50 mM Tris–HCl pH 7.5 (20°C), 0.1 mM EGTA, plus inhibitors/activators (see Section 2.2.3) as required; N.B. include 0.03% (v/v) Brij-35 when preparing dilute inhibitor solutions]
- buffer C (as for buffer A, but without BSA)
- 20% TCA (w/v)

Method

1. Add 0.1 ml of 75 mM caffeine and 0.4 ml of buffer C to one vial of [^{32}P] phosphorylase *a* (see *Protocol 2*) to give a 3 mg/ml (30 μM) solution, and store on ice. Occasionally, the redissolved phosphorylase *a* substrate shows slight cloudiness due to trace denaturation. If this occurs simply warm the vial for a minute or two at 30°C and spin out the precipitate at 12 000 *g* for 2 min in a microcentrifuge. Keep the vial at room temperature (return to storage at 4°C after use).

2. Dilute the protein phosphatase sample (crude extract or purified PP1 or PP2A) in buffer A, taking heed of the comments in Sections 3.2 and 3.3.

3. Label a set of 1.5 ml microcentrifuge tubes, duplicates for each sample and two more for blanks.[a, b]

4. Pipette 10 μl of diluted protein phosphatase into each sample tube and 10 μl of buffer A (no enzyme) into the blank tubes: place on ice.[c]

5. Add 10 μl of buffer B (containing inhibitor/activator as required ; see Section 2.2.3) to each tube and place on ice.

6. Add 10 μl of [^{32}P] phosphorylase *a* to a microcentrifuge labelled 'total'. Set aside.

7. Remove the sample and blank tubes from ice and place in a water bath at 30°C. If required, leave tubes to preincubate at 30°C to allow time-dependent inhibition of phosphatases to occur (consult Section 2.2.3). Otherwise proceed to step 8 immediately.

8. Start the phosphatase assays by adding 10 μl of [^{32}P]phosphorylase *a* at 10 or 15 sec intervals.[d] Vortex mix and incubate at 30°C for 10 min.[e]

9. Stop the reactions at 10 or 15 sec intervals by adding 100 μl of 20% TCA to each tube. Vortex mix the suspensions.

10. Centrifuge the tubes at 12 000 *g* for 2 min at room temperature in a microcentrifuge to sediment the precipitated protein.

11. Label another set of tubes exactly as in step 3.

Protocol 3. *Continued*

12. Transfer 100 μl of each clear TCA supernatant to the second set of tubes. Add 1 ml of aqueous-compatible scintillation fluid (for example, Fluoran HV or Ecoscint) to these tubes, and to the TOTAL tube from step 6, and count using a ^{32}P-program in a liquid scintillation counter.[f]

13. *Calculation* : one unit (U) of protein phosphatase activity releases 1 μmol phosphate from phosphorylase *a* per min in the standard assay. Therefore, the calculation for a 10 min assay is:

$$\text{cpm released} = \text{SAMPLE cpm} - \text{BLANK cpm} \tag{1}$$

$$\text{Activity}^g \text{ (mU/ml)} = \frac{\text{cpm release}}{\text{cpm TOTAL}} \times \frac{0.3}{10} \times 100 \times \frac{130}{100} \tag{2}$$

[a] Blanks should be included as the first and last tubes in every set of assays so that any phosphatase carryover due to a contaminated pipette tip being accidentally inserted into the vial of substrate will be noticed. Using phosphorylase *a* as substrate, the blank value should be <1% of the total counts: if the blank value rises above 5% of the total, discard the substrate and make up a fresh aliquot.

[b] Up to 40 × 10 min assays can be performed in a single experiment.

[c] The assay volume can be doubled if the specific radioactivity of the phosphorylase *a* is getting too low for accurate counting (< 30 000 cpm in 10 μl of phosphorylase *a*).

[d] Addition of phosphorylase substrate every 10 sec can be managed by using a single pipette tip for addition of substrate to the entire set of assays: however, take care not to let the tip come into contact with the phosphatase solution in the bottom of the assay tube.

[e] The assay proceeds without a lag phase and is linear with time up to 30% release of counts, because the phosphorylase concentration (10 μM) is above the K_M (≈5 μM for both PP1 and PP2A).

[f] Samples can also be counted by Cerenkov counting without added scintillant: the efficiency of Cerenkov counting is approximately half that of standard scintillation counting.

[g] 0.3 is the number of nanomoles of phosphorylase in the assay, 10 is the incubation time in minutes, 100 is to convert the results for 1 ml rather than 10 μl of enzyme and 130/100 is the fraction of the TCA supernatant that is counted.

When diluting purified PP1$_C$ or PP2A$_C$ from buffer containing 55% glycerol into a less viscous buffer, always add the concentrated enzyme to the new buffer (not the other way round) and then vortex mix very quickly: this prevents denaturation during mixing.

2.2.3 Modification to allow independent quantification of PP1 and PP2A in crude extracts

In those eukaryotic tissue extracts that have been tested, PP1-like and PP2A-like protein phosphatases (which may include novel PPP protein phosphatases, see Section 1.4) are the only enzymes with activity towards phosphorylase *a* in the absence of divalent cations (see *Table 1*). In a mixture, the two types of enzyme can be quantified individually by taking advantage of their differential sensitivity to the specific inhibitors, inhibitor-1 (I1), inhibitor-2 (I2), and okadaic acid. In *dilute* extracts, inhibitor proteins-1 and -2 are specific inhibitors of PP1, while okadaic acid is a specific inhibitor of PP2A-like enzymes at 1–2 nM. Both PP1 and PP2A are completely inhibited by okadaic acid at 5 μM. The word *dilute* is emphasized because these are tight-binding

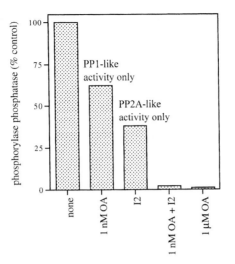

Figure 1. Effect of okadaic acid and inhibitor-2 on phosphorylase phosphatase activity in a dilute tissue extract. Assays were performed with 1 nmol/l okadaic acid (OA), after pre-incubation for 15 min with 0.2 μmol/l inhibitor-2 (I2) with or without 1 nmol/l okadaic acid, with 1 μmol/l okadaic acid, or without inhibitors.

inhibitors, so that the apparent IC_{50} values will increase as the concentration of protein phosphatase present in the assays increases. For this reason, it is crucial for accurate quantitation that the phosphorylase phosphatases activity present in the assays is ≤ 0.2 mU/ml. Under these conditions, as illustrated in *Figure 1*, PP1 is that proportion of phosphorylase phosphatase activity that is inhibited by 200 nM I1 or 200 nM I2. Alternatively, if the inhibitor proteins or peptide are not available, PP1-like activity is the activity at 1–2 nM okadaic acid (which blocks PP2A) minus the activity in the presence of 5 μM okadaic acid (which blocks both PP1- and PP2A-like enzymes). PP2A-like activity is either the phosphorylase phosphatase activity which is unaffected by the inhibitor proteins, or the activity which is blocked by 1–2 nM okadaic acid. Both methods agree within ±10% in extracts of liver, brain, and adipose tissue (13) and oilseed rape seeds (14), all of which contain similar amounts of PP1 and PP2A. If one enzyme accounts for >80–90% of the total activity, quantitation of the other becomes less accurate.

These methods can be extended to determine whether PP1-like or PP2A-like, or one of the related novel PPP protein phosphatases in a cell or tissue extract, is the most active towards any new phosphoprotein substrate. First, determine the relative activities of type 1 and 2A protein phosphatases towards phosphorylase *a*. Then, in a similar manner, use okadaic acid and/or inhibitor proteins to compare the relative contributions of PP1-like and PP2A-like activity towards the new substrate. Examples of such experiments are to be found in the following references: cardiac phospholamban (15);

carrot cell quinate dehydrogenase (16); histone H1 (17); rat liver HMG-CoA reductase (18).

The following notes provide guidelines for the preparation and use of inhibitor solutions:

- *Okadaic acid*, or its sodium salt, is sold as a concentrated solution (1–5 mg/ml) in dimethyl sulfoxide (DMSO) or as a powder which should be dissolved in DMSO (Life Technologies, Calbiochem). It is diluted into assay buffer B (*Protocol 3*) containing 0.03% Brij-35 to give a concentration of okadaic acid three times that required in the final assays. Store diluted solution at –20°C. Okadaic acid (10 μl) is added to phosphorylase phosphatase assays at step 5 of *Protocol 3*. Perform a preliminary check that the amount of okadaic acid added to the final assay is able to block PP2A activity completely without inhibiting PP1. For this purpose, either use purified PP1 and PP2A, or perform a titration experiment using a dilute crude tissue extract as source of enzyme. The end-point for PP2A inhibition at 1–2 nmol/l okadaic acid can be clearly seen (*Figure 2*), especially in extracts where the

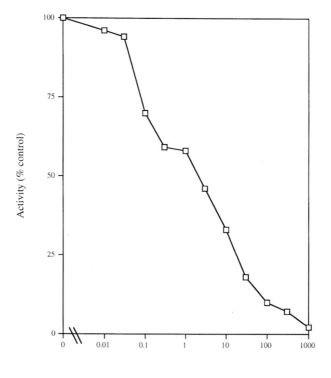

[Okadaic acid] (nM)

Figure 2. Inhibition of phosphorylase phosphatase activity by okadaic acid in a dilute oil-seed extract. The arrow points to the okadaic acid concentration (1 nmol/l) which completely inhibits PP2A (42% of total activity) without affecting PP1 activity (the remaining 58%).

PP1/PP2A phosphorylase phosphatase ratio is about 1:1. Okadaic acid in dimethylformamide (DMF) is sold by some suppliers and is suitable for use in these cell-free assays, but not for intact cell experiments (see Chapter 3).

- *Inhibitors -1 and -2* can be purified from rabbit skeletal muscle as described (19). Their molar concentration is determined by amino acid analysis. Inhibitor-1 is only active after phosphorylation by protein kinase A. Bacterially expressed mammalian inhibitor-2 is commercially available (for example, from UBI). Inhibitor proteins are added to phosphorylase phosphatase assays at step 5 of *Protocol 3*. Note that the high molecular mass complexes between the PP1 catalytic subunit and its regulatory subunits are inhibited by inhibitors-1 and -2 in a time-dependent fashion. Therefore, inhibitor proteins must be pre-incubated with diluted cell extract for 15 min at 30 °C (at step 5, *Protocol 3*).

2.2.4 Standard assay of casein phosphatase activity

Casein is a convenient substrate for routine assays of both PP2A and PP2C because it is inexpensive and free of contaminating protein phosphatase activity. This means that batches of substrate can be prepared by phosphorylation with protein kinase A and stored for up to 6 weeks with no release of $[^{32}P]$phosphate. The casein is a hydrolysed mixture, so incorporation of ^{32}P cannot be calculated exactly on a molar basis, but it is usually low (4 nmol $[^{32}P]$phosphate/mg of casein; estimated to be less than 0.1 mol $[^{32}P]$phosphate/mol of casein). For this reason, calculations are all based on the concentration of ^{32}P, rather than protein. PP1 has very low casein phosphatase activity. There is no evidence that acid and alkaline phosphatases can dephosphorylate casein that has been phosphorylated by protein kinase A, provided that the pH of the assays is close to neutral.

Protocol 4. Preparation of ^{32}P-labelled casein substrate

Materials

- casein (Sigma C 4765: an autoclaved, hydrolysed and partially dephosphorylated solution of approximately 10 mg/ml)
- catalytic subunit of protein kinase A (at least 400 mU/ml; available from Sigma and Boehringer Mannheim)
- $[\gamma$-$^{32}P]$ ATP [1 mM, $\approx 10^6$ cpm/nmol (see *Protocol 1*)]

- 5 × incubation buffer [250 mM Tris–HCl pH 7.0 (20°C), 0.5 mM EGTA, 50% (v/v) glycerol; add 2-mercaptoethanol to 0.5% (v/v) immediately before use]
- stop solution (100 mM EDTA, 500 mM NaF, 10 mM Na pyrophosphate, pH 7.0)
- Sephadex G-50 Superfine column (15 × 1 cm, equilibrated in 1 × incubation buffer)

Method[a–g]

1. Thaw the $[\gamma$-$^{32}P]$ ATP solution and determine its specific radioactivity (*Protocol 1*).

Protocol 4. *Continued*

2. In a screw-cap microcentrifuge tube, mix the following components:

[γ-^{32}P] ATP	100 μl
Mg acetate (100 mM)	100 μl
5 × incubation buffer	200 μl
water	400 μl

3. Withdraw >200 μl of casein solution from the Sigma bottle using a sterile needle, and transfer to a separate tube (the bottle has a rubber septum seal and the solution will keep for several months at 4°C if the contents are not contaminated).

4. Add 200 μl of the casein solution to the incubation mix.

5. Start the reaction by addition of protein kinase A, to give a final activity of ≈2 mU/ml.

6. Incubate at 30°C for 16 hours.

7. Add 110 μl of stop solution and incubate for 5–10 min.

8. Centrifuge at 12 000 *g* for 2 min at 4°C and discard the small pellet of denatured protein.

9. Apply the supernatant to the column of Sephadex G-50 Superfine.

10. With the aid of a Geiger counter, monitor flow of [^{32}P]casein (flow through peak) and residual [γ-^{32}P]ATP (second broader peak) through the column.

11. When the [^{32}P]casein approaches the end of the column, start collecting fractions of approximately 300 μl (we usually collect fractions manually; a fraction collector is likely to become contaminated).

12. Count a 10 μl sample of each fraction in a scintillation counter and by referring to the original specific radioactivity (step 1), calculate incorporated phosphate concentration (μM) in each fraction of the [^{32}P]casein peak. Pool the peak fractions of [^{32}P]casein and dilute to give a final concentration of 18 μM with respect to ^{32}P (usually 6 or 7 fractions need to be pooled and diluted by 1.5- to 2- fold).

[a] For assaying the casein phosphatase activity of PP2A, follow the assay procedure described in *Protocol 3* except that 10 μl of 18 μM [^{32}P]casein, as prepared in *Protocol 4*, is added to start the assay instead of phosphorylase. In the absence of added cations this method will selectively determine PP2A-like activity in most mammalian cell extracts, because PP1 has negligible casein phosphatase activity (see *Table 2*) and because acid and alkaline phosphatases do not have significant casein phosphatase activity (assayed at neutral pH) in those tissues that have been tested. However, to be confident that only PP2A is being detected, check that the activity can be completely inhibited by 1 nM okadaic acid and use the acid-molybdate method to check that the radioactivity released is due to dephosphorylation rather than proteolysis (*Protocol 5*).
[b] One unit of activity (U) is that amount of PP2A which catalyses the dephosphorylation of 1 μmol of [^{32}P]casein per min in the standard assay. The calculation of activity is therefore exactly the same as for *Protocol 3*, except that the final number should be multiplied by 0.6

because the substrate concentration in the assays is 6 μM (rather than the 10 μM used in phosphorylase phosphatase assays).

[c] Note that contamination of purified PP1 preparations by traces of PP2A is easily measured as the amount of casein phosphatase activity which is inhibited by 1 nM okadaic acid.

[d] The activity of native PP5 can be determined by its arachidonic acid-stimulated casein phosphatase activity (20).

[e] *For assay of PP2C activity* using [^{32}P]casein, perform the standard casein phosphatase assay as for PP2A but include 20 nM magnesium acetate in the assay (added as 60 mM in buffer B, *Protocol 3*). PP2C can be assayed selectively by including 5 μM okadaic acid or 40 nM microcystin in the assays to inhibit PP1 and PP2A (see *Table 1* and *Figure 1*)

[f] The calculation of activity is exactly the same as for the assay of PP2A activity using [^{32}P]casein.

[g] Note that the $A_{0.5}$ for Mg^{2+} of PP2C in skeletal muscle extracts, with casein as substrate; is 1 to 3 mM depending on whether the purified enzyme or a tissue extract is being assayed (21).

2.2.5 Procedure for checking that all TCA-soluble radioactivity released from [^{32}P]protein is inorganic phosphate

In preliminary experiments with new sources of protein phosphatase, always perform assays by both the standard and the acid-molybdate extraction methods (*Protocols 3* and *5*, respectively). You may find that the calculated phosphatase activity is lower (and more reproducible) by the molybdate extraction method. This means either that the protein substrate is not precipitated completely by TCA (check this) or that some of the TCA-soluble radioactivity in the standard assay is due to the action of proteinases releasing ^{32}P-labelled acid-soluble peptides rather than protein phosphatases. In such cases, use the acid-molybdate extraction method routinely for assays.

Protocol 5. Acid-molybdate extraction of inorganic phosphate

Materials

- acid phosphate (1.25 mM KH_2PO_4 in 0.5 M H_2SO_4)
- butan-1-ol/heptane [1:1 (v/v)]
- ammonium molybdate [5% (w/v)]

Method

1. Perform assays according to *Protocol 3*, or one of its variants, up to the transfer of 100 μl of the TCA supernatant into the second set of tubes (step 12 in *Protocol 3*).

2. Add 0.2 ml of acid phosphate to each tube.

3. Add 0.5 ml of butan-1-ol/heptane plus 100 μl of 5% (w/v) ammonium molybdate solution and vortex mix vigorously.

4. Leave at room temperature for 10 min to allow the aqueous and organic phases to separate. Meanwhile, prepare a third set of labelled microcentrifuge tubes.

5. Transfer 0.3 ml of the organic (top) phase into the fresh tubes.[a]

6. Add 0.3 ml of butan-1-ol/heptane to the TOTAL tube.

Protocol 5. *Continued*

7. Add 1 ml of scintillant (compatible with organic solvents, e.g. Fluoran-HV) to each sample and count in a scintillation counter using a ^{32}P-program.

---

[a] The calculation of activity is as for *Protocol 3* except for an extra multiplication factor of 5/3 to account for the fact that only 0.3 ml out of a possible 0.5 ml is transferred at this stage.

2.3 Measurement of protein phosphatase activity using phosphorylated synthetic peptides

Structural features other than the primary structures surrounding the phosphorylation sites are important in determining the specificities of protein phosphatases. This means that specificity towards phosphopeptides designed to mimic phosphorylation sites of proteins may bear little relevance to the specificity of protein phosphatases towards the native phosphorylated protein substrates *in vivo*. Nevertheless, ^{32}P-labelled synthetic peptides can be convenient substrates, particularly useful for PP2B whose protein substrates can be difficult to generate outside specialist laboratories. Several examples of phosphopeptide-based protein phosphatase assays are given in refs 22–26. In outline, synthetic peptides are phosphorylated using [γ-^{32}P]ATP and the relevant protein kinase, aiming for maximal phosphorylation of the residue of interest. Residual [γ-^{32}P]ATP can be removed by ion-exchange chromatography on Dowex AG1X8 columns, or by binding the peptide to a Sep-Pak C18 cartridge (Waters) equilibrated in 0.1% trifluoroacetic acid, washing extensively, and eluting the peptide in 0.1% trifluoroacetic acid in the required concentration of acetonitrile. The phosphorylated peptides are lyophilized, and may be purified further by HPLC. If the peptide contains more than one serine, threonine, and/or tyrosine residue, you may wish to ascertain which residue(s) have been phosphorylated by the kinase and dephosphorylated by the phosphatase (see Chapter 6). Assays are as described in *Protocol 3*, except that the liberated [^{32}P]phosphate is converted into its phosphate–molybdate complex, extracted with butan-1-ol/toluene and quantitated as in *Protocol 5*, because most small peptides are not precipitated by TCA.

2.4 Measurement of *p*-nitrophenylphosphate phosphatase activity

2.4.1 Introduction

Protein (tyrosine) phosphatases and acid and alkaline phosphatases can all be assayed spectrophotometrically by following the hydrolysis of *p*-nitrophenylphosphate (*p*NPP) at 400 nm [extinction coefficient varies with pH, at pH 8–9 it is 16 500 l/mol/cm (27)]. For some time there have been differing reports about whether or not PP1, PP2A, PP2C, and PP2B have *p*NPPase activity.

Recently, it was found that bacterially expressed human PP1 (available from UBI), or PP1 that had been stored for a few years, had both *p*NPPase and tyrosine phosphatase activities, but native PP1 that was freshly isolated from tissue extracts had none (7). PP2A has *p*NPPase and tyrosine phosphatase activity after it has been incubated with MgATP and a protein termed protein–tyrosine phosphatase activator (PTPA) (28). Therefore, it seems likely that the *p*NPPase activity of the protein (serine/threonine) phosphatases is only exhibited by non-native conformations.

The *p*NPP phosphatase assay does, however, warrant consideration for use with *purified* protein phosphatases because it is simple to use, and *p*NPP is cheap, readily available, and easy to use. For example, the inhibition of the *p*NPPase activity of bacterially expressed PP1 can be used as a non-radio-active assay for microcystin and okadaic acid (29). However, *p*NPP is not a suitable substrate for assaying protein (serine/threonine) phosphatases in crude cell-free extracts that may contain acid and alkaline phosphatases, and other enzymes that can hydrolyse *p*NPP.

The sensitivity of the *p*NPPase assay is generally less than for assays with ^{32}P-labelled substrates, and the specific activity will vary depending on the state of the enzyme. Note that, unlike its protein phosphatase activity, the *p*NPP phosphatase of PP2A$_C$ has been reported to be completely Mg^{2+} dependent (27). The pH optima for *p*NPP phosphatase activities are in the range 8.0–8.5.

2.4.2 Outline of assay procedure

- Prepare a 40 mM solution of *p*NPP (disodium salt) in buffer C (*Protocol 3*) just before use.
- Dilute phosphatase(s) in buffer A (*Protocol 3*).
- Start the reaction by addition of 10–100 μl enzyme to 900–990 μl substrate and record the initial velocity by recording the change in absorbance at 400 nm.
- If assaying for microcystin or okadaic acid, prepare a standard curve of *p*NPPase activity in the presence of known concentrations of toxin.
- The concentration of *p*NPP used in assays is generally 5 mM.

3. Alternative methods for identifying protein phosphatases

3.1 Microcystin as an affinity tag for identifying PPP catalytic subunits

Microcystins are a family of cyclic heptapeptides produced by certain species of freshwater blue-green algae (see *Figure 3* in Chapter 3 for structure) (8). The most common variations in microcystin occur at the X and Y positions of the ring. For example, in microcystin-LR (MC-LR), residue X is leucine and Y is arginine. In microcystin-YR (MC-YR), residue X is tyrosine and Y is arginine.

The *N*-methyl dehydroalanine (Mdha) residue of microcystin forms a covalent bond with a conserved cysteine residue in the target PPP protein phosphatases, and the covalent linkage is stable to SDS sample buffer and boiling (30–32). This means that radiolabelled microcystins can be used as affinity tags to identify protein phosphatase catalytic subunits after SDS–PAGE (*Figure 3*). Note that the microcystins will not bind to denatured or inhibited protein phosphatases (30–32).

Protocol 6. Preparation of ^{125}I-microcystin-YR and its use as an affinity tag for identifying microcystin-binding PPP catalytic subunits after SDS–PAGE

Reagents

- microcystin YR (MC-YR)
- Iodogen
- [^{125}I]sodium iodide
- 20 mM tyrosine
- 0.5 M sodium iodide

Method

1. Dissolve 20 μg of MC-YR in 20 μl methanol. Dilute to 0.1 ml with 0.12 M sodium phosphate buffer, pH 7.4. Add to a tube containing 0.01 mg dried Iodogen. Add 5 μl (~0.5 mCi) of [^{125}I]sodium iodide. Stand for 15 min at 0°C. Quench with 10 μl of 20 mM tyrosine, followed by 10 μl of 0.5 M (unlabelled) sodium iodide.

2. Pre-wet a Waters C18 Sep-Pak cartridge with methanol and equilibrate in 0.1% (v/v) trifluoroacetic acid (TFA).

3. Dilute the sample in 1 ml 0.1% (v/v) TFA. Wash with 0.1 (v/v) TFA in 10% (v/v) acetonitrile to remove sodium iodide and tyrosine. Elute the [^{125}I]MC-YR with 0.1% TFA/100% (v/v) acetonitrile.

4. Dry the sample by rotary evaporation. (Note that the equipment will inevitably become contaminated with ^{125}I.) Redissolve the [^{125}I]MC-YR in 20 μl of methanol and count in a gamma counter.

5. Add ~50–200 000 cpm of [^{125}I]MC-YR to the sample suspected to contain a microcystin-binding protein phosphatase, and incubate at room temperature for 1 h.

6. Denature the sample in SDS sample buffer and run on SDS–PAGE.

7. Identify microcystin-binding proteins by autoradiography of the SDS gel. Note that the apparent molecular weight of the protein phosphatase will include the molecular weight of the [^{125}I]MC-YR (~1000 Da).

---

4. Preparation of protein (serine/threonine) phosphatases

4.1 Precautions for measurement of protein phosphatase activities in extracts

The substrate specificities, regulatory properties, responses to divalent cations, and inhibitors of protein (serine/threonine) phosphatases can alter during purification due to proteolysis, conformational changes, and the loss of regulatory subunits (4, 31). For microcystin-binding enzymes, rapid purification of enzymes by microcystin affinity chromatography (*Protocols 7* and *8*) largely avoids these artefacts. Additional precautions are recommended:

- Use *Protocol 5* for all assays using previously unstudied extracts, or using new substrates, in case proteolysis of substrate is a problem (see comments in Section 2.2.1).

- The stability of protein phosphatase activity in different samples varies. Enzyme stability may be a particular problem when using slaughterhouse material: PP1 in mammalian heart is almost completely destroyed within 1 h of slaughter (15). Therefore, all extracts should be prepared as quickly as possible, at 0–4°C and in the presence of proteinase inhibitors. Aliquots of extracts (~100 µl, with protein concentrations >2 mg/ml) can usually be snap-frozen in liquid nitrogen and stored at –70°C for at least three months without deterioration. Thaw samples in a few seconds in a water bath at 30°C.

- It is particularly important when assaying protein phosphatase activity in extracts prepared from whole organisms such as insects or protozoa, to include sufficient trypsin inhibitor (in a protease inhibitor cocktail) to inhibit all of the trypsin released from digestive glands. Trypsin degrades the regulatory subunits of PP1 and PP2A and cleaves the C-terminal end of the 37 kDa catalytic subunits, leaving active fragments of 33/34 kDa. Proteolysis increases the phosphorylase phosphatase activity of PP1 in cell extracts by up to fivefold compared with the native forms. Disruption of lysosomes and release of their proteases can be minimized by inclusion of isotonic sucrose or mannose in homogenization buffers.

- Avoid buffers such as phosphate or glycerophosphate that inhibit protein phosphatases. Include reducing agents such as 2-mercaptoethanol (5–20 mM) or dithiothreitol (1 mM) to prevent oxidation. EDTA is required, but keep the concentration low (0.1 mM) to prevent the conversion of PP1 and PP2A to forms that are Mn^{2+} dependent and have other altered properties (these changes can also occur after prolonged storage, or after exposure of these enzymes to NaF).

- Extracts contain both high and low molecular weight substances that inhibit protein phosphatases. To avoid underestimating activities it is best to gel filter extracts through small columns of Sephadex G-50 and/or to assay at as high a dilution of the extract as possible. This is facilitated if the specific radioactivity of the protein substrate is very high ($\sim 10^6$ cpm/nmol). Perform a test series of dilutions with all new extracts. For some frequently used sources the following guidelines for final dilution on PP1/PP2A assays may be given (extracts prepared by homogenization with 2.5 ml of buffer per gram of tissue, followed by centrifugation at 10 000 *g*): rat liver, 600-fold; skeletal muscle, 300-fold; pea seedlings, 150-fold.

On the other hand, when studying the regulatory properties of a protein phosphatase it is sometimes preferable to try to mimic the physiological situation as closely as possible (33). This might mean using high concentrations of extract and substrate and short assay times to avoid missing any important regulators present in the extract.

4.2 Purification of protein (serine/threonine) phosphatases

4.2.1 Outline of strategies

The strategy for purifying protein (serine/threonine) phosphatases from eukaryotic cells and tissues depends on the intended purpose. Possible reasons for wishing to purify these enzymes, together with matching purification strategies are:

(a) *Catalytic subunits.* The catalytic subunits of PP1 and PP2A have relatively broad substrate specificities and are used as reagents for dephosphorylation of regulatory proteins in many studies. PP1 and PP2A catalytic subunits are useful in bioassays for detecting and quantifying okadaic acid and microcystin in environmental samples. PP1 and PP2C catalytic subunits can be expressed in bacteria. However, the substrate specificities and regulatory properties of the bacterially expressed PP1 are different from those of the native catalytic subunits purified from eukaryotic cells and tissues. It has not been possible to express PP2A in a heterologous system.

   (i) *PP1C and PP2AC.* The catalytic subunits of PP1 and PP2A can be purified by ethanol extraction followed by conventional chromatography (5). We have been using bovine heart muscle from the local abattoir as a source of PP2A, because PP1 is degraded by the time that the tissue reaches the laboratory (~ 2 h) (15), which makes the purification of PP2A more straightforward. PP1 and the regulatory subunits of PP2A can be isolated by microcystin affinity purification (*Protocols 7* and *8*). The PP2A catalytic subunit, however, can only be eluted from the microcystin–Sepharose using denaturing buffers.

   (ii) *PP2C.* The key steps used to purify PP2C are chromatography on thiophosphorylated myosin P-light chain [a procedure first introduced for avian smooth muscle protein phosphatases by Pato and Adelstein

(34)] and chromatography on a Mono Q FPLC column [which resolves two isoenzymes of PP2C in both rabbit skeletal muscle and liver (21)]. The myosin P-light chains used to purify PP2C can be purified in large quantities from rabbit skeletal muscle. Thiophosphorylated proteins are prepared by using adenosine 5'-*O*-(3-[^{35}S]thio)-triphosphate instead of ATP in the protein kinase reaction. The thiophosphorylated product is not hydrolysed by any of the known protein (serine/threonine) phosphatases. Thiophosphorylated proteins have also been used as affinity matrices for purification of PP2B.

(b) *Holoenzyme forms.* The distinctive substrate specificities, regulatory properties, and subcellular locations of the native forms of PP1 and PP2A, and other related enzymes, are explained by the interactions of the catalytic subunits with different regulatory subunits. Only a few of the regulatory subunits have been identified, by purifying the high molecular mass forms from mammalian cells (see *Table 1*). Various complexed forms of PP2A have been purified from several mammalian tissues. Biochemical characterization of native forms of the protein phosphatases that have been identified by cDNA and PCR cloning is only just beginning. Few protein phosphatases have been characterized by biochemical approaches from non-mammalian eukaryotes.

   (i) *Native and active forms of PP1, PP2A, and related enzymes.* For PP1, isolate the relevant subcellular fractions of interest, followed by microcystin affinity purification (*Protocols 7* and *8*). Microcystin–Sepharose chromatography has also been used to isolate native PP5 (S. Meek and C. MacKintosh, unpublished results). Methods for purifying several mammalian holoenzyme forms of PP2A are given in reference 5. The regulatory subunits of PP2A can be isolated by microcystin–Sepharose. However, we have been unable to elute the catalytic subunit of PP2A from MC–Sepharose in non-denaturing buffers.

   (ii) *PP2B.* Most purification strategies designed for PP2B from mammalian tissues rely on affinity chromatography using calmodulin–Sepharose (35). The finding that PP2B binds specifically to columns composed of immobilized cyclophilin–cyclosporin A (10) provides an alternative purification procedure.

4.2.2 Microcystin-affinity chromatography of protein phosphatases and their associated subunits

The α, β-unsaturated carbonyl group of the Mdha residue in microcystins (MCs) is arranged with the π electrons polarized so that nucleophilic addition of thiolates to the β-position is favoured at high pH. Reacting the Mdha of MC with small thiols at pH 11 gave >95% conversion to products which could no longer form covalent bonds with the PPs, but still inhibited the enzymes with high affinity (31). Therefore, the Mdha of MC is not essential for the toxin–enzyme interaction. Immobilizing microcystin via the Mdha provides a powerful affinity matrix for purification of PPs and their associated regulatory subunits (31, 32).

Buy MC–Sepharose (UBI) or prepare (*Protocol 7*) MC affinity matrices in two steps:

(a) React the Mdha with a low molecular weight molecule containing both a thiol that reacts with the Mdha and second reactive chemical group, such as amino or thiol group, that is not present elsewhere on MC. It is important to use a low molecular weight thiol that reacts sufficiently rapidly so that coupling is complete before the MC is degraded under the alkali conditions used. The resulting thiol adducts are then resistant to alkali, probably because the susceptible peptide bond between the Mdha and Glu residues is stabilized once nucleophilic addition to Mdha has been achieved.

(b) React the introduced free amino or thiol group with an activated Sepharose under mild conditions.

In principle, MC adducts could be linked to other matrices using similar chemistry. For example, ref. 36 describes the preparation and use of microcystin–biotin.

Protocol 7. Preparation of microcystin–Sepharose

Materials

- microcystin (the microcystin must contain a dehydroalanine moiety, such as MC-LR, -RR, or -YR from Calbiochem or Life Technologies Inc. Because pure MC is expensive, we use a mixture of MCs partially purified by HPLC from blue-green algae, and provided by a water company. Using a mixture means that we cannot easily monitor product formation, but this is not a concern as previous reactions with pure MCs had been reliably yielding >95% conversion to products).
- aminoethanethiol hydrochloride (Aldrich, white solid; discard if oxidized and pale green)

Method

1. Briefy purge water, DMSO, and 5N NaOH separately with N_2 gas.

2. Immediately before use, dissolve aminoethanethiol hydrochloride (0.75 g to every 0.25 ml water) which gives a 1g/ml solution. The endothermic hydration cools the solution.

3. Dissolve MC (~1–5 mg/ml) in ethanol or methanol, and add 1.5 vol. water, 2 vol. DMSO, 0.67 vol. 5N NaOH, and 1 vol. aminoethanethiol hydrochloride (1 g/ml).

4. Briefly purge the container with N_2 gas, seal, and incubate at 50°C for 30 min.

5. Cool the solution, and in a fume hood, dilute fivefold with 0.1% (v/v) trifluoroacetic acid (TFA). Take the pH to 1.5 by dropwise addition of 100% (v/v) TFA, checking the pH by spotting on to a thin slither cut from a pH indicator strip (BDH).

6. Apply the sample to a C18 Sep-pak cartridge that has been pre-wetted in methanol, and then equilibrated in 0.1% (v/v)TFA. Collect the flow through into a sealed tube for disposal. Wash the cartridge with 0.1% (v/v) TFA in 10% (v/v) acetonitrile in water. Elute the amino-ethanethiol–MCs (AET–MC) with 0.1% TFA in acetonitrile, dry by rotary evaporation, and dissolve in 0.02 ml of methanol.

7. If using a pure microcystin, run a small portion of the AET–MC on a C18 column developed with a gradient of 0.1% trifluoroacetic acid/ acetonitrile. On our Vydac system AET–MC-LR is eluted at ~33% acetonitrile, compared with ~36% acetonitrile for unmodified MC-LR.

8. React the amino group of AET–MC with either CNBr–Sepharose or CH-activated Sepharose ('activated' means that it contains an *N*-hydroxy succinamide group that reacts with amino groups) according to the manufacturer's instructions (*Affinity chromatography: principles and methods*, a booklet supplied by Pharmacia-LKB).

9. Store the prepared and washed MC–Sepharose in buffer containing a bactericide, such as sodium azide, and wash before use.

Protocol 8. Purification of hepatic glycogen-bound form of PP1 by microcystin-affinity chromatography

Materials

- MC-Sepharose
- buffer A [50 mM Tris-HCl, pH 7.0, 0.1 mM EGTA, 0.1% (v/v) 2-mercaptoethanol, 5% (v/v) glycerol plus protease inhibitors, see below]
- buffer A containing 0.5 M NaCl
- buffer A containing 3 M NaSCN

Method

1. Prepare glycogen–protein particles from one rat liver, and suspend in 3 ml of 50 mM Tris–HCl, pH 7.0, 0.1 mM EGTA, 0.1% (v/v) 2-mercapto-ethanol, 5 % glycerol (buffer A), plus the protease inhibitors 1 mM benzamidine, 0.2 mM phenylmethylsulfonyl fluoride, 0.04 mg/ml leupeptin, and 0.5 mM tosylphenylchloromethyl ketone using a hand homogenizer.

2. Add MC–Sepharose (0.5 ml, containing 0.15 mg of MC), and mix for 30 min at 4°C.

3. Pour the suspension into a small column and wash with buffer A containing 0.5 M NaCl.

Protocol 8. *Continued*

4. When the protein concentration of the eluate is less than 0.005 mg/ml, pour one void volume of buffer A containing 3 M sodium isothiocyante (NaSCN) through the column and and stop the flow for 30 min.
5. Restart the flow and elute the PP1 with buffer A containing 3M NaSCN (assay according to *Protocol 3* at high dilution >1/500 to dilute the inhibitory NaSCN).
6. Dialyse against buffer A.

The purified enzyme comprises the catalytic subunit of PP1 and the 33 kDa G_L subunit which is responsible for the distinctive properties of the complex, i.e. glycogen binding, enhanced dephosphorylation of glycogen synthase by PP1G (compared with PP1C), and allosteric inhibition by phosphorylase *a*. The G_L subunit can be detected by direct binding of digoxygenin-labelled PP1C by the overlay procedure described in *Protocol 9* and *Figure 3*. The

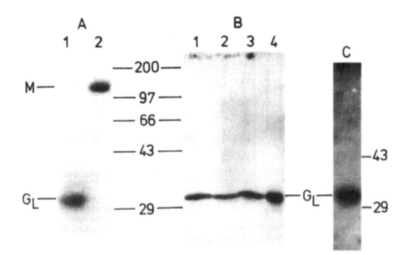

Figure 3. Identification of the G_L subunit using [^{125}I]microcystin-YR, digoxygenin (DIG)-PP1γ, and ^{32}P-labelled phosphorylase *a* as probes. The PP1-containing fractions obtained from microcystin–Sepharose chromatography of rat liver glycogen–protein particles were denatured in SDS, electrophoresed on 10% polyacrylamide gels, transferred to nitrocellulose, and blotted with the labelled PP1 probes (*Protocol 9*).(A) The microcystin column eluate from chromatography of rat hepatic glycogen–protein particles on MC–Sepharose (lane 1) and purified PP1$_M$ from chicken gizzard smooth muscle (lane 2) were blotted with [^{125}I]MC-YR-PP1γ (*Protocol 9*).(B) Glycogen-protein particles isolated from the liver of a fed rabbit (lane 1), a fasted rat (lane 2), and a fed rat (lane 3), and the microcystin column eluate from chromatography of rat hepatic glycogen–protein particles on MC–Sepharose (lane 4) were probed with DIG–PP1γ. (C) The eluate from MC–Sepharose chromatography of rat glycogen–protein particles was probed with ^{32}P-labelled phosphorylase *a* (see ref. 32 for further details). The positions of standard marker proteins myosin (205 kDa), phosphorylase (97 kDa), bovine serum albumin (66 kDa), ovalbumin (43 kDa), and carbonic anhydrase (29 kDa) are indicated.

33 kDa band was detected in isolated glycogen particles and in the purified enzyme (*Figure 3*), demonstrating that the purified protein is intact.

Microcystin–Sepharose has proved very useful for purifying the active forms of PP1 from other tissues and subcellular fractions. Microcystin-binding protein phosphatases can also be purified from soluble extracts of cells and tissues. One example is the isolation of PP5 from a cauliflower extract. However, we generally find that only ~15–50% of the total phosphorylase phosphatase activity in a soluble extract binds to the column, presumably due to inhibitors in the extracts that compete with the MC–Sepharose for binding to the protein phosphatases. For this reason 'clean-up' steps prior to MC–Sepharose might be useful.

Protocol 9. Identifying PP1C-binding regulatory subunits in overlays probed with digoxygenin (DIG)-labelled PP1 or [^{125}I]MC-YR-labelled PP1C

Materials

- purified PP1C (dialysed into buffer C, see below)
- Centricon 30 microconcentrator (Amicon, or equivalent)
- digoxygenin-carboxy-methyl-*N*-hydroxy-succinimide ester (Boehringer Mannheim, 4 mg/ml in DMSO, freshly prepared)
- anti-digoxygenin (DIG) antibodies conjugated with peroxidase-enhanced chemiluminescence detection kit (Boehringer Mannheim)
- buffer A (25 mM Tris–HCl, pH 7.5, 250 mM NaCl, and 1 mg/ ml BSA)
- buffer B (10 mM sodium phosphate, pH 8.5, 150 mM NaCl)
- buffer C (10 mM sodium phosphate, pH 7.5, 150 mM NaCl)
- buffer D [25 mM Tris–HCl, pH 7.5, 5% (w/v) skimmed milk powder (Marvel) and 500 mM NaCl]
- buffer E (25 mM Tris–HCl, pH 7.5, 500 mM NaCl)

Method (using digoxygenin-labelled PP1C)

1. Label 30 μg of PP1 with 5.3 μg digoxygenin-carboxy-methyl-*N*-hydroxy-succinimide ester in buffer B (final volume 200 μl) for 2 h at room temperature, and dialyse overnight into buffer C.

2. Run protein samples on SDS–PAGE, transfer to nitrocellulose membranes electrophoretically for 250 V h, and block non-specific sites with buffer D for 16 h at room temperature.

3. Probe nitrocellulose membranes with DIG-labelled PP1 diluted to 2.5 μg/ml in buffer A for 2 h at room temperature, then wash extensively with buffer E for 3 h.

4. Probe blots with anti-DIG antibodies conjugated with peroxidase diluted 5000-fold in buffer A for 45 min, and wash again for 3 h with buffer E.

5. Visualize with the enhanced chemiluminescence system (Amersham).

Protocol 9. *Continued*

Alternative method using [^{125}I]MC-YR-labelled PP1C

1. Label 30 µg of PP1 with [^{125}I]microcystin-YR for 16 h as in *Protocol 6*. Add BSA to 1 mg/ml and concentrate to 50 µl in a Centricon 30 micro-concentrator. Dilute to 1.5 ml and reconcentrate to 50 µl twice in buffer A to remove free microcystin.

2. Run protein samples on SDS–PAGE, transfer to nitrocellulose membranes electrophoretically for 250 V h, and block non-specific sites with buffer D for 16 h at room temperature.

3. Probe nitrocellulose membranes with labelled PP1 diluted to 2.5 µg/ml in buffer A for 2 h at room temperature, then wash extensively with buffer E for 3 h.

4. Expose membrane to X-ray film.

This procedure has been used to detect the G_L subunit (*Figure 3*), and many other PP1-binding proteins. It would be interesting to test whether synthetic peptides containing the PP1-binding motif [(R/K) (V/I)xF] block binding of PP1 to regulatory subunits in these overlays. If so, this feature would provide an additional test of binding specificity.

References

1. Cohen, P.T.W. (1997). *Trends in Biochemical Sciences*, **22**, 245.
2. Barford, D. (1996). *Trends in Biochemical. Sciences*, **21**, 407.
3. Garbers, C., DeLong, A., Deruere, J., Bernasconi, P., and Soll, D. (1996). *EMBO Journal*, **15**, 2115.
4. Cohen, P. (1994) *BioEssays*, **16**, 583
5. Cohen, P., Alemany, S., Hemmings, B.A., Resink, T.J., Strålfors, P., and Tung, H.Y.L. (1988). In *Methods in enzymology* (ed. J.D. Corbin and R.A. Johnson), Vol. 159, p. 390. Academic Press, London.
6. Alessi, D. R., Street, A.J., Cohen, P., and Cohen, P.T.W. (1993). *European Journal of Biochemistry*, **213**, 1055.
7. MacKintosh, C., Garton, A.J., McDonnell, A., Barford, D., Cohen, P.T.W., Tonks, N.K., and Cohen, P. (1996). *FEBS Letters*, **397**, 235.
8. MacKintosh, C. and MacKintosh, R.W. (1994). *Trends in Biochemical Sciences*, **19**, 444.
9. Walsh, A.H., Cheng, A., and Honkanen, R.E. (1997). *FEBS Letters*, **416**, 230.
10. Liu, J., Farmer, J.D. Jr, Lane W.S., Friedman, J., Weissman, I., and Schreiber, S.L. (1991). *Cell*, **66**, 807.
11. Fischer, E.H. and Krebs, E.G. (1958). *Journal of Biological Chemistry*, **231**, 73.
12. Cohen, P. (1983). In *Methods in enzymology* (ed. J.D. Corbin and J.G. Hardman), Vol. 99, p. 243. Academic Press, London.
13. Cohen, P. (1991). In *Methods in enzymology* (ed. T. Hunter and B.M. Sefton), Vol. 201, p. 389. Academic Press, London.

14. MacKintosh, C. and Cohen, P. (1990). *Biochemical Journal*, **262**, 335.
15. MacDougall, L.K., Jones, L.R., and Cohen, P. (1991). *European Journal of Biochemistry*, **196**, 725.
16. MacKintosh, C., Coggins, J., and Cohen, P. (1991). *Biochemical Journal*, **273**, 733.
17. Sola, M., Langan, T., and Cohen, P. (1991). *Biochmica Biophysica Acta*, **1094**, 211.
18. Clarke, P. R., Moore, F., and Hardie, D.G. (1991). *Advances in Protein Phosphatases*, **6**, 187.
19. Cohen, P., Foulkes, J.G., Holmes, C.F.B., Nimmo, G.A., and Tonks, N.K. (1988). In *Methods in enzymology* (ed. J.D. Corbin and R.A. Johnson),. Vol. 159, p. 427. Academic Press, London.
20. Chen, M.X. and Cohen, P.T.W. (1997). *FEBS Letters*, **400**, 136.
21. McGowan, C.H. and Cohen, P. (1988) In *Methods in enzymology* (ed. J.D. Corbin and R.A. Johnson), Vol. 159, p. 416. Academic Press, London.
22. Donella-Deanna, A. and Pinna, L.A. (1994). *Biochmica Biophysica Acta*, **1222**, 415.
23. Donella-Deanna, A., Krinks, M.H., Ruzzene, M., Klee, C., and Pinna, L.A. (1994). *European Journal of Biochemistry*, **219**, 109.
24. Agostinis, P., Goris, J., Pinna, L.A., Marchiori, F., Rerich, J.W., Meyer, H.E., and Merlevede, W. (1990). *European Journal of Biochemistry*, **189**, 235.
25. Rusnak, F., Beressi, A.H., Haddy, A., and Tefferi, A. (1996). *Bone Marrow Transplant*, **17**, 309.
26. Gaussin, V., Hue, L., Stalmans, W., and Bollen, M. (1996). *Biochemical Journal*, **316**, 217.
27. Takai, A. and Mieskes, G. (1991). *Biochemical Journal*, **275**, 233.
28. Cayla, X., Van Hoof, C., Bosch, M., Waelkens, E., Vandekerckhove, J., Peeters, B., Merlevede, W., and Goris, J. (1994). *Journal of Biological Chemistry*, **269**, 15668.
29. Ward, C.J., Beattie, K.A., Lee, E.Y., and Codd, G.A. (1997) *FEMS Microbiological Letters*, **153**, 465.
30. MacKintosh, R.W., Dalby, K.N., Campbell, D.G., Cohen, P.T.W., Cohen, P., and MacKintosh, C. (1995). *FEBS Letters*, **371**, 236.
31. Moorhead, G., MacKintosh, R.W., Morrice, N., Gallagher, T., and MacKintosh, C. (1994). *FEBS Letters*, **356**, 46.
32. Moorhead, G., MacKintosh, C., Morrice, N., and Cohen, P. (1995). *FEBS Letters*, **362**, 101.
33. Hubbard, M.J. and Cohen, P. (1991). In *Methods in enzymology* (ed. T. Hunter and B.M. Sefton), Vol. 201, p. 414. Academic Press, London.
34. Pato, M.D. and Adelstein, R.S. (1982). In *Methods in enzymology* (ed. D.W. Frederiksen and L.W. Cunningham), Vol. 85, p. 308. Academic Press, London.
35. Stewart, A.A. and Cohen, P. (1988). In *Methods in enzymology* (ed. J.D. Corbin and R.A. Johnson), Vol. 159, p. 409. Academic Press, London.
36. Campos, M., Fadden, P., Alms, G., Qian, Z., and Haystead, T.A.J. (1997). *Journal of Biological Chemistry*, **271**, 28478.

8

Identification of substrates for protein–tyrosine phosphatases

ANDREW J. GARTON, ANDREW J. FLINT, and
NICHOLAS K. TONKS

1. Introduction

The protein–tyrosine phosphatase (PTP) family of enzymes comprises a large number (>75) of structurally diverse enzymes involved in regulating intracellular signalling processes through the dephosphorylation of phosphotyrosine-containing substrate proteins (1, 2). Genetic studies have now clearly demonstrated that individual PTPs fulfil specific, non-redundant roles in regulating mammalian signal transduction processes (3–5). It therefore appears likely that individual PTPs recognize only a limited range of tyrosine-phosphorylated substrate proteins *in vivo*. However, genetic studies generally provide little information regarding the biochemical functions of the gene products under investigation. Thus, in order to understand fully the physiological functions of individual PTPs, it will be necessary to employ additional approaches to identify the biologically relevant substrates for a particular enzyme. This has proven to be an extremely elusive goal, which has been significantly hampered by the relatively low substrate specificity displayed by many PTPs *in vitro*. Indeed, early investigations of PTP substrate specificity using a small number of artificial substrates led to the proposal that intracellular targeting of PTPs via their non-catalytic segments may be the primary determinant of substrate specificity *in vivo*, whereas isolated PTP catalytic domains may be rather promiscuous in their substrate preferences (6). However, although subcellular location clearly plays an important role in determining the range of substrates available to many PTPs (7), evidence is now mounting that PTP catalytic domains are themselves capable of demonstrating discernible, and sometimes exquisite, substrate specificity (8).

We have recently developed new methods for identifying physiological substrates for individual PTPs and have applied these methods to the study of several members of the PTP family. These methods are generally applicable to any PTP family member and, therefore, should prove useful in delineating the substrate preferences of other PTPs, of which the physiological substrate specificity is in most cases currently completely unknown.

2. Development of substrate-trapping mutant forms of PTPs

During the last five years a substantial body of structural and kinetic data has emerged for various members of the PTP family. These studies include the elucidation of the crystal structure of PTP1B, both alone and in complex with a phosphotyrosine-containing substrate peptide (9, 10), analysis of a mutant PTP1B molecule which revealed the structure of a reaction intermediate (11), structural analysis of PTPs in complex with a variety of active site-directed inhibitors (11–14), and kinetic analysis of various forms of PTP1B and YopH (15–17). This work has allowed a highly detailed description of the PTP catalytic mechanism, including the elucidation of functional roles for the majority of the 27 invariant residues found in PTP catalytic domains (9–11, 18, 19). Our efforts to identify substrates for specific PTPs were initiated, in light of the available kinetic data, following the first structural studies performed on PTP1B (9, 10, 18). These studies provided a starting point for our attempts to generate mutant forms of PTPs with substrate-trapping properties, i.e. mutant enzymes that are capable of binding with high affinity to appropriate substrates, but which are impaired in their ability to complete the dephosphorylation reaction. Such mutants are useful reagents in identifying PTP substrates, since bound substrates become trapped in stable, dead-end complexes with the mutant PTP, allowing these enzyme–substrate complexes to be isolated and the bound substrate to be identified.

Kinetic analysis of a large number of mutant forms of PTP1B indicated two mutants that potentially display the appropriate substrate-trapping characteristics (20). Mutation of the active-site cysteine residue of PTPs (to serine or alanine) has been known for some time to result in a completely inactive enzyme, which in some cases can bind substrates with sufficiently high affinity to allow the isolation of stable complexes (10, 21–24). However, this mutation frequently yields enzymes that fail to bind substrates efficiently in stable complexes, particularly in the context of substrate-trapping experiments performed in intact cells. In contrast, mutation of an invariant aspartic acid residue (residue 181 in PTP1B) to alanine reduces the k_{cat} by a factor of 10^5, but generates an enzyme whose affinity is similar to that of the wild-type enzyme (20). Moreover, the role of this invariant aspartic acid residue in the catalytic mechanism is to catalyse the release of the substrate molecule from the active site of the enzyme via protonation of the phenolic oxygen atom of the substrate phosphotyrosine residue. Thus mutation of this residue to alanine results in an enzyme which binds substrates efficiently, but is considerably impaired in its ability to dephosphorylate and release the bound substrate molecules. An additional feature of such mutants is a reduction in the potential for electrostatic repulsion between the negatively charged side chain of the aspartic acid residue and the negatively charged substrate phosphate moiety. This electrostatic repulsion is likely to oppose the strong hydrophobic interactions between the phenol ring of the substrate tyrosine residue and hydrophobic residues of the phosphatase, which create a strongly hydrophobic environment within the PTP active-site cleft. These hydrophobic interactions are essential for the closure of the active-site cleft of the enzyme around the substrate phosphotyrosine residue during catalysis, and play a significant role in substrate dephosphorylation by ensuring that the substrate is held in position, and in the correct orientation, during catalysis. Mutation of the invariant aspartic acid residue would therefore be expected to enhance these interactions, and thereby improve the stability of complexes containing bound substrates. Indeed, we have found that this type of mutant is considerably more useful than previously described mutants in which the active-site cysteine is converted to serine or alanine. It can be used in a variety of different strategies both *in vitro* and in intact cells, and is likely to facilitate greatly the identification of physiologically relevant substrates for all PTPs.

3. Substrate trapping *in vitro*

Identification of PTP substrates by trapping performed *in vitro* involves three major steps; generation and purification of the mutant PTP, incubation of the mutant enzyme with a pool of phosphotyrosine-containing candidate substrate molecules (*Protocol 1*) and identification of the substrate molecules bound to the mutant PTP (see *Figure 1*).

Initially, the invariant aspartic acid residue within the PTP catalytic domain must be identified and mutated to alanine. This residue is found within the conserved sequence WPD (both the tryptophan and aspartic acid residues are invariant, but in some enzymes the intervening proline is substituted by a different amino acid, e.g. lysine, serine, phenylalanine) and corresponds to residue 181 in PTP1B. Following mutation, several options exist for expression and purification of the mutant enzyme.

Most experiments can be readily performed with microgram amounts of protein, which may be obtained following expression in *E. coli*, in Sf9 cells (using recombinant baculovirus), or in mammalian cells (COS or 293) following transfection. In the latter case it is important to ensure that there is no interaction with any phosphotyrosine-containing proteins present in the transfected cells, and no contamination of the mutant enzyme preparation with wild-type enzyme. A major consideration in designing substrate-trapping experiments is the precise nature of the construct to be used. Important considerations include: (i) whether to include non-catalytic sequences; (ii) whether to add epitopes or other tags (e.g. GST) to facilitate purification of the expressed protein; (iii) in the case of transmembrane PTPs containing two copies of the PTP domain sequence, whether to include both domains in the

Andrew J. Garton et al.

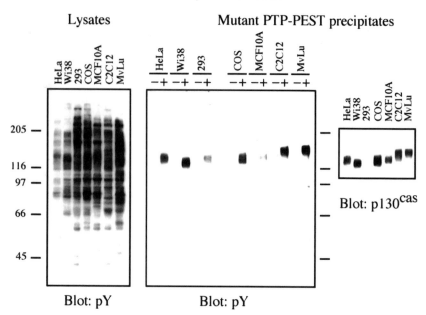

Figure 1. Identification of p130^{cas} as a substrate for PTP–PEST by substrate trapping *in vitro*. The indicated cell lines were treated with pervanadate and lysed as described in *Protocol 1*. Aliquots (30 μg of protein) were analysed by SDS–PAGE and antiphosphotyrosine immunoblotting (left panel). Substrate trapping was performed using 2 μg of baculovirus-expressed purified mutant (D199A) PTP–PEST protein, which was immobilized on protein A–Sepharose beads via a monoclonal PTP–PEST antibody which had been covalently coupled to the matrix using dimethyl pimelimidate (lanes marked +); samples in control (–) lanes were incubated with the same antibody beads, excluding only the mutant PTP–PEST protein. Proteins associating with the mutant PTP–PEST were immunoblotted with antibodies to phosphotyrosine (middle panel) or p130^{cas} (right panel). p130^{cas} was selected as a possible candidate substrate for PTP–PEST based on the apparent molecular mass and relative abundance of the tyrosine-phosphorylated protein associated with the mutant PTP–PEST protein (middle panel). The left portion of the middle panel (HeLa, Wi38 and 293 samples) was exposed for 5 min, whereas the right portion was exposed for 30 sec. Note that a single major phosphotyrosine-containing protein was precipitated from each lysate by the mutant PTP–PEST protein (middle panel), and that this tyrosine-phosphorylated protein in each case corresponds to p130^{cas} (right panel). [Data from Garton *et al.* (1996). *Mol. Cell. Biol.* **16**, 6408.]

same construct or analyse each domain in isolation. Regardless of the nature of the construct to be utilized, it is essential also to generate a control construct containing the wild-type form of the PTP catalytic domain, in order to distinguish between substrates bound to the mutant PTP catalytic site (i.e. via a substrate-trapping interaction) and tyrosine-phosphorylated proteins which bind via other mechanisms (particularly via non-catalytic segments present in the expressed construct). Similarly, it is also essential to verify that substrate-trapping can be blocked by pre-incubation of the mutant PTP with the active site-directed inhibitor vanadate, again to ensure that any apparent

186

substrate-trapping observed truly results from interactions of the tyrosine phosphorylated protein with the active site of the mutant PTP.

Once expressed, the mutant PTP protein must be purified and then attached to a matrix before use in substrate-trapping experiments. The two most convenient methods for achieving these goals are to use glutathione–Sepharose beads following expression of the mutant protein as a GST fusion protein, or to use antibodies (recognizing either the PTP itself or epitopes added to the expression construct) coupled to protein A–Sepharose beads. If the latter approach is used, it is often useful to couple the antibody molecules covalently to the affinity matrix using dimethyl pimelimidate (25), since these antibody proteins will otherwise obscure two portions of the final SDS–PAGE gel (with apparent molecular weights of 50 kDa and 20 kDa, corresponding to the heavy and light chain components of the antibody molecule, respectively) and thus prevent detection of any tyrosine-phosphorylated substrate proteins which migrate at these positions in the gel. Also, if purified by immuno-precipitation with a PTP-specific antibody, it is essential to ensure that the isolated mutant PTP is not contaminated with any endogenous active enzyme molecules. This potential problem may be avoided by producing the mutant PTP in cell lines which normally do not express the PTP, by isolating the mutant enzyme using appropriate epitope tag antibodies, or by expression of the mutant as a GST fusion protein followed by isolation using glutathione–Sepharose beads.

Generation of a pool of tyrosine-phosphorylated potential substrate molecules is most conveniently achieved by pervanadate treatment of cells growing in culture (*Protocol 1*). Pervanadate treatment can be used for any cell type growing in culture; therefore, the starting material for these experi-ments should be a cell line that normally expresses high levels of the PTP being investigated. Pervanadate stimulates tyrosine phosphorylation within the cell by non-selective inhibition of the endogenous PTPs, resulting in a dramatic increase in the phosphotyrosine content of a large number of cellu-lar proteins (26). It is important to point out that the level of phosphorylation achieved by this treatment is far greater than that observed normally within the cell, even in cells overexpressing oncogenic protein tyrosine kinases. Thus, it cannot be assumed that the nature and phosphorylation state of the phos-photyrosine-containing proteins generated by this procedure is necessarily an accurate representation of the potential physiological substrates normally found within the cell. Regardless of these considerations, this procedure generates a wide variety of highly tyrosine-phosphorylated cellular proteins, and therefore provides a good starting point for analysis of PTP substrate specificity.

Other approaches for the generation of complex mixtures of potential substrates may also be considered, such as the use of phosphotyrosine-containing peptide libraries similar to those used to delineate the recognition motifs of SH2 domains (27). In this approach, the entire complement of substrate phosphopeptides capable of binding with high affinity to an immobilized mutant PTP could be sequenced directly, permitting the elucidation of strong amino acid preferences of the PTP in the immediate vicinity of the phosphorylated tyrosine residue of the substrate molecule. An additional, somewhat similar, approach for PTP substrate identification was recently described, in which a library of peptides containing the non-hydrolysable phosphotyrosine analogue difluorophosphonomethyl-phenylalanine was screened for peptides capable of binding to an affinity matrix containing immobilized wild-type PTP. Bound candidate substrate peptides were identified directly by mass spectrometry (28).

The buffer composition and the steps involved in preparing the cell lysate in *Protocol 1* are designed to generate a population of tyrosine-phosphorylated potential substrate proteins which are stable with respect to their phosphorylation state, while allowing for substrate trapping to occur efficiently upon addition of the mutant PTP. These two somewhat incompatible requirements are reconciled by the inclusion of the irreversible active site-directed PTP inhibitor iodoacetic acid in the lysis buffer (rendering the endogenous PTPs inactive by covalent modification of the active-site cysteine residue), followed by the addition of a fourfold excess of reducing agent (DTT) to inactivate any remaining iodoacetic acid (preventing modification of the mutant PTP added thereafter). An additional consideration is the potential inhibition of substrate-trapping interactions by the pervanadate used to stimulate the cells. Although the concentration of pervanadate in the lysate is likely to be considerably lower than that used in treatment of the cells (0.1 mM), this compound is an extremely potent PTP inhibitor (IC_{50} approximately 10 nM), which could therefore potentially interfere with subsequent substrate-trapping events by interacting with the active site of the mutant PTP. The inclusion of EDTA in the lysis buffer alleviates this problem owing to a well-documented chelating effect (8, 29) that lowers the effective concentration of pervanadate in the lysate.

In addition to their use in substrate-trapping experiments, lysates prepared in this way can also be used to study the specificity of substrate dephosphorylation by wild-type PTPs, by incubating the lysate with the active PTP and immunoblotting the phosphatase-treated lysate with an antibody specific for phosphotyrosine (8). Such experiments can provide useful supporting data regarding the nature of the substrates that are efficiently recognized by the active enzyme. The major consideration in performing these studies is to use limiting amounts of the PTP, and perform the incubation at a low temperature (e.g. on ice), to allow subtle substrate preferences to be observed. Indeed, even in the case of a highly selective PTP such as PTP–PEST, which preferentially dephosphorylates p130[cas] over a wide range of enzyme concentrations (8), prolonged incubations at high enzyme concentrations result in the complete dephosphorylation of all phosphotyrosine-containing proteins present in the lysate.

Identification of tyrosine-phosphorylated potential substrates observed to associate with the mutant PTP may be achieved by several methods. If the potential substrate is sufficiently abundant, it may be purified to allow its identification directly by protein sequencing. Since the affinities of inter-actions between mutant PTPs and their substrates are relatively high (in some cases allowing quantitative precipitation of the substrate from the lysate), the substrate purification can frequently be achieved in a single step using the mutant PTP as an affinity column, followed by isolation of the substrate for sequencing by preparative-scale SDS–PAGE. One important consideration in using aspartic acid mutant forms of PTPs as affinity purification reagents is that these mutants, although catalytically impaired, still possess measurable PTP activity [equivalent to 1 catalytic cycle per hour at 30°C in the case of PTP1B (20)]. Thus it is essential to perform the purification as rapidly as possible, at 4°C and in the absence of reducing agents (which are required by PTPs to maintain catalytic activity), in order to prevent substrate dephos-phorylation and the resultant release of bound substrate molecules from the mutant PTP.

In cases where the abundance of the substrate molecule is relatively low, prohibiting large-scale purification and sequencing, it is often necessary to attempt to identify the substrate based on various characteristics, followed by immunoblotting analysis using an appropriate antibody (see *Figure 1*). Useful properties to examine include the apparent molecular weight of the protein on SDS–PAGE gels (compared with proteins of similar molecular weight which are known to be tyrosine phosphorylated), subcellular fractionation properties, and expression pattern in a range of cell lines (e.g. haematopoietic, epithelial, fibroblast, neuronal cells). In addition to these methods of substrate identification, it is likely that the rapidly emerging technology of proteomics will soon become sufficiently sophisticated to allow the direct identification of unknown substrate proteins following their analysis by two-dimensional gel electrophoresis (30, 31).

Protocol 1. Trapping of PTP substrates by incubation with mutant PTPs *in vitro*

Reagents

- purified mutant and wild-type PTP proteins
- affinity matrix [glutathione–Sepharose (Pharmacia) for mutant PTP–GST fusion proteins, or protein A–Sepharose (Pharmacia) with pre-coupled antibody]
- pervanadate solution [prepare immediately before use: add 0.5 ml of 0.1 M sodium orthovanadate, and 5.7 μl of 30 % (v/v) hydrogen peroxide solution to 0.5 ml water]
- buffer A [20 mM Tris–HCl, pH 7.4, 1 mM EDTA, 1 mM benzamidine, 1 μg/ml leupeptin, 1 μg/ml aprotinin, 10 % (v/v) glycerol, 1 % (v/v) Triton X-100, 100 mM NaCl, 5 mM iodoacetic acid]
- anti-phosphotyrosine antibody [e.g. 4G10 (UBI), PY20 (Transduction Laboratories)]

Protocol 1. *Continued*

Method

1. Grow the cells in a 10 cm diameter tissue culture plate until they reach 80–90% confluence, then add 0.002 vol. of pervanadate solution (final concentration 0.1 mM), and incubate for 30 min at 37 °C.

2. Rinse the cells twice in PBS at 4 °C, then lyse on ice by the addition of 0.75 ml buffer A. Scrape the cells from the plate and incubate the lysate in a 1.5 ml Eppendorf tube on a rocking platform at 4 °C for 30 min.

3. Add 0.01 vol. of 1M DTT (to inactivate any remaining unreacted iodoacetic acid), then clarify the lysate by centrifugation at 10 000 *g* for 10 min at 4 °C. The resulting supernatant can then either be stored at –70 °C or used immediately in substrate-trapping experiments.

4. Couple 2 µg of the mutant PTP to an appropriate bead matrix (e.g. use 10 µl of protein A–Sepharose with an appropriate bound PTP antibody or 10 µl of glutathione–Sepharose beads for GST-mutant PTP fusion proteins).

5. Incubate the lysate (500 µg of total protein) with the mutant PTP matrix at 4 °C for 30–45 min on a rocking platform.

6. Collect the precipitates by brief centrifugation (10 sec at 1000 *g*), remove the supernatant, and rapidly wash the pelleted beads three times with 1 ml of ice-cold buffer A (omit iodoacetic acid). Add 50 µl of SDS–PAGE sample buffer to the beads, heat at 100°C for 5 min, and remove the beads from the samples by centrifugation prior to loading on to an SDS–PAGE gel.

7. Analyse the proteins associated with the mutant PTP by SDS–PAGE and immunoblotting using a phosphotyrosine-specific antibody.

4. Substrate trapping in intact cells

Many PTPs are restricted to specific subcellular locations which limits the spectrum of tyrosine-phosphorylated potential substrates accessible to them. It is therefore beneficial to study the substrate specificities of such enzymes within intact cells, to ensure that only physiologically relevant substrates will be identified. In general, substrate-trapping experiments performed within intact cells are subject to similar considerations to those described above for experiments performed *in vitro*. However, several additional factors need to be considered. In order to ensure that only appropriate substrates are available to the mutant PTP, it is essential that the expressed enzyme is correctly localized within the cell. Therefore, the expressed protein should include the entire protein sequence and, if it also includes additional sequences (e.g.

antibody epitopes, GST), the subcellular location of the fusion protein should be compared with that of the endogenous protein (or overexpressed non-fusion protein), to confirm that such sequences do not affect the intracellular targetting of the PTP. This type of analysis should ideally be performed using a wild-type construct, in addition to the substrate-trapping mutant form of the enzyme, since this approach can often yield much useful information regarding the subcellular site of action of the enzyme being investigated. Frequently, expressed substrate-trapping mutant forms of PTPs can be visualized in regions of the cell somewhat different from the subcellular locations in which the wild-type enzymes are normally observed, owing to strong interactions with tyrosine-phosphorylated substrates whose locations overlap with, but are not identical to, the observable steady-state intracellular location of the wild-type PTP (7, 20, 21). The most striking example of this phenomenon is the dramatic, EGF-induced redistribution of a substrate-trapping mutant form of the 45 kDa splice variant of TCPTP. Under basal conditions, both wild-type and mutant transfected forms of this enzyme are localized predominantly in the nucleus. However, EGF stimulation results in a rapid redistribution of the mutant PTP to a cytoplasmic location owing to its stable association with tyrosine-phosphorylated substrates such as the EGF receptor (7). This effect is not observed with the wild-type enzyme which remains largely nuclear upon EGF stimulation, presumably because the normal interactions of TCPTP with tyrosine-phosphorylated EGF receptor molecules are highly transient in nature. Thus, the use of substrate-trapping forms of PTPs may allow the visualization of mutant PTPs that have been retained at physiologically relevant sites within the cell by prolonged interaction with their substrates. Such observations are clearly very useful in establishing the function of a PTP and may not be achievable by any other approach.

A second major factor determining the success of an *in vivo* substrate-trapping experiment is the level of expression of the mutant PTP. In most cases, in order to achieve significant trapping of tyrosine-phosphorylated substrates it is necessary to overcome the effects of the endogenous, wild-type protein. Therefore, one would anticipate that considerably higher levels of expression of the trapping mutant, compared with endogenous PTP, would be required. The reason for this is apparent when one considers that substrate trapping is at best a stoichiometric phenomenon, whereas the wild-type endogenous enzyme interacts with tyrosine-phosphorylated substrate molecules in a transient and catalytic manner, and is therefore capable of interacting with (and rapidly dephosphorylating) many substrate molecules, thereby preventing association with the mutant PTP protein. For these reasons, *in vivo* substrate-trapping experiments are frequently most successful when carried out in COS or 293 cells, where very high levels of expression can readily be achieved using appropriate expression vectors. An alternative strategy for avoiding the actions of the endogenous enzyme may be applicable to PTPs that display rapid transient changes in expression levels under appropriate conditions. This would involve the use of an appropriately regulated expression system to induce expression of the mutant PTP under conditions where the endogenous enzyme is present at a very low level.

In order to achieve efficient substrate trapping by mutant PTPs within intact cells, it may first be necessary to stimulate tyrosine phosphorylation within the transfected cells. In most cells under normal growth conditions, the steady-state level of tyrosine phosphorylation of the majority of potential substrate proteins will be very low, and the turnover rate may also be relatively slow. However, it may also be informative to attempt substrate-trapping experiments in serum-starved cells in the absence of stimuli. Under these conditions, proteins which undergo rapid constitutive turnover of phosphotyrosine can become trapped in a highly tyrosine-phosphorylated state due to protection, by bound mutant PTP molecules, of the phosphoryl-ated residues from dephosphorylation by endogenous active PTPs (*Figure 2*). For example, expression of the substrate-trapping mutant form of PTP1B in

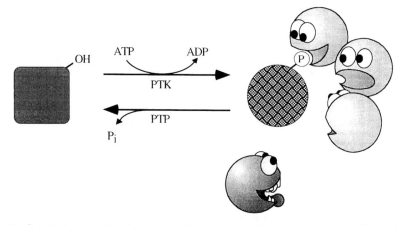

Figure 2. Substrate trapping in intact cells promotes the accumulation of tyrosine-phosphorylated substrate proteins. The figure shows a substrate protein in equilibrium between tyrosine phosphorylated and non-phosphorylated states. Normally, the opposing actions of protein tyrosine kinases (PTKs) and phosphatases (PTPs) maintain a relatively low level of phosphorylation of the protein. The balance is upset by the introduction into the cell of a large excess of a mutant PTP which is catalytically impaired, but can still bind to appropriate substrates with high affinity (i.e. a substrate-trapping mutant PTP, shown here as a 'toothless' enzyme to indicate its ability to bind substrates but not remove the phosphate group). Under appropriate conditions, this binding will block the interaction of the phosphorylated substrate molecule with the endogenous active PTP (enzyme with teeth), allowing the tyrosine-phosphorylated form of the substrate to accumulate. Note that the biological consequences of this type of accumulation may be minimal, since the phosphorylated tyrosine residues are bound tightly by the mutant PTP and therefore are not accessible to their downstream targets (SH2 domains, PTB domains, etc.). Thus, even though the substrate may exhibit very high levels of tyrosine phosphorylation, its biological properties may more closely resemble those of the non-phosphorylated protein.

serum-starved COS cells is sufficient in itself to increase the steady-state level of tyrosine phosphorylation of endogenous EGF receptor molecules in the endoplasmic reticulum (20). In fact, the EGF receptor constitutes the major tyrosine-phosphorylated protein in lysates from COS cells expressing mutant PTP1B. These observations strongly suggest that one physiological function of the endoplasmic reticulum-localized enzyme PTP1B is to negate any ligand-independent phosphorylation of EGF receptor molecules occurring during transit from the endoplasmic reticulum to the cell surface, thereby preventing inappropriate activation of intracellular signalling pathways in the absence of ligand.

In contrast to the above observations for PTP1B, merely overexpressing mutant forms of PTPs sometimes has a negligible effect on intracellular phosphotyrosine levels, and does not, in itself, result in efficient substrate trapping (see Figure 6 in ref. 20). This observation indicates that the substrates of such PTPs may be subject to inducible tyrosine phosphorylation. In such cases, an additional stimulus is required to generate a sufficient level of substrate tyrosine phosphorylation to allow stable association with the mutant PTP (7, 8). Since the nature of the physiological substrate for the PTP under investigation will initially be unknown, several different phosphotyrosine-inducing cell treatments should be tested, e.g. growth factors, cytokines, attachment of cells to extracellular matrix proteins. Alternatively, the mutant PTP can be co-transfected with activated forms of protein tyrosine kinases such as v-src (32). Another approach that may prove useful for certain PTPs is to stimulate tyrosine phosphorylation within cells expressing the mutant PTP, using pervanadate. If this approach is taken, efficient substrate trapping will only be possible if the active-site cysteine of the PTP catalytic domain is mutated. This is because inhibition of PTPs by vanadate arises from a direct interaction of the active-site cysteine thiol group with the vanadium atom, which results in the formation of a covalent bond between the inhibitor and enzyme molecules (11). In addition, studies performed *in vitro* have demonstrated that, under appropriate conditions, pervanadate treatment of PTPs results in the irreversible oxidation of the active-site cysteine residue to cysteic acid (29). Therefore, the pervanadate used to inhibit endogenous PTP activity would also inhibit interactions of substrate molecules with aspartic acid mutant forms of PTPs over a similar concentration range (8). When these interactions are prevented by mutation of the cysteine residue, vanadate and pervanadate are far less efficient PTP inhibitors, which permits binding of appropriate substrates to the catalytic site of the mutant enzyme even in the presence of these inhibitors. It is important to point out that this particular approach may not be appropriate for all PTPs. A major disadvantage is the relatively low efficiency of substrate trapping generally displayed by mutant PTPs harbouring a mutated active-site cysteine residue (see Section 2). This problem may be partially overcome if very high levels of overexpression of the mutant PTP can be achieved in an appropriate cell line (e.g. COS or 293 cells). In addition, the efficiency of substrate trapping by such mutants may be significantly enhanced in certain cases by distinct high-affinity interactions between the substrate protein and non-catalytic segments of the PTP (33). A potentially useful modification to this approach might be to utilize a double mutant form of the enzyme in which both the cysteine and aspartic acid residues are altered. Preliminary results obtained for PTP1B indicate that this type of mutant displays similar substrate-trapping properties to the aspartic acid mutant form of the enzyme, in trapping experiments performed in intact cells (A.J. Flint and N.K. Tonks, unpublished data). It is therefore conceivable that such a mutant, by retaining the key properties of both of the singly mutated derivatives, may prove significantly more efficient at trapping substrates within pervanadate-treated cells than either of the singly mutated PTPs.

Protocol 2.   Trapping of PTP substrates in intact cells

Reagents

- 2 × HBS (50 mM Hepes, pH 7.1, 1.5 mM Na$_2$HPO$_4$, 10 mM KCl, 280 mM NaCl)
- buffer A [Tris–HCl, 20 mM, pH 7.4, 1 mM EDTA, 1 mM benzamidine, 1 μg/ml leupeptin, 1 μg/ml aprotinin, 10 % (v/v) glycerol, 1 % (v/v) Triton X-100, 100 mM NaCl, 5 mM iodoacetic acid]
- protein A–Sepharose (with pre-coupled PTP antibody), or glutathione–Sepharose, both from Pharmacia
- anti-phosphotyrosine antibody [e.g. 4G10 (UBI), PY20 (Transduction Laboratories)]

Method

1. Grow the cells to be transfected to 50–70 % confluence, add fresh medium to the cells prior to transfection (10 ml per 10 cm diameter plate).

2. Mix 20 μg of plasmid DNA encoding the mutant PTP, 50 μl of 1.25 M CaCl$_2$, and 200 μl of deionized water, then add the solution dropwise with mixing to 250 μl of 2 × HBS. Allow the precipitate to form at room temperature for 30 min, then add dropwise to the cells.

3. Allow the cells to grow for 12–15 h, then replace the medium and allow the cells to grow for a further 24 h.

4. Stimulate the cells (if necessary) to increase the cellular phosphotyrosine content (potentially useful stimuli may include growth factors, cytokines, or pervanadate).

5. Rinse the cells twice with 10 ml of PBS at 4°C, then lyse on ice by the addition of 0.75 ml of buffer A. Scrape the cells from the plate and incubate in a 1.5 ml Eppendorf tube, on a rocking platform, at 4°C for 30 min. Clarify the lysate by centrifugation at 10 000 *g* for 10 min at 4°C.

6. Couple an appropriate PTP antibody to 10 μl of protein A–Sepharose beads.

7. Precipitate the mutant PTP and associated substrate proteins by incubating the lysate (500 μg of total protein) with the PTP antibody at 4°C for 1 h on a rocking platform. Alternatively, if the PTP is expressed as a GST fusion protein, incubate the lysate for 30 min with 10 μl of glutathione–Sepharose beads.

8. Collect the precipitates by brief centrifugation (10 sec at 3000 *g*), remove the supernatant, and rapidly wash the pelleted beads three times with 1 ml of buffer A (omit iodoacetic acid). Add 50 μl of SDS–PAGE sample buffer to the beads, heat at 100°C for 5 min, and remove the beads from the samples by centrifugation prior to loading on to an SDS–PAGE gel.

9. Analyse the proteins associated with the mutant PTP by SDS–PAGE and immunoblotting using a phosphotyrosine-specific antibody.

5. Future applications

In addition to the approaches described here, it is likely that novel alternative methods will soon become available for utilizing substrate-trapping mutant forms of PTPs. One approach that could prove useful in future studies is a modified yeast two-hybrid screen based on a method previously used for identification of proteins recognized, in a phosphotyrosine-dependent manner, by the PTB domains of the SHC and Numb adaptor proteins (34, 35). In this approach, cDNA encoding a tyrosine kinase catalytic domain would be incorporated into the same vector as that used for expression of a LexA fusion protein containing the substrate-trapping mutant PTP catalytic domain, with expression of the two proteins being driven by different promoters. Co-expression in yeast of this 'bait' vector with a second vector containing a library of VP16 fusion proteins would then allow the tyrosine phosphorylation of the library proteins, and their subsequent association with the mutant PTP fusion protein, an interaction that is potentially sufficiently stable to allow detection by standard two-hybrid methods. Similarly, a two-hybrid screen could be performed using yeast that constitutively express a tyrosine kinase. In such approaches it is important to choose the tyrosine kinase to be expressed carefully, since these enzymes may be toxic when constitutively expressed in yeast cells (36, 37).

Another potentially useful screening approach would be a modified expression cloning strategy, in which a library of target proteins is expressed in a strain of *E. coli* that constitutively expresses a tyrosine kinase. The colonies are then screened by standard filter overlay methods for their ability to associate with a purified mutant PTP protein. Somewhat similar approaches have been used previously to identify active protein–tyrosine phosphatases (38), and substrates for protein kinases (39).

The ability of catalytically inactive forms of PTPs to promote tyrosine phosphorylation of appropriate substrates within the cell (20) has led to the suggestion that these mutant PTPs may function as 'dominant negative' reagents for studying signalling pathways regulated by PTPs. However, in many cases the primary function of substrate tyrosine phosphorylation is to initiate the formation of multiprotein signalling complexes by providing a binding site for association of other signalling proteins via phosphotyrosine recognition domains, including SH2 and PTB domains (40). Since the enhanced phosphorylation induced by expression of the mutant PTP results from the stable association of the mutant PTP catalytic domain with the phosphorylated tyrosine residue, these phosphorylated residues remain unavailable for binding to their appropriate downstream effector molecules (*Figure 2*). Therefore, the trapping mutants may exert effects *in vivo* that are analogous to those of the wild-type PTP. Thus it is apparent that any results obtained using such mutant PTPs to study signalling processes must be interpreted with great caution. However, two alternative mutants, which display reduced catalytic activity but lack substrate-trapping properties, have been described; these mutants are expected to prove more useful as dominant negative reagents since they potentially represent catalytic domains that completely lack biological activity. Mutation (to methionine) of an invariant arginine residue which stabilizes the transition state structure of the enzyme–substrate complex (residue 221 in PTP1B), results in an enzyme with considerably reduced catalytic activity and much lower affinity for substrate (20). Alternatively, deletion of the PTP signature motif surrounding the active-site cysteine residue (as described in ref. 41) results in an enzyme without catalytic activity. This mutant displays considerably reduced substrate-binding affinity, since many residues within this motif form hydrogen bonds with the phosphate group of the substrate, and therefore are essential for efficient substrate binding.

6. Perspectives

It is now well established that, whereas the membrane proximal catalytic domains of transmembrane PTPs generally display catalytic activity, the membrane distal domains of most transmembrane PTPs are inactive due to the absence of key catalytic residues. Therefore, it has been suggested that these inactive PTP domains may serve a binding function, perhaps recruiting target substrate proteins for dephosphorylation by the active membrane proximal catalytic domain. However, this has yet to be verified and in some cases, such as domain II of CD45 (42), the presence of bulky hydrophobic residues in the PTP signature motif renders such a binding function unlikely. Similarly, genes have now been identified that encode PTPs predicted to be entirely lacking in enzymatic activity since they lack the invariant aspartic acid residue (43, 44). Therefore, these proteins resemble naturally occurring substrate-trapping mutant PTPs, and again may serve a binding function rather than a catalytic one. These observations have led to the proposal of a 'dominant negative' role for such proteins, involving promotion of phosphorylation-dependent signalling processes by protecting phosphorylated substrate proteins from dephosphorylation by active cellular phosphatases. However, as outlined earlier (see Section 5 and *Figure 2*) it is unclear whether these proteins truly function in this manner or whether their function is merely to bind to appropriate targets, perhaps directly affecting the properties of the target protein. Alternatively, such a binding function may be required to reduce the susceptibility of the target protein to dephosphorylation, perhaps during trafficking of the protein between intracellular compartments, in order to allow subsequent interactions with appropriate downstream effectors.

The three-dimensional structures of tyrosine-specific PTPs (9, 12, 45, 46) share a similar overall fold with the dual specificity phosphatases (47), which catalyse the dephosphorylation of phosphoserine/phosphothreonine in addition to phosphotyrosine residues; however, the structures of the two enzyme subfamilies differ in several key aspects. The catalytic mechanisms of PTPs and dual-specificity phosphatases are also similar; both types of enzyme utilize a nucleophilic cysteine residue to catalyse the dephosphorylation reaction, and both also utilize an essential invariant aspartic acid residue as a general acid during catalysis (10, 16, 48). In spite of these similarities, it is currently unclear whether aspartic acid mutant forms of dual-specificity PTPs would function as efficient substrate-trapping molecules. This is because substrate trapping by tyrosine-specific PTPs is largely driven by hydrophobic interactions with the phosphotyrosine residue buried within the active-site cleft of the enzyme, primarily involving an invariant tyrosine residue (Y46 in PTP1B) and the residue immediately C-terminal to the invariant aspartic acid (F182 in PTP1B) (10). In contrast, the active-site cleft of dual-specificity phosphatases is relatively shallow, which is likely to result in considerably reduced substrate-trapping properties. Furthermore, it is unclear whether these enzymes undergo a conformational change akin to that observed with tyrosine-specific PTPs upon substrate binding. In the case of the tyrosine-specific PTPs this loop movement, involving the residues surrounding the catalytically essential aspartic acid residue (179–187 in PTP1B), is essential for substrate binding (and also for substrate trapping) as it brings a bulky hydrophobic residue (F182 in PTP1B) into position to interact with the phenol ring of the substrate tyrosine residue. Indeed, this hydrophobic interaction plays a key role in driving the movement of this loop towards the bound substrate. However, the analogous residue in dual-specificity PTPs is not hydrophobic in nature, suggesting that substrate binding to dual-specificity PTPs does not depend on analogous hydrophobic interactions, and that substrate binding is unlikely to initiate a similar conformational change in the structures of these enzymes. Although these considerations appear to indicate that dual-specificity phosphatases may generally have less potential in substrate-trapping experiments, recent data suggests that substrate recognition by dual-specificity phosphatases may sometimes be enhanced by additional interactions via non-catalytic segments of the enzymes (49, 50), potentially permitting stable interactions to be observed in the context of mutant forms of these enzymes.

An additional intriguing development is the recent identification of two members of the dual-specificity phosphatase family which lack an active-site cysteine residue, and which have been demonstrated to be devoid of catalytic activity (51, 52). These proteins (STYX and Sbf-1) constitute a new class of naturally occurring, potential substrate-trapping phosphatases, and it has been suggested that they may exert their biological effects through their ability to associate in stable complexes with appropriate phosphorylated proteins *in vivo*. However, it is currently not clear whether this is indeed the case, and no physiological binding partners for these inactive phosphatase catalytic domains have yet been identified. Furthermore, the Sbf-1 protein lacks both the active-site cysteine residue and the catalytically essential arginine residue within the highly conserved PTP signature motif (51). As mentioned in Section 5, mutation of the analogous arginine residue in PTP1B (Arg-221) significantly reduces the affinity of the enzyme for substrates (20), and the absence of this residue in the context of Sbf-1 would therefore be expected to result in this protein exhibiting limited binding properties. In addition, it is not clear that the biological functions of Sbf-1 require the phosphatase-related domain of the protein, since Sbf-1 is a large protein (1693 amino acids) which also contains at least one other functional protein–protein interaction domain, termed SID (SET domain interacting domain). In fact, when expressed by itself, the SID domain has been shown to exhibit biological properties identical to those of the full-length Sbf-1 protein, in both fibroblast transformation and myoblast differentiation assays performed *in vitro* (51).

The methods outlined in this chapter are proving invaluable for the identification of physiological substrates for PTPs. They are applicable to any tyrosine-specific PTP, since the residues mutated to generate substrate-trapping mutants are present in all active PTP catalytic domains. It is therefore likely that the physiological substrates for a large number of PTPs will now be identified relatively rapidly, paving the way for future investigations into the biological roles of individual members of this important family of signal-transducing enzymes.

References

1. Tonks, N.K. and Neel, B.G. (1996) *Cell* **87**, 365.
2. Neel, B.G. and Tonks, N.K. (1997) *Curr. Opin. Cell Biol.* **9** 193.
3. Shultz, L.D., Schweitzer, P.A., Rajam, T.V., Yi, T., Ihle, J.N., Matthews, R.J., Thomas, M.L., and Beier, D.R. (1993) *Cell* **73**, 1445.

4. Schaapveld, R.Q., Schepens, J.T., Robinson, G.W., Attema, J., Oerlemans, F.T., Fransen, J.A., Streuli, M., Wieringa, B., Nennighausen, L., and Hendriks, W.J. (1997) *Dev. Biol.* **188**, 134.

5. You-Ten, K.E., Muise, E.S., Itie, A., Mixhaliszyn, E., Wagner, J., Jothy, S., Lapp, W.S., and Tremblay, M.L. (1997) *J. Exp. Med.* **186**, 683.

6. Mauro, L.J. and Dixon, J.E. (1994) *Trends Biochem. Sci.* **19**, 151.

7. Tiganis, T., Bennett, A.M., Ravichandran, K.S., and Tonks, N.K., (1998) *Mol. Cell. Biol.* **18**, 1622.

8. Garton, A.J., Flint, A.J., and Tonks, N.K. (1996) *Mol. Cell. Biol.* **16**, 6408.

9. Barford, D., Flint, A.J., and Tonks, N.K. (1994) *Science* **263**, 1397.

10. Jia, Z., Barford, D., Flint, A.J., and Tonks, N.K. (1995) *Science* **268**, 1754.

11. Pannifer, A.D.B., Flint, A.J., Tonks, N.K., and Barford, D. (1998) *J. Biol. Chem.* **273**, 10454.

12. Stuckey, J.A., Schubert, H.L., Fauman, E.B., Zhang, Z.Y., Dixon, J.E., and Saper, M.A. (1994) *Nature* **370**, 571.

13. Fauman, E.B., Yuvaniyama, C., Schubert, H. L., Stucket, J. A., and Saper, M.A. (1996) *J. Biol. Chem.* **271**, 18780.

14. Schubert, H.L., Fauman, E.B., Stuckey, J.A., Dixon, J.E., and Saper, M.A. (1995) *Protein Sci.* **4**, 1904.

15. Lohse, D.L., Denu, J.M., Santoro, N., and Dixon, J.E. (1997) *Biochemistry* **36**, 4568.

16. Zhang, Z.Y., Wang, Y., and Dixon, J.E. (1994) *Proc. Natl. Acad. Sci. USA* **91**, 1624.

17. Zhang, Z.Y., Palfey, B.A., Wu, L., and Zhao, Y. (1995) *Biochemistry* **34**, 16389.

18. Barford, D., Jia, Z., and Tonks, N.K. (1995) *Nat. Struct. Biol.* **2**, 1043.

19. Zhang Z.Y. (1998) *Crit. Rev. Biochem. Mol. Biol.* **33**, 1.

20. Flint, A.J., Tiganis, T., Barford, D., and Tonks, N.K. (1997) *Proc. Natl. Acad. Sci. USA* **94**, 1680.

21. Black, D.S. and Bliska, J.B. (1997) *EMBO J.* **16**, 2730.

22. Furukawa, T., Itoh, M., Krueger, N.X., Streuli, M., and Saito, H. (1994) *Proc. Natl. Acad. Sci. USA* **91**, 10928.

23. Sun, H., Charles, C.H., Lau, L.F., and Tonks, N.K. (1993) *Cell* **75**, 487.

24. Xu, X and Burke, S.P. (1996) *J. Biol. Chem.* **271**, 5118.

25. Schneider, C., Newman, R.A., Sutherland, D.R., Asser, U., and Greaves, M.F. (1982) *J. Biol. Chem.* **257**, 10766.

26. Heffetz, D., Bushkin, I., Dror, R., and Zick, Y. (1990) *J. Biol. Chem.* **265**, 2896.

27. Songyang, Z., Shoelson, S.E., Chaudhuri, M., Gish, G., Pawson, T., Haser, W.G., King, F., Roberts, T., Ratnofsky, S., Lechleider, R.J., Neel, B.G., Birge, R.B., Fajardo, J.E., Chou, M.M., Hanafusa, H., Schaffhausen, B., and Cantley, L.C. (1993) *Cell* **72**, 767.

28. Huyer, G., Kelly, J., Moffat, J., Zamboni, R., Jia, Z., Gresser, M.J., and Ramachandran, C. (1998) *Anal. Biochem.* **258**, 19.

29. Huyer, G., Liu, S., Kelly, J., Moffat, J., Payette, P., Kennedy, B., Tsaprailis, G., Gresser, M.J., and Ramachandran, C. (1997) *J. Biol. Chem.* **272**, 843.

30. Geisow, M.J. (1998) *Nat. Biotechnol.* **16**, 206.

31. Humphery-Smith, I and Blackstock, W. (1997) *J. Protein Chem.* **16**, 537.

32. Spencer, S., Dowbenko, D., Cheng, J., Li, W., Brush, J., Utzig, S., Simanis, V., and Lasky, L.A. (1997) *J. Cell Biol.* **138**, 845.

33. Garton, A.J., Burnham, M.R., Bouton, A.H., and Tonks, N.K. (1997) *Oncogene* **15**, 877.
34. Dho, S.E., Jacob, S., Wolting, C.D., French, M.B., Rohrschneider, L.R., and McGlade, C.J. (1998) *J. Biol. Chem.* **273**, 9179.
35. Lioubin, M.N., Algate, P.A., Tsai, S., Carlberg, K., Aebersold, A., and Rohrschneider, L.R. (1996) *Genes Dev.* **10**, 1084.
36. Kornbluth, S., Jove, R., and Hanafusa, H. (1987) *Proc. Natl. Acad. Sci. USA* **84**, 4455.
37. Brugge, J.S., Jarosik, G., Andersen, J., Queral-Lustig, A., Fedor-Chaiken, M., and Broach, J.R. (1987) *Mol. Cell. Biol.* **7**, 2180.
38. Ishibashi, T., Bottaro, D.P., Chan, A., Miki, T., and Aaronson, S.A. (1992) *Proc. Natl. Acad. Sci. USA* **89**, 12170.
39. Fukunaga, R. and Hunter, T. (1997) *EMBO J.* **16**, 1921.
40. Pawson, T. (1995) *Nature* **373**, 573.
41. Tang, T.L., Freeman, R.M., Jr, O'Reilly, A.M., Neel, B.G., and Sokol, S.Y. (1995) *Cell* **80**, 473.
42. Charbonneau, H., Tonks, N.K., Kumar, S., Diltz, C.D., Harrylock, M., Cool, D.E., Krebs, E.G., Fischer, E.H., and Walsh, K.A. (1989) *Proc. Natl. Acad. Sci. USA* **86**, 5252.
43. Lu, J., Notkins, A.L., and Lan M.S. (1994) *Biochem. Biophys. Res. Commun.* **204**, 930.
44. Cao, L., Zhang, L., Ruiz0Lozano, P., Yang, Q., Chien, K.R., Graham, R.M., and Zhou, M. (1998) *J. Biol. Chem.* **273**, 21077.
45. Hoffmann, K.M., Tonks, N.K., and Barford, D. (1997) *J. Biol. Chem.* **272**, 27505.
46. Bilwes, A.M., den Hertog, J., Hunter, T., and Noel, J.P. (1996) *Nature* **382**, 555.
47. Yuvaniyama, J., Denu, J.M., Dixon, J.E., and Saper, M.A. (1996) *Science* **272**, 1328.
48. Denu, J.M., Zhou, G., Guo, Y., and Dixon, J.E. (1995) *Biochemistry* **34**, 3396.
49. Muda, M., Theodosiou, A., Gillieron, C., Smith, A., Chabert, C., Camps, M., Boschert, U., Rodrigues, N., Davies, K., Ashworth, A., and Arkinstall, S. (1998) *J. Biol. Chem.* **273**, 932.
50. Doi, K., Gartner, A., Ammerer, G., Errede, B., Shinkawa, H., Sugimoto, K., and Matsumoto, K. (1994) *EMBO J.* **13**, 61.
51. Cui, X., De Vivo, I., Slany, R., Miyamotot, A., Firestein, R., and Cleary, M.L. (1998) *Nat. Genet.* **18**, 331.
52. Wishart, M.J., Denu, J.M., Williams, J.A., and Dixon, J.E. (1995) *J. Biol. Chem.* **270**, 26782.

9

Assay and purification of protein–serine/threonine kinases

D. GRAHAME HARDIE, TIMOTHY A.J. HAYSTEAD,
IAN P. SALT, and STEPHEN P. DAVIES

1. Introduction

It is now very clear that activation of protein kinase cascades (ably assisted by appropriate regulation of protein phosphatases) is the major mechanism by which eukaryotic cells respond to changes in the extracellular environment, such as the presence of extracellular signal molecules or adverse environmental conditions. It also appears to be the major mechanism through which cellular processes which occur discontinuously (e.g. DNA replication, mitosis) are switched on and off. Although there are exceptions, protein kinases are usually expressed constitutively, and the key event is activation of pre-existing kinase protein, through phosphorylation, dephosphorylation, binding of an allosteric activator (e.g. cyclic AMP, 5'-AMP, Ca^{2+}, diacylglycerol), or binding of a regulatory protein (e.g. a cyclin). It is therefore very important to develop methods to assay protein kinase *activity*. In some cases (e.g. MAP kinases) the activating phosphorylation produces a mobility shift on SDS–polyacrylamide gel electrophoresis (PAGE), and the activation can be assessed by Western blotting. However, non-activating phosphorylation events can also produce mobility shifts, and this method can be unreliable or misleading. There is, therefore, no real substitute for the development of direct assays of protein kinase activity. This is the first main topic of this chapter (see Section 2).

Traditionally, protein kinase projects commenced with purification of the protein kinase, followed later by cloning of the encoding DNA. With the advent of PCR using redundant primers to clone homologous DNAs, and the cloning of expressed sequence tags (ESTs) and whole genomes, this process is increasingly being done in reverse (see Chapter3). Biochemical studies of protein kinases are often now performed on expressed recombinant protein that has been isolated by virtue of its fusion to a purification tag such as glutathione-S-transferase or polyhistidine. However, purification of the native, endogenous protein kinase has not been completely superceded for a number of reasons:

- native, endogenous protein kinases may have covalent modifications which may not be faithfully reproduced in the chosen expression system
- many protein kinases have regulatory subunits in addition to the catalytic subunit: in some cases these are essential for activity and, even if they are not, much information about the regulation of the complex may be missed if they are not present
- from a cloned sequence alone it remains difficult to identify target proteins (although see Chapter 13), so many protein kinases are still characterized by starting with an activity assay involving phosphorylation of a target protein

Protein kinases can be purified by any regular method of protein purification, but there are a small number of specific affinity methods that can be attempted, and these form the second major topic of this chapter (see Section 3).

2. Assay of protein kinases

In this section we will review the many different variants of protein kinase assays, and give typical protocols for some of them. The activity of protein kinases can be assayed in a number of ways:

(a) Many protein kinases can be assayed by incorporation of phosphate from ATP into a synthetic peptide (typically, 10–15 residues) based on the sequences of phosphorylation sites on target proteins. This is probably the most convenient type of assay for routine use, and for high-through-put screening. Many protein kinases, especially those which have multiple targets *in vivo*, phosphorylate short peptides with kinetic parameters similar to those of the native target proteins: these kinases appear to recognize short primary sequence motifs, and peptide assays are convenient and useful. Section 2.1 discusses this type of assay and *Protocol 1* describes a typical example.

(b) Some protein kinases phosphorylate synthetic peptides poorly, or not at all. These are usually kinases that are highly selective for a single target, such as the MAP kinase kinases (MEKs). They may be recognizing an aspect of three-dimensional structure of their target in addition to primary sequence. Alternatively, they may be binding to additional determinants on their target protein outside of the immediate area of the phosphorylation site. For example, JNK1 phosphorylates the transcription factor c-Jun at Ser-63 and Ser-73, but a region between residues 30–57 also appears to be required for recognition, and this interaction is still functional even if a 55 residue spacer is inserted between this recognition domain and the phosphorylation site (1). Similarly, chimeras constructed between the p44 (ERK1) and p38 (SAPK2) members of the MAP kinase subfamily show that recognition by the upstream pathway (i.e. the MAP kinase kinase) depended on the identity of the first 40 residues, and not on the identity of the 'T loop' which contains the actual phosphorylation sites, Thr-202 and Tyr-204 (2). The first 40 residues include the αC helix that is close to the T loop in the folded structure, but remote from it in the linear sequence. This is analogous to the recognition of a 'conformational epitope' by an antibody. Kinases that fall into this class will have to be assayed using the native protein target as substrate, or at least an expressed domain derived from it that contains the recognition features. Section 2.2 discusses this type of assay, and *Protocols 3* and *4* describe typical examples.

If the kinase changes the functional properties of its target, it can be assayed by monitoring that change in function. Section 2.3 discusses this type of assay.

2.1 Synthetic peptide kinase assays

See Chapter 10 for a more comprehensive treatment of synthetic peptide kinase assays.

2.1.1 Choice of synthetic peptide

Synthetic peptides suitable for kinase assays could be derived from several different routes:

1. *Peptides based on sequences around phosphorylation sites on natural target proteins.* The 'classical' example of this is the *kemptide* substrate for cyclic-AMP-dependent protein kinase, which was based on the phosphorylation site on the L- isoform of pyruvate kinase (3). Another example is the *SAMS* peptide (HMRSAMSGLHLVKRR) which was developed for the assay of AMP-activated protein kinase (4). The peptide was based on the sequence from residues 73 to 85 (HMRSSMSGLHLVK) on rat acetyl–CoA carboxylase, containing the primary phosphorylation site (Ser-79). However, two arginines were added at the C-terminus to ensure binding to phosphocellulose paper (see below), and the serine at position 5 was replaced by alanine to eliminate a site for cyclic-AMP-dependent protein kinase. Once the key residues required for recognition by the AMP-activated protein kinase had been established using variants of the *SAMS* peptide (5), the *AMARA* peptide (AMARAASAAALARRR) was developed subsequently (6). This peptide has the key residues required for recognition, but others (apart from the basic tail) are alanine. *AMARA* is a more potent and selective substrate for the AMP-activated protein kinase than *SAMS*.

2. *Pseudosubstrate sequences.* Many ligand-activated protein kinases are

maintained in the inactive state in the absence of their activator because the active site is occupied by a *pseudosubstrate* sequence within a regulatory domain of the kinase. Pseudosubstrate sequences have all of the determinants required to be a substrate, but lack a phosphorylatable residue in the appropriate position. House and Kemp (7) found that a synthetic peptide based on the pseudosubstrate sequence of protein kinase C-α made an excellent substrate if the non-phosphorylatable alanine residue was changed to serine. This approach was utilized by Schaap *et al.* (8) to devise a peptide substrate for protein kinase C-ε, for which a protein substrate was not known.

3. *Peptide libraries.* Where a protein substrate is not known, a recognition motif can in some cases be established using a peptide library approach such as that described by Songyang and Cantley in Chapter 16. Lam *et al.* (9) used a peptide library method to devise a novel peptide substrate for Src family protein–tyrosine kinases.

2.1.2 Method for separation of phosphorylated peptide and unreacted ATP

Although non-radioactive kinase assays are now being developed, most kinase assays still utilize [γ-^{32}P]ATP, and the key to a successful assay is the method which is used to separate the radioactive, phosphorylated peptide from unreacted ATP. The most widely used method for peptide substrates remains the phosphocellulose paper assay (10). Peptides containing basic residues bind to the negative charges on this paper, while ATP does not. The reaction mixture is pipetted on to a square of the paper, which is then washed in a large volume of diluted phosphoric acid. An advantage of this method is that a large number of paper squares can be washed simultaneously in the same solution.

This method is only possible if the peptide contains two or more basic residues that mediate the binding to phosphocellulose paper. Toomik *et al.* (11) pointed out that even peptides with two basic residues do not always bind quantitatively, particularly because the phosphorylation itself introduces (at neutral pH) two negative charges which inhibit binding. These problems can usually be overcome by adding extra basic residues at the N- or C-terminus of the peptide, which is why the *AMARA* peptide (AMARAASAAALARRR) used for assay of AMP-activated protein kinase contains three arginines at the C-terminus, even though these are not essential for phosphorylation. Clearly, when a peptide assay utilizing the phosphocellulose paper method is being developed, it is essential to confirm that the phosphorylated peptide binds quantitatively to the paper (see ref. 11).

In order to get round the problem of peptides not binding quantitatively to phosphocellulose paper, Toomik *et al.* (12) developed an alternative assay in which the peptide was tritium labelled, and was separated from unreacted

ATP by binding to ferric adsorbent paper, which binds phosphorylated peptides selectively. This method also replaces the use of ^{32}P with the safer tritium isotope. However. it does require the synthesis of tritiated peptide, and does not appear to have been widely adopted.

If the phosphocellulose paper method is not suitable (e.g. when utilizing short acidic peptides as substrates), numerous other methods to separate the phosphorylated peptide from ATP have been used, e.g. thin layer chromatography methods (13). Alternative protocols for peptide kinase assays are also discussed in more detail in Chapter 10 of this volume. *Protocol 1* is for the phosphocellulose paper assay of the AMP-activated protein kinase: it can be readily adapted for the assay of other peptide kinases. In order to convert the results of a kinase assay from radioactivity into more meaningful units (i.e. moles phosphate), it is necessary to know the specific radioactivity of the [γ-^{32}P]ATP. While it is possible to calculate the specific radioactivity using the data provided by the radioisotope supplier, it is much better to measure it oneself, especially because the [γ-^{32}P]ATP is usually diluted with cold ATP before use. *Protocol 2* provides a simple and accurate method for doing this. Every time a new mix of labelled and cold ATP is made, ATP standards should be prepared as in *Protocol 2*. If these are counted (and hence the specific radioactivity recalculated) every time an unknown is counted, radioactive decay is automatically corrected.

Protocol 1. Peptide kinase assay using phosphocellulose paper

Equipment and reagents

- phosphocellulose paper (Whatman P81, cut into 1 cm squares, numbered with a hard pencil. If one corner is folded over, it makes the squares easier to pick up using forceps.)
- assay buffer (50 mM Na Hepes, pH 7.0, 1 mM DTT, 0.02% Brij-35)
- synthetic peptide substrate (e.g. the *SAMS* peptide for AMP-activated protein kinase assay, HMRSAMSGLHLVKRR, 1 mM in assay buffer)

- AMP (1 mM; the allosteric activator of AMPK, not required for other protein kinases)
- Mg–ATP (1 mM [γ-^{32}P]ATP, diluted to 200–500 cpm/pmol with unlabelled ATP, plus 25 mM MgCl$_2$^{a})
- kinase, diluted appropriately in assay buffer
- phosphoric acid (1%, v/v)
- scintillation counter
- non-aqueous scintillation fluid (e.g. Optiscint HiSafe, Wallac)

Method

1. Mix the following components in 1.5 ml microcentrifuge tubes (Eppendorf type):

   peptide substrate   5 µl

   kinase   5 µl

   AMP   5 µl

   assay buffer   5 µl

   Start the reactions at 15 or 20 sec intervals by adding 5 µl of Mg–ATP,

---

Protocol 1. *Continued*

vortex, and incubate at 30°C for 10 min. Also perform blanks where the peptide is omitted.

2. Stop the reactions at 15 or 20 sec intervals by removing 15 μl and pipetting it on to a P81 paper square held in forceps. When it has soaked in, drop the square into a beaker containing 500 ml of phosphoric acid. Up to 50 squares may be dropped into the same beaker.

3. When all the reactions have been stopped, stir the beaker containing the paper squares gently[b] on a magnetic strirrer for 5 min at room temperature.

4. Pour off the phosphoric acid to radioactive waste (filters may be retained in the beaker using forceps or a small sieve). Add 500 ml of phosphoric acid, stir again for 5 min, and pour off the acid again. Repeat this washing procedure once more.[c, d]

5. Rinse the paper squares with acetone and allow to dry in the air on a sheet of aluminium foil. Sort them into their numbered order, and place each square into a scintillation vial containing 10 ml of scintillation fluid.[e] Count them using a ^{32}P programme. Count paper squares containing ATP standards (see *Protocol 2*) at the same time.

---

[a] ATP is a weak acid, and unlabelled ATP is normally supplied in an acid form; it is important to neutralize it with NaOH or KOH (e.g. as a 100 mM stock) before adding $MgCl_2$, otherwise an insoluble Mg–ATP complex may precipitate.
[b] Stirring the paper squares vigorously may cause them to disintegrate, and make the pencilled numbers hard to read!
[c] Some people wash the filters using an inner plastic beaker punctured with holes, which sits inside the main beaker; the holes allow liquid to exchange but retain the paper squares.
[d] The efficiency of your washing procedure should be tested using blank reactions lacking kinase.
[e] With the exception of those used for the ATP standards, the vials of scintillation fluid can usually be re-used after removing the paper squares; count them to check that the radioactivity is back to background levels. If re-using them, be careful not to get fluid on the outside of the vials, as this can cause them to stick in the counter.

Protocol 2. Determining the specific radioactivity of ATP

Equipment and reagents
- UV spectrophotometer
- quartz cuvette, semi-micro size (500 μl)

Method

1. Pipette 490 μl of water into the cuvette and zero the spectro-photometer on it at 260 nm.

2. Add 10 μl of stock Mg–ATP solution, mix with a plastic stirrer bar, and read the absorbance (A_{260}). If the original Mg–ATP solution was 1 mM,

the absorbance should be ~0.3 (extinction coefficient of ATP at 260 nm = 15 l mmol^{-1} cm^{-1}).

3. Pipette 20 μl aliquots from the cuvette onto P81 squares, allow to dry, and count as in *Protocol 1*. Do this in duplicate or triplicate and determine the average. The specific radioactivity of ATP in cpm/nmol = cpm ÷ [20 × (A_{260} ÷ 15)].

2.1.3 High-throughput peptide kinase assays

Protocol 2 can be adapted to a 96-well format for high-throughput assays. The assays are conducted in polypropylene 96-well plates with round-bottom wells. Assays are started using an eight-channel pipette, and the plate is then mounted on a small orbital shaker in a bench-top air incubator. Equipment suitable for this is described in *Protocol 5*. Assays are stopped by adding 5 μl of 6% (v/v) phosphoric acid from an eight-channel pipette. Aliquots (20 μl) of the stopped assays are then pipetted on to a Wallac P30 Filtermat, which is marked with 96 squares. The Filtermat is washed with 1% (v/v) phosphoric acid and acetone as in *Protocol 1*, air-dried, and counted using the Cerenkov effect in a Wallac 1450 Microbeta counter.

2.2 Protein substrate kinase assays

A possible complication when using a protein, rather than a synthetic peptide, as a substrate in a kinase assay is that the preparation of the protein may itself contain protein kinases. The extent of this problem will of course depend on the source and purity of the target protein. It is not usually a serious problem when using proteins expressed in *Escherichia coli*, especially when isolated by means of a purification 'tag' such as polyhistidine or a fused glutathione-*S*-transferase (GST) domain. Proteins purified in that manner are usually very pure, and in any case *E. coli* does not appear to express any broad specificity protein kinases. Contamination of the target protein by kinases is, however, an issue which should be considered when the target protein has been purified from a eukaryotic source, including Sf9 cells when utilizing the baculovirus system. Unless the presence of contaminating kinases has already been ruled out, blanks should be performed where the target protein preparation is incubated with [γ-^{32}P]ATP in the absence of added kinase, as well as blanks without substrate (see *Protocol 3*).

As with a peptide kinase assay, the key here is the method used to separate the ^{32}P-labelled protein from unreacted [γ-^{32}P]ATP. With the exception of very basic proteins such as histones, most proteins do not bind quantitatively to phosphocellulose paper. Although a number of different methods have been used, we will describe two which are rather general in their application:

(a) Almost all proteins are insoluble in 25% trichloroacetic acid (TCA), while ATP remains soluble. *Protocol 3* describes a protein kinase assay using TCA precipitation. Although somewhat more tedious to perform than *Protocol 1*, it is straightforward, gives accurate results, and is widely applicable.

(b) *Protocol 3* may not give accurate results if the target protein is seriously contaminated with other protein kinases, or the protein kinase preparation is contaminated with other kinase substrates. In these cases the 'protein substrate alone' or 'kinase alone' blanks may give incorporation almost as high as in the full reaction, and the correction for these blanks will be inaccurate. In such cases, kinase assays may still be possible if the proteins and [γ-^{32}P]ATP are separated by SDS–PAGE (*Protocol 4*). This has the advantage over *Protocol 3* that contaminating ^{32}P-labelled proteins will usually be resolved from the target protein by the electrophoresis. It is more time consuming to perform than the TCA precipitation assay, and is probably less accurate for determining the absolute stoichiometry of phosphorylation in mole phosphate per mole of target protein.

Protocol 3. Protein kinase assay by TCA precipitation

Equipment and reagents

- microcentrifuge suitable for 1.5 ml tubes (several models are available which hold 12 or more tubes; the centrifuge should be capable of at least 10 000 *g*; refrigeration is not necessary)
- microcentrifuge tubes (1.5 ml polypropylene, Eppendorf type)
- TCA (25%, w/v)
- BSA (10 mg/ml)
- protein substrate (typically between 0.1 and 1 mg/ml)
- other reagents (buffer, kinase, activators, Mg–ATP) as for *Protocol 1*

Method

1. Make up 25 μl assays[a] as in *Protocol 1* but using the protein in place of the peptide substrate. As well as the full reaction (protein + kinase + buffer + ATP) perform a 'kinase blank' (kinase + buffer + ATP), a 'protein substrate blank' (protein + buffer + ATP), and a 'buffer blank' (buffer + ATP).[b, c]

2. Incubate for an appropriate time (e.g. 30 min at 30°C).

3. Stop the reaction by adding 1 ml of TCA.[d]

4. When all reactions have been stopped, add 30 μl of BSA,[e] cap the tube, vortex mix, and leave on ice for at least 2 min.

5. Centrifuge the tubes for 2 min with the tags holding the caps facing radially outwards.[f]

6. Suck off the supernatant[g] by sliding a glass Pasteur pipette down the face of the tube that was facing radially inwards[h] (i.e. opposite to the tag holding the cap). The protein pellet may not be visible, so be careful not to scrape the tube bottom with the pipette tip. A few μl of supernatant will be left behind, but this will be removed in steps 7 and 8.

7. Add 1 ml of TCA, vortex mix, centrifuge for 2 min, and remove the supernatant as before.

8. Repeat step 7 twice more.

9. Count the tubes, using the Cerenkov effect, by placing the whole tube into a scintillation vial and counting using a tritium programme[i] on a scintillation counter.

10. To calculate the results, subtract the counts obtained in the 'kinase blank' from that in the full reaction. This gives the incorporation into the protein itself. Subtract the counts obtained in the 'buffer blank' from those in the 'protein substrate blank'. This gives the phosphorylation of the protein due to contaminating endogenous kinases.[c] In each case, divide by the specific radioactivity of the ATP to convert cpm into moles of phosphate.

11. To determine the specific radioactivity of ATP, use *Protocol 2*, but instead of pipetting the 20 μl aliquots from the cuvette on to P81 paper, place them into empty microcentrifuge tubes and count them at the same time as the TCA pellets.

[a] If performing a time course rather than a single time point, you can take several aliquots out of a single reaction tube and add them to pre-prepared microcentrifuge tubes containing 1 ml of TCA.

[b] The buffer blank corrects for non-covalently bound radioactivity trapped in the protein pellet, and for background radiation.

[c] The 'protein substrate alone' and 'buffer' blanks can be omitted if you are sure there are no endogenous kinases in the protein substrate preparation

[d] The most likely source of error (as well as radioactive contamination) in this protocol is getting liquid containing [γ-^{32}P]ATP trapped around the tube lid. Never open a tube containing [γ-^{32}P]ATP which you have agitated in any way without centrifuging for a few sec first. This brings all of the radioactive liquid down to the bottom of the tube.

[e] The BSA acts as a carrier for precipitation, and also ensures that all of the protein pellets are the same size, and trap the same amount of non-covalently bound radioactivity

[f] Although the protein pellet is not usually visible, in a fixed angle microcentrifuge rotor it forms a streak near the bottom of the tube on the tube face that was radially outwards. The tag on the tube cap provides orientation for this.

[g] The initial supernatants and the appropriate washes will be radioactive: handle with appropriate safety precautions.

[h] Sliding the pipette down this face ensures that the protein pellet will not be disturbed.

[i] Photons produced by the Cerenkov effect from ^{32}P have lower energy than those produced by liquid scintillation, and a programme designed for scintillation counting of ^{32}P may give lower efficiency.

Protocol 4. Protein kinase assay using SDS–PAGE

Equipment and reagents

- polyacrylamide gels (typically 10% acrylamide): we utilize the Laemmli buffer system (14)
- equipment for SDS–PAGE (e.g. BioRad Protean II)

Protocol 4. *Continued*

Method

1. Carry out the kinase reaction as for *Protocol 3*, but instead of adding TCA, add 0.25 volumes of 5 × SDS sample buffer. As well as the full reaction, carry out a 'kinase blank' and a 'protein substrate blank':[a] a 'buffer blank' is not necessary.

2. Incubate the sample (use a screw-capped tube) in a boiling water bath for 2 min.

3. Apply the samples to the wells of a stacking gel and carry out electrophoresis.[b]

4. Remove the gel, stain, and destain by your usual method and dry on a gel dryer between sheets of cellophane.

5. Analyse the dried gel by autoradiography or on a phosphorimager.[c]

---

[a] The 'protein substrate blank' may be omitted if you are sure there are no kinases in the protein substrate preparation.

[b] The unreacted [γ-^{32}P]ATP runs with the bromophenol blue dye front: one option is to run this off the bottom of the gel, when the lower electrophoresis buffer will become contaminated. If this is not done, the initial staining and destaining solutions will become radioactive. In either case, handle with the appropriate safety precautions.

[c] Quantification can be achieved either by densitometry of the autoradiogram, or analysis using the phosphorimager software. However, unless [^{32}P]protein standards of known radioactive content are run on the same gel, the results will be relative rather than absolute. Absolute radioactivity can be obtained by excising the radioactive bands from the gel and counting in a scintillation counter. Either way, the absolute accuracy is probably less than when using *Protocol 3*.

2.3 Kinase assays based on the function of the target protein

Since these involve assays of the function of the target protein, detailed protocols are not appropriate here. Functional assays were used to detect the first protein kinase to be discovered [phosphorylase kinase (15)]. They are still very useful in some cases, especially when synthetic peptides are poor substrates, and where a protein target is phosphorylated at multiple sites that a conventional protein kinase assay (*Protocol 3*) would not distinguish. Functional assays are used, for example, to assay MAP kinase kinase kinases (e.g. Raf-1) and MAP kinase kinases, the eventual readout being a MAP kinase assay. Since the function of proteins can be affected by many processes other than phosphorylation, it is important to show that the effect is both time and Mg–ATP dependent. Another useful control is to show that the effect is reversed by adding a purified protein phosphatase. This is a particularly convincing result if the effect is blocked by a protein phosphatase inhibitor such as okadaic acid (see Chapter 3).

2.4 Immunoprecipitate kinase assays

The mammalian genome probably encodes somewhere between 1000 and 2000 protein kinases (16) and it is probably wishful thinking to imagine that any single substrate will be specific for just one of them. The question as to whether the assay used is measuring only the kinase it was designed for should therefore always be considered. For example, the *SAMS* peptide (see *Protocol 1*) is a relatively specific substrate for the AMP-activated protein kinase (4). However, if one measures *SAMS* kinase activity in crude lysates of some cells it is clear that other kinases can slowly phosphorylate the peptide (17). In addition, both isoforms of the kinase catalytic subunit [α1 and α2 (18)] phosphorylate the peptide equally well and are not distinguished by this assay. A way around these problems is to immunoprecipitate the kinase using a specific antibody, and to assay it using the resuspended immunoprecipitate (IP). This depends on the availability of a specific immunoprecipitating antibody that does not interfere with the kinase activity, but these are increasingly becoming commercially available. *Protocol 5* describes an immunoprecipitate kinase assay for the AMP-activated protein kinase: this can be adapted for other protein kinases.

Protocol 5. Immunoprecipitate kinase assay for AMP-activated protein kinase

Equipment and reagents

- bench-top refrigerated centrifuge
- roller mixer
- bench-top air incubator to hold orbital shaker (e.g. Stuart Scientific SI60)
- IP buffer [50 mM Tris–HCl, pH 7.4 at 4°C, 150 mM NaCl, 50 mM NaF, 5 mM Na pyrophosphate, 1 mM EDTA, 1 mM EGTA: prepare 1 × stock; before use add 1 mM DTT, 0.1 mM benzamidine, 0.1 mM PMSF, 5 μg/ml soybean trypsin inhibitor, 1% (v/v) Triton X-100 (final concentrations)]

- small orbital shaker [IKA-VIBRAX-VXR, Janke & Kunkel, fitted with a holder (type VX 2E) which takes microcentrifuge tubes]
- protein G–Sepharose (Pharmacia Biotech)
- anti-AMPK α-subunit antibodies [affinity-purified sheep anti-AMPK antibodies as described in Woods *et al.* (19); either α1- or α2-specific antibodies can be used, or both isoforms can be precipitated using a mixture of the two]

Method

1. Wash 40 μl (packed volume) of protein G–Sepharose[a] with 5 × 1 ml of IP buffer, and resuspend in 120 μl IP buffer. This should be enough for 30 assays.

2. Add 40 μg of sheep anti-AMPK antibody to the protein G–Sepharose slurry and mix on a roller mixer for 45 min at 4°C.

3. Centrifuge the slurry (18 000 *g*, 1 min, 4°C) and wash the pellet five times with 1 ml ice-cold IP buffer. Resuspend the final pellet with another 120 μl IP buffer and divide the slurry into 5 μl aliquots.

Protocol 5. *Continued*

4. To each 5 μl aliquot add 50–100 μl of kinase (e.g. crude cell lysate) and mix for 2 h at 4°C on a roller mixer.

5. Centrifuge the mixture (18 000 *g*, 1 min, 4°C) and wash the pellet with 5 × 1 ml of ice-cold IP buffer containing 1 M NaCl (to remove non-specifically bound protein). Wash the pellet with 3 × 1 ml of lysate assay buffer and resuspend in 30 μl Hepes–Brij buffer prior to assay. It is best to assay immunoprecipitates immediately, but they can be stored at –20°C for a few days.

6. Assay the immunoprecipitates using either *Protocol 1* or *3*, except that the reactions are performed on an orbital shaker mounted in the bench-top air incubator.[b]

[a] When pipetting any solution containing Sepharose beads, use pipette tips with the last 5 mm cut off to avoid trapping the beads in the tip.
[b] This arrangement keeps the Sepharose beads suspended throughout the assay: if the tubes are not shaken in this way the results will not be reproducible.

3. Purification of protein kinases

Protein kinases are typical globular enzyme proteins, and can be purified using any standard method for protein purification. However, they are not usually highly abundant proteins, and at least 10 000-fold purification from crude extracts will typically be required to achieve homogeneity. Affinity chromatography methods are therefore highly desirable, and this section will focus on such methods.

3.1 Affinity methods based on the kinase ATP-binding site

Affinity chromatography matrices based on ATP have been available for many years. However, most of them involve attachment of ATP via the adenine or ribose moieties, and these are not successful in affinity purification of protein kinases. The reason for this became apparent when the first protein kinase catalytic domain structure [cyclic-AMP-dependent protein kinase (20)] was determined by X-ray crystallography. Residues in the nucleotide-binding site make intimate contact with Mg–ATP, and orientates the molecule such that its γ-phosphate is exposed at the lips of the catalytic cleft (21). The precise orientation of ATP in a protein kinase functions to bring its γ-phosphate close to the hydroxyl side chains of the receiving tyrosine, threonine, or serine residue upon substrate binding. Transfer of the phosphate to the hydroxyl is direct, involving both the local amino acids at the lips of the catalytic cleft and the receiving amino acid.

Understanding the mechanisms by which proteins bind small molecules has often proved to be a powerful means for designing ligands for their affinity purification. Because of the unique orientation by which protein kinases bind ATP, it became clear that only the γ-phosphate could provide a suitable point at which to immobilize ATP: attachment of ATP through any other moiety (e.g. the ribose or purine rings) would block binding to the catalytic cleft. This was the rationale for the development of γ-phosphate-linked ATP–Sepharose by Haystead et al. (22). It has been used subsequently by several laboratories to purify protein kinases to homogeneity (22–26). However, it should be remembered that protein kinases share structural similarities with other adenine nucleotide-binding proteins, in particular those belonging to the dinucleotide or dehydrogenase class. Indeed, the degree of structural conservation within the nucleotide-binding pockets suggests that these protein classes may have been derived from common ancestral proteins. γ-Phosphate-linked ATP–Sepharose therefore binds NAD-linked dehydrogenases and ATP-binding proteins other than protein kinases, such as HSP90 (27), and can be used to generate a fraction enriched in protein kinases, dehydrogenases, and ATP-binding proteins from crude cell extracts. These different classes of nucleotide-binding proteins can then often be specifically eluted from the column utilizing specific nucleotides, e.g. NAD^+, ADP, or ATP (*Figure 1*).

Protocol 6. Affinity purification using γ-phosphate-linked ATP–Sepharose

Equipment and materials

- γ-phosphate-linked ATP–Sepharose (Upstate Biotechnology Inc, NY, cat. no. 16–106)
- nucleotide solutions (prepare each separately in equilibration buffer at 10 mM: NADH, NADPH, AMP, ADP, ATP: *check pH after adding nucleotides*)
- equilibration buffer [25mM Tris–HCl pH 7.4, 60 mM $MgCl_2$ 150 mM NaCl, 1mM DTT, 0.15 mM sodium orthovanadate, 1mM benzamidine, 1mM phenylmethane sulphonyl fluoride (PMSF)].

Method[a]

1. Apply the extract or the partially purified protein fraction to a pre-column of Sepharose 4B in equilibration buffer (this step removes proteins that bind to Sepharose non-specifically).

2. Wash the γ-phosphate-linked ATP–Sepharose column extensively with equilibration buffer.

3. Apply the sample to ATP–Sepharose in equilibration buffer, and wash extensively until no further protein is eluted from the column.

4. Wash the column with equilibration buffer containing 1 M NaCl until no further protein is eluted.

5. Wash the column with equilibration buffer without NaCl.

6. Wash the column extensively with equilibration buffer containing 10 mM NADH or NADPH (this step elutes proteins belonging to the

---

213

Protocol 6. *Continued*

dinucleotide-binding class, e.g. glyceraldehyde phosphate dehydro-genase, but not most protein kinases[b]).

7. Wash the column extensively with equilibration buffer containing 10 mM AMP (this step elutes AMP-binding proteins such as glycogen phosphorylase[b]).

8. Wash the column extensively with equilibration buffer containing 10 mM ADP (this step elutes ADP-binding proteins, e.g. HSP90).

9. Wash the column extensively with equilibration buffer containing 10 mM ATP[c] (this step elutes protein kinases[d] and other proteins that bind ATP in a similar manner, e.g. arginine deiminase).

[a] It is important to include $MgCl_2$ in all buffers throughout the described procedure.
[b] Some protein kinases, e.g. AMP-activated protein kinase, have nucleotide-binding sites other than the catalytic site and can be eluted using NADH or AMP.
[c] Remember that a protein kinase eluted from the column will be in 10 mM ATP: this will decrease the specific radioactivity of [γ-^{32}P]ATP, and extensive dilution or dialysis may be required before attempting a kinase assay.
[d] If one is only interested in obtaining a fraction enriched in protein kinases, we suggest either washing the column with a cocktail of AMP, NADH/NADPH ,and ADP, or including these nucleotides (at 10 mM) in the equilibration buffer.

3.2 Affinity methods based on the kinase protein/peptide-binding site

Another obvious method to purify protein kinases is to use affinity matrices based on their peptide or protein substrates. However, a drawback with this approach is that all proteins, and to a lesser extent peptides, contain multiple charged side chains and can act as ion-exchange matrices. Thus, while the protein kinase may bind to a column matrix based on the substrate protein, so will many other proteins, through non-specific charge interactions. Ion-exchange effects can be minimized by running the separation at high ionic strength, but this often also results in failure of the protein kinase to bind. This approach may be successful for protein kinases that interact with their substrates through regions other than the phosphoacceptor site. For example,

Figure 1. A (top panel): SDS–PAGE analysis of fractions derived using specific elution steps (EGTA, NADH, AMP, ADP, and ATP) after proteins in a crude 100 000 *g* supernatant fraction from rabbit skeletal muscle had been bound to a large γ-phosphate-linked ATP–Sepharose column. The major proteins in the NADH, AMP, ADP, and ATP eluates could be identified directly by mixed peptide sequencing (see Chapter 1) and were glyceraldehyde phosphate dehydrogenase (GAPDH), glycogen phosphorylase, heat shock protein 90, and MAP kinase (MAPK), respectively. The middle (B) and lower (C) panels show that the EGTA and ATP washes from the column eluates could be further resolved, and additional polypeptides identified, by anion-exchange chromatography subsequent to ATP–Sepharose.

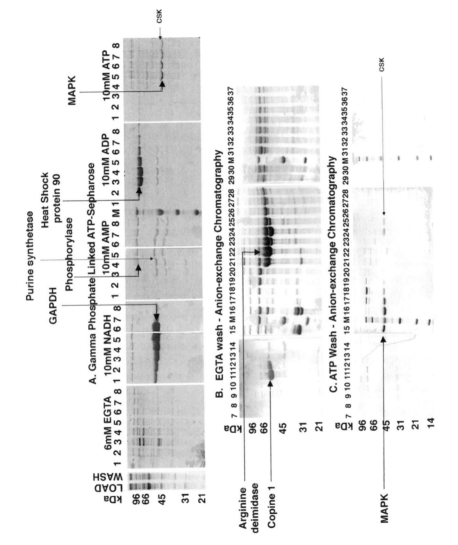

Purine synthetase

Heat Shock
protein 90

GAPDH

Phosphorylase

A. Gamma Phosphate Linked ATP-Sepharose

MAPK

kDa
96
66
45
31
21

WASH
LOAD

6mM EGTA
1 2 3 4 5 6 7 8

10mM NADH
1 2 3 4 5 6 7 8

10mM AMP
1 2 3 4 5 6 7 8 M

10mM ADP
1 2 3 4 5 6 7 8

10mM ATP
1 2 3 4 5 6 7 8

CSK

B. EGTA wash - Anion-exchange Chromatography

Arginine
deimidase

Copine 1

kDa
96
66
45
31
21

7 8 9 10 11 12 13 14 15 M 16 17 18 19 20 21 22 23 24 25 26 27 28 29 30 M 31 32 33 34 35 36 37

C. ATP Wash - Anion-exchange Chromatography

MAPK

kDa
96
66
45
31
21
14

7 8 9 10 11 12 13 14 15 M 16 17 18 19 20 21 22 23 24 25 26 27 28 29 30 M 31 32 33 34 35 36 37

CSK

the JNK protein kinase was originally partially purified via its ability to bind to a glutathione-*S*-transferase fusion with the N-terminal domain of c-Jun. The complex was removed from solution by binding to glutathione-agarose (28).

The problem of non-specific ionic interactions is reduced but not eliminated by the use of short peptide substrates. A good example of kinase purification utilizing a peptide substrate (29) involved a peptide (YRRAAVPPSPSLSRH-SSPHQSEDEE, phosphorylated serines underlined) derived from a sequence on rabbit muscle glycogen synthase containing one site (the most C-terminal serine) for casein kinase II (CKII) and three sites for glycogen synthase kinase-3 (GSK3). The peptide was phosphorylated on Ser-21 with an apparent Km of 10 μM by CKII, and a substantial purification of CKII was achieved using a column where the peptide was coupled to agarose via the N-terminal tyrosine. However, since the kinase was eluted by conditions of moderate ionic strength (200 mM NaCl), it was unclear whether this was a true affinity interaction, rather than an ion-exchange effect. A more convincing case for an affinity interaction was obtained using GSK3. With some substrates, GSK3 requires a prior phosphorylation at the P+4 position (i.e. 4 residues C-terminal to the phosphorylated serine) (30). As expected, the peptide was only phosphorylated by GSK3 (apparent K_m 5 μM) if it was pre-phosphorylated on Ser-21 by CKII. GSK3 passed straight through a column comprising the unphosphorylated peptide, but other proteins that bound non-specifically were retained. However, a substantial proportion of the GSK3 which passed through the dephosphopeptide column did bind to a column which was prephosphorylated on Ser-21 using CKII, and this could be recovered with a salt wash. This two-column procedure resulted in a 30-fold purification of GSK3 in good yield, and the protein was ~60% pure (29). Although the requirement of GSK3 for a prior phosphorylation represents something of a special case, this two-step approach, in which the initial column is used to remove proteins that bind through non-specific ionic interactions, may perhaps be adaptable to other systems.

Another report of a successful substrate peptide affinity column was the system developed by Michell *et al.* (31) to purify AMP-activated protein kinase (AMPK). The laboratory of one of us (D.G.H.) had previously failed (S.P. Davies and D.G. Hardie, unpublished results) to achieve specific binding of AMPK to columns based on the *SAMS* peptide (HMRSAMSGLHLVKRR) which is a well-established AMPK substrate with an apparent K_m of ~30 μM (4). By screening a collection of synthetic peptides, Michell *et al.* (31) found a peptide (LKKLTRRPSFSAQ) which had a somewhat lower K_m (4 μM) than *SAMS*. When coupled to an *N*-hydroxysuccinamide ester-activated Sepharose column, this peptide bound AMPK complexes containing the α1-subunit, but not the α2-subunit. Remarkably, the complex bound with very high avidity, and 2 M NaCl and 30% (by vol.) ethylene glycol were required for elution. This is a much tighter binding than one would expect based on an apparent

K_m of 4 μM, and the authors speculated that the aminohexanoic acid linker between the peptide and the resin may have contributed to the high avidity, by binding to a hydrophobic pocket on the kinase surface. The necessity for ethylene glycol in the elution buffer may support this view, but the molecular basis for this very tight interaction remains unclear, as is the basis for the selectivity of the column for complexes containing the α1-isoform of the catalytic subunit. This selectivity is surprising given that the specificity of α1 and α2 complexes are very similar (18). This reinforces the view that the design of peptide substrate affinity columns to some extent remains an art rather than a science.

Protocol 7 describes the preparation of the peptide column, and *Protocol 8* describes the purification of AMPK complexes containing the α1-subunit by chromatography on the peptide column.

Protocol 7.   Preparation of a peptide affinity column for AMPK-α1 complexes

Equipment and reagents

- HiTrap NHS-activated Sepharose column (5 ml, Pharmacia Biotech, cat. no. 17–0717–01)
- peptide (LKKLTRRPSFSAQ) for AMPK-α1 specific column
- coupling buffer (0.2 M NaHCO$_3$, 0.5 M NaCl, pH 8.3)
- HCl (1 mM, ice cold)
- buffer A (0.5 M ethanolamine, 0.5 M NaCl, pH 8.3)
- buffer B (0.1 M acetate, 0.5 M NaCl, pH 4.0)
- storage buffer (50 mM Na phosphate, 0.1% NaN$_3$, pH 7.0)

Method

1. Dissolve peptide in coupling buffer. For a 5 ml column, 5 ml of a 2 mg/ml solution is suitable.
2. The HiTrap column is supplied in 100% isopropanol. This is removed by washing with 3 × 10 ml of ice-cold 1 mM HCl, using a 10 ml syringe.[a]
3. Immediately inject 5 ml of ligand solution on to the column. Seal the column and let it stand for 30 min at room temperature.
4. Excess active groups are deactivated, and non-covalently bound peptide washed out by sequential washes as follows:
5. 3 × 10 ml buffer A, 3 × 10 ml buffer B, 3 × 10 ml buffer A.
6. Let the column stand for 30 min.
7. 3 × 10 ml buffer B, 3 × 10 ml buffer A, 3 × 10 ml buffer B.
8. Wash into buffer for kinase purification, or storage buffer if the column is not to be used immediately.

[a] Take care not to introduce air bubbles at this stage, and do not exceed a flow rate of ~2 drops/sec, otherwise the gel can become irreversibly compressed.

D. *Grahame Hardie* et al.

Protocol 8. Purification of AMPK-α1 complexes on peptide affinity column

Equipment and reagents

- peptide affinity column (see *Protocol 7*)
- column buffer [0.05 *l* Tris–HCl, pH 7.4 at 4°C, 50 mM NaF, 5 mM Na pyrophosphate, 1 mM EDTA, 1 mM EGTA, 0.02% (w/v) Brij-35, 10% (v/v) glycerol, 1 mM DTT, 0.1 mM PMSF, 1 μg/ml soybean trypsin inhibitor, 1 mM benzamidine]

- buffer A (column buffer plus 0.5 M NaCl plus 0.1% Triton X-100)
- buffer B (column buffer plus 2 M NaCl plus 0.1% Triton X-100)
- buffer C [column buffer plus 0.5 M NaCl plus 30% (w/v) ethylene glycol]

Method

1. Equilibrate the column at 4°C in column buffer plus 0.5 M NaCl plus 0.1% Triton X-100 using a 10 ml syringe.

2. Dilute the AMPK preparation to 2 ml in buffer A and apply to the column.

3. Allow the column to stand for 30 min at 4°C.

4. Wash with 4 × 10 ml of buffer A.

5. Wash with 4 × 10 ml of buffer B.

6. Elute kinase with 4 × 10 ml of buffer C.

References

1. Kallunki, T., Deng, T., Hibi, M., and Karin, M., (1996). *Cell*, **87**, 929.
2. Brunet, A. and Pouyssegur, J. (1996). *Science*, **272**, 1652.
3. Kemp, B.E., Graves, D.J., Benjamini, E., and Krebs, E.G. (1977). *J. Biol. Chem.*, **252**, 4888.
4. Davies, S.P., Carling, D., and Hardie, D.G. (1989). *Eur. J. Biochem.*, **186**, 123.
5. Weekes, J., Ball, K.L., Caudwell, F.B., and Hardie, D.G. (1993). *FEBS Lett.*, **334**, 335.
6. Dale, S., Wilson, W.A., Edelman, A.M., and Hardie, D.G. (1995). *FEBS Lett.*, **361**, 191.
7. House, C. and Kemp, B.E. (1987). *Science*, **238**, 1726.
8. Schaap, D., Parker, P.J., Bristol, A., Kriz, R., and Knopf, J. (1989). *FEBS Lett.*, **243**, 351.
9. Lam, K.S., Wu, J., and Lou, Q. (1995). *Int. J. Pept. Protein Res.*, **45**, 587.
10. Roskoski, R., Jr, (1983). In *Methods in enzymology* (ed. Corbin, J.D., Hardman, J.G.), Vol. 99, p. 3. Academic Press, New York.
11. Toomik, R., Ekman, P., and Engström, l. (1992). *Anal. Biochem.*, **204**, 311.
12. Toomik, R., Ekman, P., Eller, M., Jarv, J., Zaitsev, D., Myasoedov, N., Ragnarsson, U., and Engström, L (1993). *Anal. Biochem.*, **209**, 348.
13. Lou, Q., Wu, J., and Lam, K.S. (1996). *Anal. Biochem.*, **235**, 107.
14. Laemmli, U.K. (1970). *Nature*, **227**, 680.

15. Krebs, E.G. and Fischer, E.H. (1956). *Biochim. Biophys. Acta*, **20**, 150.
16. Hunter, T. (1994). *Sem. Cell Biol.*, **5**, 367–376.
17. Salt, I.P., Johnson, G., Ashcroft, S.J.H., and Hardie, D.G. (1998). *Biochem. J.*, **335**, 533–539.
18. Salt, I.P., Celler, J.W., Hawley, S.A., Prescott, A., Woods, A., Carling, D., and Hardie, D.G. (1998). *Biochem. J.*, **334**, 177.
19. Woods, A., Salt, I., Scott, J., Hardie, D.G., and Carling, D. (1996). *FEBS Lett.*, **397**. 347.
20. Knighton, D.R., Zheng, J.H., Ten Eyck, L.F., Ashford, V.A., Xuong, N.H., Taylor, S.S., and Sowadski, J.M. (1991). *Science*, **253**, 407.
21. Taylor, S.S., Radzio-Andzelm, E., and Hunter, T. (1995). *FASEB J.*, **9**, 1255.
22. Haystead, C.M.M., Gregory, P., Sturgli, T.W., and Haystead, T.A.J. (1993). *Eur. J. Biochem.*, **214**, 459.
23. Davies, S.P., Hawley, S.A., Woods, A., Carling, D., Haystead, T.A.J., and Hardie, D.G. (1994). *Eur. J. Biochem.*, **223**, 351.
24. Yu, H., Li, X., Marchetto, G.S., Dy, R., Hunter, D., Calvo, B., Dawson, T.L., Wilm, M., Andregg, R.J., Graves, L.M., and Earp, H.S. (1996). *J. Biol. Chem.*, **271**, 29993.
25. Earp, H.S., Huckle, W.R., Dawson, T.L., Li, X., Graves, L.M., and Dy, R. (1995). *J. Biol. Chem.*, **270**, 28440.
26. Sicheri, F., Moarefi, I., and Kuriyan, J. (1997). *Nature*, **385**, 602.
27. Grenert, J.P., Sullivan, W.P., Fadden, P., Haystead, T.A.J., Clark, J., Minnaugh, E., Krutzsch, H., Ochel, H.J., Schulle, T.W., Sausville, E., Neckers, L.M., and Toft, D.O. (1997). *J. Biol. Chem.*, **272**, 23843.
28. Hibi, M., Lin, A.N., Smeal, T., Minden, A., and Karin, M. (1993). *Genes Dev.*, **7**, 2135.
29. Woodgett, J.R. (1998). *Anal. Biochem.*, **180**, 237.
30. Fiol, C.J., Mahrenholz, A.M., Wang, Y., Roeske, R.W., and Roach, P.J. (1987). *J. Biol. Chem.*, **262**, 14042.
31. Michell, B.J., Stapleton, D., Mitchelhill, K.I., House, C.M., Katsis, F., Witters, L.A., and Kemp, B.E. (1996). *J. Biol. Chem.*, **271**, 28445.

10

Assay of protein kinases and phosphatases using specific peptide substrates

MARIA RUZZENE and LORENZO A. PINNA

1. Introduction

The successful and widespread use of peptide substrates for the assay of protein kinases is grounded on the observation that the great majority of these enzymes are able to recognize phosphoacceptor sites defined by local structural features within their target proteins. Although this local sequence specificity is not the only tool ensuring the selectivity of protein kinases (indeed, in some cases, notably protein–tyrosine kinases, it may not even be a major one) many members of the protein kinase family phosphorylate short peptides reproducing their natural phosphoacceptor sites. The advantage of synthetic peptides, compared with protein substrates, for kinase assays and specificity studies include:

- they can be readily obtained in large amounts
- they have a well-defined chemical composition
- they are readily amenable to chemical modifications aimed at rendering their phosphorylation more efficient, more specific, and easier to test

However, the kinetics of peptide phosphorylation compared with those of the intact protein, and the relevance of local structural features, are quite variable for different protein kinases. The majority of protein–serine/threonine kinases (PSKs) are markedly site specific, and consequently display high catalytic efficiency towards relatively short peptides, provided that the peptides include the required specificity determinants. This approach provides a tool for tailoring peptides that are specific for a given Ser/Thr kinase, as well as allowing a sensitive assay of its activity. The site specificity of most protein–tyrosine kinases (PTKs) is not as clear-cut as that of Ser/Thr kinases. This is probably because factors other than the sequence around the phosphoacceptor sites are more important for tyrosine kinases' specificity, especially the presence of protein–protein recognition modules outside the catalytic domain, such as SH2, SH3, PTB, and PH domains (reviewed in ref. 1). This may frustrate attempts to develop *specific* peptide substrates for individual PTKs. However, it does not prevent the sensitive assay of their activity using peptide substrates, whose kinetic parameters are sometimes comparable to the best peptide substrates of PSKs. The only shortcoming in these cases would be that the peptide substrates of PTKs are, generally speaking, more promiscuous than those of PSKs. This does not rule out the possibility of discrimination between different PTKs using more than one peptide substrate. While it is highly unlikely that an individual peptide will be completely specific for a given PTK, the activity ratio towards two distinct peptides can still be drastically different for two different PTKs.

A small set of highly specific protein kinases display phosphorylation rates with peptide substrates that are negligible compared with those with the parent protein, and, consequently, peptide substrates are of little use in these cases. Some calmodulin-dependent protein kinases (e.g. calmodulin-dependent protein kinase IV), the eIF2α kinases, the G-protein-coupled receptor kinases, Raf, and the dual specificity MAP kinase kinases (MEKs) fall into this category. It is conceivable that, in the future, effective peptide substrates for these dedicated protein kinases could be developed, albeit possibly not related to the sequence of the natural protein substrates. Pertinent to this are observations that, on the one hand, the mutation of protein phosphoacceptor sites may improve their phosphorylation over that of the wild type, and, on the other hand, that crucial determinants established with peptide substrates may prove to be of lesser importance with the intact protein substrate. An example is provided by C-terminal Src protein kinase (Csk) a dedicated PTK very poorly active towards even large peptides reproducing its natural target, but which phosphorylates unrelated peptides with high efficiency (2). As pointed out in the next section, the peptide library approach (see Chapter 16) could prove especially helpful to select peptide substrates suitable for assaying protein kinases in this class, as well as protein kinases whose natural targets are still unknown.

2. The residue and sequence specificity of protein kinases

Efficient phosphorylation of amino acids by protein kinases requires their incorporation into a discrete peptide sequence. During the catalytic event, the residue undergoing phosphorylation will bind to the active site, whereas the surrounding residues will interact with elements of the kinase outside the *sensu stricto* catalytic site. These latter have been identified in some instances by site-directed mutagenesis (e.g. 3–6) or by crystal structure analysis (7, 8). Both the nature of the phosphorylatable amino acid, and of the residues surrounding it, will therefore influence the catalytic parameters, giving rise to 'residue specificity' and 'sequence specificity', respectively (reviewed in ref. 1). From a practical standpoint, the residue specificity gives rise to just two main types of residue selection, corresponding to the two major classes of protein kinases, i.e. serine/threonine-specific (PSKs) or tyrosine-specific (PTKs). In most instances the border between PSKs and PTKs, which can be predicted on the basis of primary structure motifs, is well defined and can be empirically drawn using peptide substrates of the two sorts, exclusively phosphorylated by either PSKs or PTKs. In contrast, the features that under-lie the preference for serine or threonine within the PSK class remain unclear. As a general rule, serine residues are preferred over threonine residues, in some cases to such an extent that it would be tempting to postulate a subclass of serine-specific protein kinases inactive on threonine (e. g. the Golgi casein kinase and p90^{S6K}). In contrast, other PSKs (e.g. AMP-activated protein kinase and Cdc2) tolerate Ser↔Thr substitutions quite well. These differences have sometimes been exploited to improve the selectivity of peptide substrates (e.g. see ref. 9).

The selectivity of most Ser/Thr kinases, however, relies mainly on *sequence specificity*, often expressed by a so-called 'consensus sequence'. The elements of the sequence that are especially required to ensure efficient phosphoryl-ation are referred to as 'specificity determinants', whose nature (generally either basic, hydrophobic or acidic residues, or proline) and position depend on the kinase considered. The relevance of individual determinants can also vary, indicating that some are more dispensable than others. The indispens-able residues are highlighted in the common representations of the consensus sequence, while additional features that are not strictly required (albeit that they may substantially improve the phosphorylation efficiency) are often neglected. Likewise, not enough emphasis is always placed on *negative deter-minants*, whose presence can compromise the phosphorylation of otherwise suitable sites, and which can be exploited to improve the selectivity of peptides, as they can prevent phosphorylation by kinases other than the desired one(s).

The consensus sequences of a number of representative protein kinases are shown in *Table 1*. Most of them have been drawn from the analysis of natural phosphoacceptor sites, in conjunction with kinetic studies with peptides re-producing these, either exactly or with suitable modifications. This approach proved especially rewarding with many PSKs, while with PTKs *sensu stricto* consensus sequences consistent with the structure of natural phosphoacceptor sites were established only in a few cases. It should be stressed, however, that even in the case of PSKs, consensus sequences do not represent an absolute rule: atypical sites, where the specificity determinants are located at positions different from those highlighted in the consensus, are often phosphorylated with an efficiency comparable to 'canonical' sites.

Table 1. Consensus sequences of some protein kinases[a]

Basophilic PSKs	Consensus sequences[b]
PKA	R–(R/K)–X–(**S/T**)–B
PKG	R–(R/K)–X–(**S/T**)–B
PKC	(R/K)–(R/K)–X–(**S/T**)–B–(R/K)–(R/K)
p70[S6K]	(K/R)–X–R–X–X–(**S/T**)–B
p90[S6K]/MAPKAPK1	X–X–(R/K)–X–R–X–X–**S**–X–X
	R–R–R–X–**S**–X–X
MAPKAPK2	X–X–B–X–R–X–X–**S**–X–X
CaMK I	B–X–R–X–X–(**S/T**)–X–X–X–B
CaMK II	B–X–(R/K)–X–X–(**S/T**)–X–X
Phosphorylase kinase	K–R–K–Q–I–**S**–V–R
AMPK	B–(X,R/K/H)–X–X–(**S/T**)–X–X–X–B
HSV-PK and PRV-PK	R–R–R–R–X–(**S/T**)–X
Proline–directed PSKs	
ERK (MAP kinases)	P–X–(**S/T**)–P–P
Cdc2 (and other cyclin-dependent PKs)	X–(**S/T**)–P–X–(K/R)
Acidophilic PSKs	
CK2	X–(**S/T**)–X–(E/D/Sp/Yp)–X
G-CK	X–**S**–X–(E/Sp)–X
CK1	(Sp/Tp)–X–X–(**S/T**)–B
	(D/E)$_n$–X–X–(**S/T**)–X–X (n ≥ 4)
GSK-3	(**S/T**)–X–X–X–Sp–X
PSKs	
Src family/Abl/c-Fps	I–**Y**–G/E
Syk	(E/D)–**Y**–E
EGF/PDGF/IGF1/Insulin receptor PKs	E–**Y**–M/F/V

[a] Abbreviations: PKA, cAMP-dependent protein kinase; PKG, cGMP-dependent protein kinase; MAPKAPK1, 2, MAP kinase-activated protein kinase-1, -2; CaMK, calmodulin-dependent protein kinases; AMPK, AMP-activated protein kinase; HSV-PK, herpes simplex virus protein kinase; PRV-PK, pseudorabies virus protein kinase; ERK, extracellular signal-regulated kinase; CK2, protein kinase CK2 (casein kinase 2); G-CK, Golgi casein kinase; CK1, protein kinase CK1 (casein kinase 1); GSK3, glycogen synthase kinase-3.
[b] Amino acids are indicated by the one-letter code; B stands for any hydrophobic amino acid and X for any residue; Sp and Tp denote phosphoserine and phosphothreonine, respectively. Interchangeable residues at a given position are grouped between parentheses, and separated by slashes. The target residues are in bold type. For other sequences and/or additional information see ref. 1.

3. Design of synthetic peptide substrates

The 'ideal' peptide substrate for a given protein kinase should be:

- readily phosphorylated by the desired kinase, with favourable kinetic constants
- poorly phosphorylated, if at all, by other kinases
- suitable for a fast, simple and sensitive assay

Peptides that are efficiently phosphorylated and suitable for convenient assays can be prepared for the majority of known protein kinases, either PSKs or PTKs. In contrast, strict selectivity is a property of only a small minority of peptide substrates. Most peptides are phosphorylated by a variety of more or less related protein kinases, albeit with different efficiencies, such a promiscuity being especially pronounced among peptide substrates of PTKs.

3.1 Efficiency of peptide substrates

In general, the design of a peptide substrate firstly requires the identification of phosphoacceptor site(s) affected by the kinase of interest in its protein targets. Often, this analysis also discloses conserved features that may represent specificity determinants. The actual relevance of these structural elements is then checked by synthesizing peptides where these residues are either conserved or substituted, and by using them as substrates for the kinase of interest. The minimum length of the peptide, required for efficient phosphorylation should also be checked. Although very short peptides (6–7 residues long) may occasionally prove excellent substrates, in general a more extended sequence is required, composed of 10–15 residues. A 'good' peptide substrate is expected to display a K_m in the low μM range (<100 μM) and a V_{max} value comparable to that of the protein substrate. One should also bear in mind that the sequence derived from the natural protein substrate is not necessarily the optimal one. Often the phosphorylation of a peptide substrate can be substantially improved by the inclusion of additional positive determinants that may be lacking in the original site, or by the elimination of negative determinants that may be present. Useful hints about substitutions or modifications likely to improve the phosphorylation efficiency can be provided by the comparative analysis of many natural phosphoacceptor sites, and by systematic studies with numerous peptide substrate derivatives.

A short-cut towards the generation of optimal peptide substrates has been recently made possible by the development of peptide library approaches. These methods, dealt with in Chapter 16, are intended to provide through a single experiment (at least in principle) an amount of information that would otherwise require hundreds of experiments with a wide variety of successively designed peptide substrates. They may also prove helpful for designing peptide substrates for kinases whose physiological phosphoacceptor sites are still unknown, and perhaps even for protein kinases which are inactive on peptides reproducing their natural substrate but might still act on unrelated short sequences. The most successful of these methods are based on oriented libraries (10, 11; see Chapter 16) consisting of peptides including either a serine or a tyrosine residue embedded in a sequence of degenerate residues. After phosphorylation by the kinase and removal of the dephosphorylated peptides, the degenerate phosphorylated peptides are sequenced by Edman degradation as a mixture. If certain amino acids are prevalent in a particular sequencing cycle, this means that the kinase had selectively phosphorylated peptides containing that amino acid at that position. In principle, one might imagine that the sequence reconstructed from the residues most highly selected at each position would represent the optimal sequence recognized by the kinase. However, it should be remembered that this represents a *virtual* sequence. Even if the library contained a peptide with this actual sequence, the signal from a single peptide would be far too low to be detectable. The peptide library approach can have shortcomings if the importance of a residue at a given position depends on the nature of the surrounding sequence.

The potential and limitations of this approach are discussed further in Chapter 16. From the standpoint of those searching for an optimal peptide substrate, it is crucial to check if a *real* peptide designed using the *virtual* optimal sequence selected from the library is efficiently phosphorylated by the kinase. In many instances this has proved to be the case. Biases can result, however, from the tendency of oriented libraries to select specificity determinants at positions where they are not really needed, as discussed in ref. 1 (this could even worsen selectivity, see below). Other problems include the constraints imposed by the fixed scaffold of all the library peptides, notably the invariant position of the phosphoacceptor residue in the centre of the sequence, and a basic triplet, usually KKK, inserted at the C-terminal end for technical reasons (10).

3.2 Specificity of peptide substrates

Conferring absolute specificity to a peptide substrate for an individual kinase, to such an extent that no other kinase will affect it, is in most cases an un-achievable goal. In many instances, however, it is possible to generate relatively selective peptides which are phosphorylated by the desired kinase much more efficiently than by the majority of other kinases, including those belonging to the same specificity class (see *Table 1*). In this connection, one can consider the example of the three types of casein kinases, all equally active on casein (hence their name) but now routinely monitored using highly specific peptide substrates (see *Table 2*). These kinases are all *acidophilic*, so-called because the specificity determinants are acidic residues surrounding the phosphoacceptor site. The design of specific peptide substrates was made possible by detailed knowledge of the specificity determinants of CK1 (requiring at least four Asp residues between n–3 and n–6), CK2 (requiring either Asp or Glu at position n+3 and additional acids nearby) and G-CK (needing Glu, but not Asp, at n+2). The n–X (n+X) nomenclature [P–X (P+X) is used by some workers] refers to the position X residues N-terminal (C-terminal) to the phosphorylated amino acid. The detailed knowledge of the specificity of the casein kinases allowed the design of peptides that fulfil the requirements of each of them, without being appreciably phosphorylated by the others, nor by basophilic or proline-directed protein kinases.

To some extent the attainment of specific peptide substrates is facilitated by incompatibilities between the specificity determinants of different kinases.

Table 2. A list of synthetic peptide substrates used for assaying protein kinases[a]

Peptide[b]	Parent protein[c]	PK[d]	K_M (μM)[e]	Ref.
LRRA**S**LG	Pyruvate kinase	PKA	6	45, 46
		PKG	210	47
RRRA**S**VA	Pyruvate kinase	PKA	6	48
		PKC	140	48
		PRV-PK	20	48
		HSV-PK	n.r.	48
GRTGRRN**S**I	PKI	PKA	0.1	47
		PKG	2	47
GSRRRRRY	Galine	PKC	12	49
		HCR	60	50
		dsI	300	50
QKRP**S**QRSKYL	MBP	PKC	7	51
KRAKRK**T**AKKR	MLC	PKC	0.5	52
RFAVRDRMQ**T**VAVGVIKAVDKK	eEF-1α	PKCδ	23	53
AKRKRKG**S**FFYGG	(library)	PKCδ	1	54
AALVRQM**S**VAFFK	(library)	PKCμ	9	54
RPRTS**S**F		PKBα	5	55
		p90s6k	12	55
		p70s6k	125	55
RPRAA**T**F		PKBα	25	55
		p90s6k	>500	55
		p70s6k	>500	55
HMRSAM**S**GLHLVKRR	ACoAC	AMPK	26	56
		SNF1	108	56
AMARAA**S**AAALARRR		AMPK	10	56
		SNF1	650	56
		CaMK I	n.r.	56
LKKLTRRP**S**FSAQ	ADR1	AMPK	4	57
		PKA	n.r.	57

Table 2. *Continued*

Peptide[b]	Parent protein[c]	PK[d]	K_M (μM)[e]	Ref.
LRRRLSDANF	Synapsin I	CaMK I	4	58
		CaMK II a	45	59
		CaMK IV	0.2	59
KKRAARATSNVFS	MLC	sm MLCK	12	60
KKRARAATSNVFA	MLC	Twitchin	4	61
		sm MLCK	20	61
		CaMK II	54	61
PLARTLSVAGLPGKK	GS	CaMK IV	1	59
		CaMK II	9	59
KSDGGVKKRKSSSS	CaM-K II	CaMK IV	8	62
KKPLNRTLSVASLPGL-amide	GS	MAPAKAP-Kin 2	12	63
		CaMK II	8	63
		$p90^{S6k}$	36	63
KKFNRTLSVA	GS	MAPAKAPK2	9	63
		CaMK II	n.r.	63
		$p90^{S6k}$	n.r.	63
KKLNRTLTVA	GS	CaMK II	n.r.	63
KKKNRTLSVA	GS	$p90^{S6k}$	0.2	63
		$p70^{S6k}$	3	9
KKRNRTLSVA	GS	$p70^{S6k}$	1	9
		$p90^{S6k}$	0.7	9
KKRNRTLTVA	GS	$p70^{S6k}$	5	9
		$p90^{S6k}$	40	9
PLRRTLSVAA	GS	CaMK II	4	9
AVAAKKSPKKAKKPA		Cdc2	3	64
NFKTPVKTIR	CK2 β sub	Cdc2	40	65
AcetylISPGRRRRKamide		Cdc2	1	66
VTPRTPPRR	MBP	Cdc2	n.r.	67
		Cdk2	n.r.	68

Peptide	Protein source	Kinase	K_m (μM)	Ref
APRTPGGRC-polylys		ERK	74	69
KRELVEPKTPSGEA	EGF-R	ERK1	190	70
		ERK2	270	70
		Cer.act.PK	2200	71
KKKKEHQVLMKTVCGTPGY	CaM-K IV	CaMKK α	263	72
KKKDPGSVLSTACGTPGY	CaM-K I	CaMKK α	n.r.	72
KRELVEPLTLPSGEA?	EGF-R	Cer.act.PK	400	71
KYHGHTMSDPGVSYR	PDH	PDHK	460	73
PLSRTLSVRSL?	GS	PhK	810	74
		PKA	n.r.	74
KRKQISVRGSL	Ph	PhK	900	64
		PKA	n.r.	64
RRREEEEESAAA		βARK	1100	75
RRRAAAAASEEE		RhK	2000	75
		RhK	1800	75
		CK2	n.r.	75
PRRRRSSRPVR	Thynnine	PRV-PK	20	76
KRREILSRRPSpYR	CREB	GSK3 α	140	77
		GSK3 β	200	77
PRPASVPPSPSLSRHSSPHQSpEDEEEP?	GS	GSK3	2	28
PLSpRTLSVASLPGL-amide*	GS	CK1	7	78
SpSpSpEESIT*	β-Casein	CK1	14	79
IGDDDDAYSITA*	Inhibitor-2	CK1	58	18
RRKHAAIGDDDDAYSITA	Inhibitor-2	CK1	250	18
RRKDLHDDEEDEAMSITA	Inhibitor-2	CK1	172	16
RRRADDSDDDDD		CK2	19	16
		G-CK	3062	17
RRRAEESEEEEE		CK2	19	16
		G-CK	15	17
KKIEKFQSEEQQQ	β-Casein	G-CK	663	17
YIYGSFK*	(library)	c-Src	55	80
		Lyn	n.r.	80
		Fyn	n.r.	80

Table 2. *Continued*

Peptide[b]	Parent protein[c]	PK[d]	K_M (μM)[e]	Ref.
AEEEIYGEFEAKKKK	(library)	c-Src	33	11
KVEKIGEGTYGVVYK	cdc2	c-Src	101	81
		Lyn	277	81
		Lck	133	81
		Fyn	487	81
		c-Fgr	80	12
		Syk	1500	12
EDNEYTA *	c-Src	Syk	58	82
		c-Fgr	430	82
		Lyn	3800	82
RRLIEDAEYAARRG	c-Src	c-Src	2800	83
c(EDNEYTA) *	c-Src	Lyn	20	84
FGEFGEFGEYGEFGEFGD *		c-Src	73	83
KKKEEEEYMPMEDL	mT	v-Src	97	85
		v-Abl	97	85
EAIYAAPFAKKK	(library)	c-Abl	4	11
EQEDEPEGDYEEVLE *	HS1	Syk	4	12
		Csk	625	12
		Lyn	5000	12
EDENLYEGLNLDDCSMYEDI *	ARAM Igα	Syk	25	12
		c-Fgr	33	12
		Lyn	98	12
		Csk	211	12
EEEPQYEEIPIYLELLP *	mT	Csk	34	2
		c-Fgr	185	2
EEEPQFEEIPIYLELLP *	mT	Csk	63	2
		c-Fgr	473	2
KKSRGDYMTMQIG	IRS-1	IR	30	86
		IGF1 R	178	86

Peptide	Protein	Kinase		
KKKSPGEYVNIEFG	IRS-1	v-Src	300	85
		v-Abl	520	85
		IGF1 R	26	86
		IR	52	86
DRVYIHPF *	Angiotensin II	Lyn	1200	82
		several PTKs		

[a] A selection of peptide substrates with K_m values in the lowest range for the indicated kinases are reported. In some cases, the peptides have been chosen for their more favourable V_{max}/K_m ratio, even if their K_m values are not the lowest reported. For a detailed analysis and comparison with other peptides, see the indicated references.

[b] Sequences are indicated with the one-letter code for amino acids; the target residues are bold type: Sp, phosphoserine; an asterisk (*) or a question mark (?) denote peptides whose phosphorylation assay by the phosphocellulose paper method is inapplicable or dubious, respectively.

[c] The parent protein is indicated, when applicable, even if the peptide sequence does not correspond exactly to the original one because of substitutions or additions, introduced to improve phosphorylation and/or specificity, or to allow phosphocellulose paper assay. Abbreviations: ACoAC, acetyl-CoA carboxylase; ADR1, yeast transcriptional activator of the *ADH2* gene; ARAM, antigen recognition activation motif; CREB, cAMP response element binding protein; GS, glycogen synthase; HS1, haematopoietic lineage-cell specific protein; IRS-1, insulin receptor substrate-1; MBP, myelin basic protein; MLC, myosin light chain; mT, middle T; PDH, pyruvate dehydrogenase; Ph, phosphorylase; PKI, heat stable inhibitor protein of PKA.

[d] Some protein kinases (denoted in *italics*) are highly specific enzymes whose peptide substrates are phosphorylated very slowly despite their sometimes favourable K_m values.

Abbreviations: AMPK, AMP-activated protein kinase; βARK, β adrenergic receptor kinase; CaMK, calmodulin-dependent protein kinases; Cdc2 kinase, protein kinase expressed by *CDC2* gene; cdk, cyclin-dependent kinase; Cer. Act. PK, ceramide activated protein kinase; CK1, protein kinase casein kinase 1; CK2, protein kinase CK2 or casein kinase 2; Csk, C-terminal Src protein kinase; dsI, double strain RNA inhibitor = eIF2α kinase; ERK, extracellular signal-regulated kinase; G-CK, Golgi casein kinase; GSK3, glycogen synthase kinase-3; HCR, heme controlled repressor = eIF2α kinase; IGF1 R, insulin-like growth factor-1 receptor; I-R, insulin receptor; MAPKAP, MAP kinase activated protein kinase; PDHK, pyruvate dehydrogenase kinase; PhK, phosphorylase kinase; PKA, cAMP-dependent protein kinase; PKB = RAC, PK related to PKA and PKC; PKG, cGMP-dependent protein kinase; PRV-PK, pseudorabies virus protein kinase; RhK, rhodopsin kinase; sm MLCK, smooth muscle myosin light chain kinase; SNF1, sucrose non-fermenting protein kinase.

Thus, a proline at n+1, which is the *sine qua non* for of all proline-directed kinases, acts as a powerful negative determinant for the majority of other kinases (1). Similarly, the basic residues that are specificity determinants for *basophilic* kinases, and also several proline-directed kinases, are detrimental to acidophilic kinases like CK2.

Another tool to improve the selectivity of peptide substrates is replacement of the phosphoacceptor residue: this may prove helpful when two kinases with otherwise similar consensus sequences display a markedly different tolerance for threonine versus serine (9).

In the case of PTKs, due to their promiscuity and lack of stringent site specificity, the generation of selective peptide substrates is particularly difficult. It appears that the positions where specificity determinants act more potently in the case of PTKs are those flanking the tyrosine at n–1 and n+1 (see *Table 1*). A bulky hydrophobic residue at n–1 (generally Ile) is strongly selected in an oriented library by all the Src family kinases tested so far, as well as by Abl, Csk ,and c-Fps (2, 11), a predilection that has been confirmed with conventional peptide studies (2). In contrast, receptor kinases (and Syk) display a marked preference for acidic residues (generally Glu) at n–1 (11, 12). The nature of the preferred residue at n+1 appears to be dictated by the composition of the so-called 'P+1 loop'. In cAMP-dependent protein kinase (PKA) this loop forms a hydrophobic pocket, with the last residue of the pocket, Leu-205, making the main contact with the determinant at position n+1. In PTKs the residue equivalent to Leu-205 in PKA is either methionine, indicating a preference for a hydrophobic residue at n+1 (e.g. receptor PTKs and Csk), or threonine, which is compatible with the accommodation of glycine or another hydrophilic residue (e.g. the Src family and other non-receptor kinases). An exception is provided by Syk, where the equivalent residue to Leu-205 in PKA is a tyrosine. Syk displays a striking preference for peptides where tyrosine is embedded between several acidic residues. Based on this principle, excellent peptide substrates for Syk could be designed (12) (see *Table 2*). It is not known how efficiently these peptides are phosphorylated by the Syk-related ZAP kinase. Although some of these highly acidic peptides display poor kinetics with other non-receptor PTKs (e.g. 12), they are still appreciably phosphorylated by the latter, so these peptides are not entirely specific for Syk. If, however, they are used in combination with different peptide substrates (e.g. the Cdc2-derived peptide, see *Table 2*) the activity ratio can be used as a reliable criterion to identify Syk activity.

A list of peptide substrates that have been used for the assay of a variety of protein kinases is provided in *Table 2*, with the indication of K_m values, whenever available. A number of points should be made. First, the number of protein kinases matching a given peptide does not allow any direct inference about actual specificity, since the list is incomplete and does not include in any case those kinases that have been shown to be *inactive* on the peptide considered. For a deeper insight into peptide selectivity, it is advisable to consult the pertinent reference(s). Secondly, in many instances differences in K_m values can be exploited to improve the selectivity of peptide substrates towards certain kinases, by performing the assay with appropriate low peptide concentrations. Thirdly, a low K_m does not always indicate a selective substrate, and *vice versa*. The outstanding affinity of the peptide GRTGRRNSI for PKA ($K_m = 0.1\ \mu M$) does not make it more specific than the traditional 'kemptide', LRRASLG ($K_m = 6\ \mu M$); both peptides are phosphorylated by several basophilic protein kinases. By contrast, the β-casein-derived peptide KKIEKFQSEEQQ is a very specific substrate for the Golgi apparatus casein kinase (G-CK), despite its high K_m (663 μM), since it is not appreciably acted upon by other protein kinases. Sometimes, an increase in K_m value is acceptable if it is the consequence of a modification (such as an addition of basic residues) aimed at rendering the peptide suitable for a fast and reliable assay technique (see Section 3). In this connection it should be noted that the majority of the peptides listed in *Table 2* either are (or have been made) suitable for the phosphocellulose paper assay; exceptions are indicated by an asterisk or question mark.

Clearly, a better knowledge of negative determinants variably perceived by different kinases will prove helpful for tailoring peptides that display higher selectivity. Another strategy that might deserve attention is to exploit the *active-site* specificity with artificial phosphoacceptor derivatives different from the natural hydroxyl amino acids, i.e. serine, threonine, and tyrosine. Especially promising in this respect appear to be studies with alcohol-bearing compounds that can serve as protein kinase substrates, and which are disclosing unexpected differences in the active-site specificity of otherwise closely related kinases, both PSKs and PTKs (13, 14). Likewise, the use of tyrosine analogues could provide a key for monitoring individual members of the PTK family. Pertinent to this is the observation that the replacement of tyrosine by its analogue, 7-hydroxy-1,2,3,4-tetrahydroisoquinoline-3-carboxylic acid, in a cyclic peptide derived from the activation loop of c-Src fully prevents phosphorylation by the Src-related kinase c-Fgr, while it is tolerated fairly well by Syk (Ruzza, P., Donella-Deana, A., Calderan, A., Filippi, B., Cesaro, L., Pinna, L.A., and Borin, G., unpublished data).

3.3 Measurement of peptide phosphorylation

Even the best and most specific peptide substrate would be of little use unless a reliable, sensitive, and simple method for assaying its phosphorylation was available. If phosphorylation is performed by radiolabelling with $[\gamma\text{-}^{32}P]ATP$, it is normally sufficient to achieve a fast and complete separation of the radiolabelled phosphopeptide from the large excess of radioactive ATP, and its hydrolysis product $[^{32}P]$phosphate. In an assay where both radioactive ATP and the phosphoacceptor peptide substrate are present at 500 μM, and assuming that only 5% of the peptide is converted into its phosphorylated form, at the end of incubation there will be an ~20-fold excess of radioactive

ATP over radioactive phosphopeptide. In order for the level of residual radioactivity due to ATP to be acceptable (e.g. <10% of the radioactivity incorporated into the phosphopeptide), no more than 0.5% of the original ATP would remain in the sample after the separation procedure. This is an optimistic scenario when the kinase is poorly active, or when the kinetics are performed with very low concentrations of peptide. In these cases it is desirable that <0.01% of the unreacted ATP remains to contaminate the phosphopeptide product, in order to make the assay feasible and reliable.

This performance can be achieved using the phosphocellulose paper procedure, based on the principle that at low pH values ATP (and inorganic phosphate) does not bind to phosphocellulose, whereas any singly phosphorylated peptide including two or three basic residues will bind. This method, first developed by Glass *et al.* (15), can be directly applied to most peptide substrates of basophilic protein kinases, which already include basic residues as specificity determinants (see *Table 2*). It can also be adapted to peptide substrates for kinases that, while not requiring basic residues, nevertheless tolerate them. This includes proline-directed kinases, and many PTKs. Although it is commonly stated that the latter require acidic determinants, in reality they are quite promiscuous as far as peptide substrates are concerned and, with a few exceptions, will phosphorylate peptides to which basic residues have been added quite well.

To ensure that phosphopeptides quantitatively bind to phosphocellulose paper, it is advisable to include at least three basic residues rather than two (see also Chapter 9 for a discussion of this topic). An example of this is provided by angiotensin II, an octapeptide widely used for the assay of PTKs, which has only two basic residues (DRVYIHPF). In our experience, angiotensin II itself is not suitable for reliable assays, since 30–40% of its phosphorylated derivative can be lost upon washing of phosphocellulose paper (see *Table 3*). To make binding quantitative (> 95% recovery after washing) another arginine can be added at the C-terminal end. This additional arginine did not appear to impair phosphorylation, at least by Src-related kinases.

Even three basic residues may not be sufficient if the peptide is phosphorylated at more than one site. This may be the case if more than one phosphoacceptor residue is present, or if additional phosphorylated residue(s) are included from the outset to prime phosphorylation by 'phosphate-directed' kinases (e.g. GSK3, see *Table 3*).

Whenever the basic residues are not present in the 'natural' sequence which formed the basis for the peptide substrate, they must be inserted at position(s) where they do not compromise phosphorylation efficiency or selectivity [e.g. by rendering the peptide a substrate for undesired basophilic kinase(s)]. This is exemplified by the design of peptide substrates suitable for the phosphocellulose paper assay of casein kinases (16, 17). These kinases are acidophilic and often perceive basic residues as negative determinants. In the case of CK2, whose specificity determinants are mostly located C-terminal to the

Table 3. Variable recovery of phosphopeptides by different procedures[a]

Peptides	Recovery of phosphopeptide by:[b]		
	P-cellulose paper	AG 1-X8 anion exchange	DEAE-paper
DRVYIHPF (angiotensin II)	45	100	n.t.
DRVYIHPFR	95	100	n.t.
EDNEYTA	< 1	90	n.t.
RRLIEDNEYTARG	98	98	n.t.
SAEEEDQYN	< 1	82	n.t.
RRRADDSDDDD	97	30	80
ESEEEEE	< 1	30	82
RKMKDTDSEEEIR	98	85	n.t.
RRASVA	98	98	n.t.
SpSpSpEESIT	< 1	< 1	n.t.
EQEDEPEGDYEEVLE	n.t.	60	88
DEDADIYDEEDYDL	n.t.	20	97
PEGDYEEVLE	n.t.	70	70
PEGDYAAVLE	n.t.	98	20

[a]Unpublished data provided by A. Donella Deana and F. Meggio.
[b]Values are expressed as a percentage relative to the recovery of phosphopeptide by the acid hydrolysis method, assumed to be 100%; n.t., not tested.

serine, the problem could be circumvented by placing the basic residues upstream at positions n–4 to n–6, where they are reasonably well tolerated. At position n–1 and anywhere downstream they are not tolerated. The resulting peptide, <u>RRR</u>ADDSDDDDD (*Table 2*), besides being an excellent substrate for CK2 by virtue of its seven Asp residues, and suitable for the phosphocellulose paper method by virtue of its three N-terminal arginines, is also quite specific. The lack of basic residues at n–2 and n–3 prevents its phosphorylation by most basophilic kinases, whereas the alanine at n–3 (instead of another Asp) prevents phosphorylation by another casein kinase, CK1. The Golgi apparatus casein kinase (G-CK), whose consensus sequence is S-X-(E/Sp)-X, does not phosphorylate this peptide because it contains Asp rather than Glu at n+2 (see *Table 1*). Conversely, a highly specific peptide substrate for G-CK, suitable for the phosphocellulose assay, was derived from the sequence surrounding serine-35 in β-casein (KKIEKFQSEEQQQ) exploiting the basic residues present in the N-terminal moiety of the natural sequence.

In the case of CK1, which requires either a phosphorylated residue at n–3, or an acidic cluster upstream from n–2, placing the basic residues downstream was not successful since the resulting peptide (DDEEDEEMSETA<u>RRR</u>) was a very poor substrate (18). This problem was circumvented by placing the basic triplet upstream, quite far from the acidic cluster. The resulting peptide (<u>RRK</u>DLHDDEEDEAMSITA), although having a K_m higher than that of the peptide devoid of basic residues, is still a fairly good substrate (16).

If the phosphocellulose paper method is not applicable, alternative procedures (described below) based on AG 1-X8 anion-exchange assay and DEAE-cellulose paper are available. The former is not successful with acidic peptides, which tend to be retarded and elute more or less together with the radioactive ATP. It would be advisable in these cases to replace acidic residues with neutral ones if they are not required as specificity determinants. The opposite strategy applies to the DEAE-cellulose paper method, which exploits the fact that at neutral pH (where aspartic and glutamic acid side chains are completely dissociated) very acidic peptides bind much more tightly than ATP, which can be eluted prior to the peptide. This method is quantitative only if the peptide contains at least seven net negative charges (see *Table 3*), so its applicability can be improved by attaching additional acidic residues, provided that they are not detrimental to phosphorylation efficiency. In this respect, DEAE-cellulose paper is complementary to phosphocellulose paper, and can be applied to peptide substrates for kinases that do not tolerate basic residues, while accepting a large number of acidic residues. It has been successfully applied to CK2 assays (19) and to the PTK Syk (see *Table 3*), both of which are acidophilic. This method may also prove useful with other acidophilic kinases, such as CK1 and GSK3.

There is also a procedure that works with any kind of phosphopeptide, irrespective of the amino acid composition and the identity of the phospho-amino acid. This is based on acid hydrolysis of ATP under conditions where the phosphoester bonds of the peptides are not affected. Radioactive in-organic phosphate is removed by molybdate treatment and extraction of the phosphomolybdate complex with an organic solvent (see below). A limitation of this 'universal' method, as well as the fact that it is more complicated than the paper methods, is the high background. This is due to incomplete removal of $[^{32}P]$phosphate generated from $[\gamma\text{-}^{32}P]ATP$ hydrolysis, which is a particular problem if the kinase activity to be measured is low. If this is not the case, the method may be the first choice procedure to compare the phosphorylation of peptides of markedly different composition (e.g. basic, acidic, and neutral), which could not be compared using a single ion-exchange procedure. The method can also be used to check the applicability of the other methods (as in *Table 3*).

4. Phosphorylation of peptides

4.1 The phosphorylation reaction

4.1.1 Assay conditions

Assays are usually performed at 30°C in a total volume of 20–50 μl. Typical final concentrations in the assay are:

- 20–50 mM buffer (usually Tris–HCl or Hepes) at pH 7.0–8.0
- 10–500 μM $[\gamma\text{-}^{32}P]ATP$ (specific radioactivity 500–5000 cpm/pmol)
- 5–10 mM $MgCl_2$ or $MnCl_2$, or both

Other components to be added are the peptide substrate, the kinase, and any essential co-factor such as an allosteric activator of the kinase. The optimal conditions depend on the specific enzyme being used.

ATP is utilized by all protein kinases as a phosphate donor, but some (e.g. CK2, and to a lesser extent Cdc2) can use GTP as well. This can form the basis of a more specific assay for these kinases. An ATP concentration of 10 μM is usually sufficient for straightforward activity assays, but higher concentrations are required for quantitative analysis, especially the determination of K_m values for peptides. In this case, the ATP concentration should exceed the K_m value for ATP (for most kinases this is in the range 1–150 μM) by several-fold. The desired specific radioactivity of [γ-^{32}P]ATP can be obtained by mixing non-radioactive ATP with the required amounts of [γ-^{32}P]ATP from a 'carrier free' stock (for example, the solution in 50% ethanol supplied by Amersham at a specific radioactivity of 3000 Ci/mmol). ^{32}P has a half-life of only 14.3 days; decay tables are available to calculate the specific activity on the day of use. A precise assessment of [γ-^{32}P]ATP specific activity is essential for calculating reaction velocities and the stoichiometry of phosphorylation (see *Protocol 2* in Chapter 9).

Mg^{2+} or Mn^{2+} ions are required both to form the $Mg–ATP^{2-}/Mn–ATP^{2-}$ complex which is the true substrate of all protein kinases, and because binding of divalent metals may also be required at other sites on the enzyme. Usually Mg^{2+} is preferred by PSKs, and Mn^{2+} by PTKs. However, many exceptions to this are known, and a mixture of both ions is often used. In principle, the use of either Mg^{2+} or Mn^{2+} could improve the selectivity of the assay.

Protocol 1.   Standard peptide phosphorylation reaction

Equipment and reagents

- thermostatic water bath or incubator
- vortex mixer
- microcentrifuge
- equipment for radioactive safety (e.g. Perspex shields, Geiger counter)
- assay buffer (0.5 M Tris–HCl or Hepes–NaOH at pH 7.0–8.0)
- [γ-^{32}P]ATP (3000 Ci/mmol; Amersham supplies a stock in 50% ethanol)

- $MgCl_2$ and/or $MnCl_2$ (100 mM)
- unlabelled ATP (50 mM; make up in water and neutralize with Tris base or dilute NaOH)
- peptide solution (10 mg/ml; most peptides are soluble in water, but neutralize with Tris base or ammonia if necessary)
- kinase
- appropriate co-factors/activators

Method

1. A mix of the reagents which are common to all of the assays can be prepared in advance. Add the reagents (buffer, [γ-^{32}P]ATP, cold ATP, $MnCl_2$, and/or $MgCl_2$, peptide substrate, co-factors) to polypropylene tubes. Adjust to the desired final volume with water. Vortex and centrifuge for few seconds.

2. Dilute the kinase in the appropriate buffer. Start the reactions by

Protocol 1. *Continued*

adding the enzyme to each tube at fixed intervals of time; mix and place the tubes in the incubator at 30°C.

3. Stop the reactions by one of the methods described in Section 4.2.

4. Blank reactions should contain all reagents except the kinase; if a significant autophosphorylation of the kinase occurs, an additional blank should contain all components except the peptide substrate.

The peptide substrate can be initially used at a wide range of concentrations (0.1–2 mM); a kinetic analysis can then be performed to determine the most appropriate concentration to be used in routine assays. In some instances, 2-mercaptoethanol (10 mM) or dithiothreitol (0.1 mM) is added to maintain proteins in the reduced state, while BSA (1 mg/ml) and/or a detergent [such as Triton X-100, or Tween 80 (0.02–0.1% v/v)] are added to stabilize the kinase. The detergent may also help to avoid binding of the peptide to the assay tubes. The appropriate ionic strength should be determined by initial trials, since many protein kinases are inhibited by high salt, while others require a defined salt concentration for optimal activity. Reactions are typically carried out for 5–15 minutes, after which they are stopped in a manner depending on the method of analysis (to be discussed in later sections). *Protocol 1* describes a typical kinase assay reaction. See *Protocol 1* in Chapter 9 for a variation on the same theme.

4.2 Techniques for post-assay separation of phosphopeptide and ATP

To detect peptide phosphorylation, it is necessary to separate the radioactive phosphopeptide from unhydrolysed [γ-^{32}P]ATP, and [^{32}P]phosphate which may also be present due to the presence of ATPases or phosphatases (Section 3.3). The technique employed to this aim depends on the nature of the peptide being tested. Several different methods are available, as described in the following sections.

4.2.1 Phosphocellulose paper assay

This technique (15) is the most convenient for basic peptides; it exploits the binding of the peptide to phosphocellulose cation-exchanger paper, under conditions of low pH which protonate the negative charges due to carboxylic side chains and prevent binding of [γ-^{32}P]ATP (see *Protocol 2*). A version of this technique is also given as *Protocol 1* in Chapter 9.

The criteria necessary for a peptide to bind quantitatively to phospho-cellulose paper under these conditions are discussed in Section 3.3. Whenever assays are conducted for the first time with a new peptide, trials should be performed using different techniques of separation of peptide and ATP, as shown in *Table 3*.

Protocol 2. Kinase assay using phosphocellulose cation-exchange paper

Materials
- phosphocellulose cation-exchange paper (P81 Whatman; cut into 2x2 cm squares)
- phosphoric acid (0.5 %, v/v)
- hair dryer
- liquid scintillation counter
- 500 ml plastic beaker (for washing papers; this beaker should have inserted into it a slightly smaller plastic beaker, in which holes of about 0.5 cm diameter have been made, to retain the papers during washing)

Method

1. After the desired incubation time (see *Protocol 1*), stop the phosphorylation reaction by spotting the reaction mixture (or an aliquot of it) on to the P81 paper squares.

2. When the liquid is adsorbed, drop the papers into the 500 ml beaker, which contains about 200 ml of phosphoric acid. Place the beaker on to a magnetic stirrer for about 5 min; discard the liquid as radioactive waste and replace it with 200 ml of fresh phosphoric acid.

3. Wash three times with phosphoric acid (200 ml per wash for ~5 min each time).

4. Wash for a few seconds with acetone and dry papers with a hair dryer.[a]

5. Place the paper squares into scintillation vials and count them in scintillation fluid.

[a] This step is not essential, but the paper squares are easier to handle when dry.

4.2.2 AG 1-X8 anion-exchange chromatography assay

For phosphopeptides which do not contain enough basic residues to allow binding to phosphocellulose paper, separation of $[\gamma\text{-}^{32}P]ATP$ can be achieved by chromatography on AG 1-X8 anion-exchange resin (20). In 30 % (v/v) acetic acid, ATP binds quite strongly to the resin by virtue of its three negative charges, while phosphopeptides bearing just one phosphate group usually do not. They pass through the column and can be used for further analysis. They can be directly counted for incorporated radioactivity by Cerenkov counting (without addition of scintillant), and can be recovered by lyophilization for further use (e.g. as a radioactive substrate for protein phosphatase, see Section 5.1.1 and *Protocol 7*).

More often, a combination of AG 1-X8 chromatography and acidic hydrolysis followed by phosphate extraction (see Section 4.2.3) is a good device to ensure that the measured radioactivity is incorporated into the phosphopeptide and not due to contaminating $[^{32}P]$phosphate or $[\gamma\text{-}^{32}P]ATP$ (see *Protocol 4*). If the peptide contains acidic residues, it will bind to the column more tightly and may be difficult to resolve completely from [γ-^{32}P]ATP. For this reason, the AG 1-X8 anion-exchange chromatography assay is not suitable for very acidic peptides, especially multiply phosphorylated peptides (see *Table 3*).

4.2.3 Assay by acidic hydrolysis of [γ-^{32}P]ATP and separation of ^{32}Pi from phosphopeptide

If the phosphorylated peptide contains a similar number of net negative charges to ATP in 30% acetic acid, it is not suitable for analysis by AG 1-X8 chromatography. In these cases, the first choice method is the hydrolysis of [γ-^{32}P]ATP, followed by extraction of the phosphate (as a phosphomolybdate complex) in an organic phase (21). Treatment at 100°C for 15 minutes in the presence of 1 N HCl is sufficient to hydrolyse the acid anhydride bonds of ATP, while preserving the ester bonds of phosphoamino acids. The reaction is neutralized by addition of NaOH, and the formation of a phosphomolybdate complex is induced by the presence of ammonium molybdate. The phospho-molybdate complex is soluble in organic solvents such as isobutanol/toluene (1:1) (22), allowing its extraction from the phosphopeptide, which remains in the aqueous phase. The radiolabelled phosphopeptide can be quantified by counting an aliquot of the aqueous phase by scintillation counting.

This assay method is appropriate for SerP, ThrP, and TyrP-containing pep-tides, since very little hydrolysis of any of these phosphoamino acids occurs under the conditions used. The method works with all peptides, no matter how acidic or basic, with the possible exception of markedly amphipathic peptides which may concentrate at the water/organic phase interface. This is, therefore, the first choice method whenever a comparison of phosphorylation of a variety of peptides of markedly different composition is being under-taken. It can be applied directly after the phosphorylation reaction (see *Protocol 3*) or, for suitable peptides, after chromatography on AG 1-X8 resin (see *Protocol 4*) in order to reduce the background, which is the main drawback of this method (see Section 3.3).

Protocol 3.   Kinase assay using acid hydrolysis and phosphomolybdate extraction

Material

- HCl (2 M and 1 M)
- ammonium molybdate [5% (w/v); to 60 ml H$_2$O, add 22.4 ml concentrated H$_2$SO$_4$ (18 M) and 10 g ammonium molybdate (add in small amounts, waiting for solubilization before adding more); adjust volume to 200 ml with water]
- solution A^a (mix 50 ml of 1 M NaOH, 50 ml of 5% ammonium molybdate, 38 ml of 50% TCA, and water to a final volume of 190 ml)

- NaOH (1 M)
- TCA (trichloroacetic acid, 50% w/v)
- water-saturated isobutanol/toluene (1:1) (mix 1 part isobutanol to 1 part toluene and about 0.5 part H$_2$O; wait for complete separation of aqueous and organic phases and take the organic phase)
- incubator at 100°C, or boiling water-bath
- vortex mixer or rotational shaker
- liquid scintillation counter

10: Assay of protein kinases

Method

1. Stop the phosphorylation reaction[b] by adding an equal volume of 2 M HCl, in order to obtain a final concentration of 1 M HCl.

2. Add 1 M HCl to a final volume of 0.5 ml.

3. Boil for 15 min.

4. Stop the hydrolysis by placing the tubes into ice and adding 1.9 ml of solution A.

5. Allow to cool, then add 5 ml of isobutanol/toluene (1:1). Cap the tubes and shake for about 30 sec.[c]

6. Wait for physical separation of the two phases; remove the upper hydrophobic layer and discard as radioactive waste.

7. Repeat steps 5 and 6 twice more.

8. Withdraw 2 ml of the lower aqueous phase and count in at least 8 ml of scintillation liquid, after vigorous shaking.

9. Calculate the total radioactivity incorporated into the phosphopeptide, multiplying the cpm detected in 2 ml of the aqueous phase by the factor 1.2.

[a] Stable for at least two weeks at room temperature.
[b] Incubations should be carried out in 10 ml tubes with caps resistant to boiling.
[c] Mixing of the aqueous and organic phases can be obtained by vortexing, or using a rotational shaker.

Protocol 4. Kinase assay by combining AG 1-X8 chromatography and acid hydrolysis

Materials

- disposable columns (e.g. Pasteur pipettes plugged with glass wool), and a suitable rack to hold them
- AG 1-X8 anion-exchange resin [equilibrated in 30% (v/v) acetic acid]
- acetic acid [30% (v/v)]
- HCl (6 M)
- ammonium molybdate [5% (w/v), prepared as in Protocol 3]

- solution B[a] (mix 15 ml of 12 M NaOH, 50 ml of 5% ammonium molybdate and 30 ml of 50% TCA)
- water-saturated isobutanol/toluene (1:1) (prepared as in Protocol 3)
- incubator at 100°C, or boiling water-bath
- vortex mixer or rotational shaker
- liquid scintillation counter

Method

1. Stop the phosphorylation reactions by adding 0.5 ml of 30% (v/v) acetic acid.[b]

2. Prepare a number of columns corresponding to the number of samples, by pouring about 0.5 ml AG 1-X8 resin into each column. Wash the resin extensively with 30% acetic acid.

Protocol 4. *Continued*

3. Load the samples on to the columns and collect the flow through fraction.

4. Wash three times with 1 ml 30% (v/v) acetic acid, and collect the eluate in the same tubes as the flow through, to a total volume of 3.5 ml.[c]

5. Withdraw 1 ml from the eluate, transfer it into 10 ml polypropylene tubes, and add 0.6 ml 6 N HCl.

6. Boil for 15 min.

7. Stop the hydrolysis by placing the tubes into ice and adding 1.9 ml of solution B.

8. Follow *Protocol 3* from step 5 to step 8.[d]

9. Calculate the total radioactivity incorporated into the phosphopeptide, multiplying the cpm detected in 2 ml of aqueous phase by the factor 6.125.

---

[a] Stable for at least 2 weeks at room temperature.
[b] If the incubation volume is higher than 50 μl, make sure the final concentration of acetic acid is 30% (v/v) by adding an appropriate amount of a more concentrated solution.
[c] The assay can be terminated at this step by directly counting the eluate, as described in Section 4.2.2. If high background values are observed (in the absence of enzyme or peptide), possibly due to contamination with [γ-^{32}P]ATP or the presence of other acid-labile compounds, continue the assay from step 5.
[d] Since only small amounts of acid-labile compounds are usually present after AG 1-X8 chromatography (compared with conditions described in *Protocol 3* where all of the [γ-^{32}P]ATP is hydrolysed), step 7 of *Protocol 3* can often be skipped.

4.2.4 Assay by detection of radiolabelled phosphoamino acids

Detection and quantitation of a peptide phosphorylation reaction can also be performed by the hydrolysis of peptide bonds (6 M HCl, 4 h at 110°C) followed by separation of phosphoamino acids. The technique can be successfully employed for phosphoserine- (SerP) and phosphothreonine- (ThrP) containing peptides, but it is not useful for phosphotyrosine- (TyrP) containing peptides, since the hydrolysis of this phosphoamino acid is too high under the conditions required for peptide hydrolysis. Compared with the other methods for phosphorylation assay, this one has the advantage that it allows separation between SerP and ThrP, thus giving useful information in the case of peptides containing both kinds of amino acids. The separation can be achieved by high voltage paper electrophoresis (23) (*Protocol 5*). At pH 1.9, SerP and ThrP are equally negatively charged and both migrate towards the anode, but migration of SerP is faster than that of ThrP due to lower steric hindrance. [^{32}P]Phosphate, derived from ATP hydrolysis, separates from both phosphoamino acids due to its much higher mobility. Quantitation can be achieved by means of autoradiography of the paper and scintillation counting of the spots corresponding to the radiolabelled phosphoamino acids. In quantitative (or comparative) studies, corrections must be made for hydro-lytic losses of SerP (48%) and ThrP (14%) (24). The general applicability of this procedure was established by assaying the ^{32}P incorporated by different protein kinases into a series of peptides with dissimilar structures, and demonstrating that they underwent comparable loss of inorganic phosphate (25).

Alternatively, two-dimensional separation of phosphoamino acids on thin-layer cellulose plate can be performed (26) (see *Protocol 11* in Chapter 5). This is convenient when a qualitative analysis of TyrP-containing peptides is required, since TyrP is not resolved from ThrP under the conditions described in *Protocol 5*. In order to avoid complete release of free phosphate from phosphotyrosine, only a partial hydrolysis of the peptide is performed in this case (6 M HCl at 110°C for 1 h instead of 4 h).

Protocol 5. Kinase assay by peptide hydrolysis and separation of phosphoamino acids

Materials

- HCl (6 M)
- sealable glass vials
- 110°C oven
- drying device (centrifugal vacuum concentrator, e.g. Savant SpeedVac equipped with an NaOH trap to collect acid, or an air-flow device)
- pH 1.9 buffer [mix 50 ml formic acid (98%), 156 ml glacial acetic acid, and water to 2 litres]

- phosphoamino acid standards (20 mM SerP, 20 mM ThrP)
- ninhydrin solution [0.2% (w/v); dissolve 100 mg ninhydrin in 49 ml acetone and 1 ml glacial acetic acid]
- Whatman paper (3MM Chr)
- hair dryer
- high voltage electrophoresis apparatus
- equipment for autoradiography and scintillation counting

Method

1. Stop the phosphorylation reaction by transferring the mixture into sealable glass vials containing about 3 ml of 6 M HCl.

2. Seal the vials and place them in the 110°C oven for 4 h.

3. Cool the vials on ice, open them, and completely evaporate the liquid (with centrifugal vacuum concentrator or air flow).

4. Wash twice, adding about 1 ml of water and taking to dryness each time.

5. Add 50 μl of buffer at pH 1.9 and vortex thoroughly to allow complete solubilization of the dried hydrolysis products.[a]

6. Add 5 μl of cold phosphoamino acid standard solution to each sample.

7. Prepare a strip of Whatman 3MM paper with the desired width;[b] mark the origin with a pencil line a few cm from the cathode end. Mark the positions to load each sample 1–2 cm apart on the origin.

Protocol 5. *Continued*

8. Load the samples a small amount at a time, allowing to drying between each application.[c]

9. Wet the paper with pH 1.9 buffer. Avoid directly wetting the origin line, but allow the buffer to soak into it from both sides by capillary action. Absorb excess buffer from the paper surface using a blotter.

10. Place the paper strip on the high voltage electrophoresis apparatus, whose buffer tanks have been filled with pH 1.9 buffer. If necessary, connect the strip to the tanks by means of paper bridges, wetted with pH 1.9 buffer. Apply a voltage (150 V/cm of paper width[d]) for 150 min, using a suitable cooling system.

11. Stop the electrophoresis, dry the strip, and spray it with 0.2% (w/v) ninhydrin solution. Place the paper into the oven for few min. Purple spots will appear where the phosphoamino acid standards are present.

12. Detect the radiolabelled phosphoamino acids by autoradiography. Cut out the radioactive spots and count in a liquid scintillant.[e] Apply a correction for hydrolytic losses of phosphoamino acids (see Section 4.2.4).

---

[a] It is often necessary to repeat this step after the first loading of the sample on the paper (step 8), in order to totally remove radioactivity from the vial.

[b] Note that using a paper strip which is too wide can induce overheating. At the usual 3000 V used for electrophoresis, the maximal width is about 15 cm, but it can vary according to the cooling device used.

[c] This step can be speeded up by using a hair dryer.

[d] 3000 V is usually suitable. If resolution of SerP and ThrP is not necessary, the electrophoretic time can be reduced to 1 h, since this will allow adequate resolution from [^{32}P]phosphate.

[e] Alternatively, use a phosphorimaging system (e.g. InstantImager, Packard).

4.2.5 DEAE-cellulose paper assay

This technique (19) (*Protocol 6*) is complementary to the phosphocellulose paper assay, being suitable for assays with acidic rather than basic peptides. It exploits the binding of the peptide to DEAE- (diethylaminoethyl-) cellulose anion-exchange paper, under conditions of pH and ionic strength which do not allow binding of [γ-^{32}P]ATP. Phosphopeptides with at least six acidic amino acids are detectable by this method, but the binding is quantitative only for peptides containing eight or more acidic residues. In this latter case the results are comparable with those obtained by total acid hydrolysis (Section 4.2.3 and *Table 3*), but provide a much faster method of analysis.

Protocol 6. Protein kinase assay using DEAE-cellulose anion-exchange paper

Materials

- 0.5 M EDTA
- DEAE-cellulose paper (NA 45, Schleicher & Schuell, NH, cut into 2 cm × 2 cm squares)
- washing buffer (100 mM NaCl, 50 mM Tris–HCl, pH 7.4)
- 500 ml beaker, with a smaller plastic beaker insert (see *Protocol 2*)
- hair dryer
- liquid scintillation counter

Methods

1. Stop the phosphorylation reaction by addition of 0.5 M EDTA to obtain a final concentration of 0.14 M.

2. Spot the reaction mixture (or an aliquot of it) onto the DEAE-cellulose paper squares.

3. When the liquid is adsorbed, drop the papers into the 500 ml beaker containing about 200 ml of washing buffer. Place the beaker on to a magnetic stirrer for about 5 min; discard the liquid as radioactive waste and replace it with 200 ml of fresh washing buffer.

4. Repeat washing three times with washing buffer (200 ml per wash for at least 5 min).

5. Wash for a few seconds with acetone and dry the papers with a hair dryer.[a]

6. Place the paper squares into scintillation vials and count them in scintillant.

[a] This step is not essential, but the paper squares are easier to handle when dry.

4.2.6 Alternative methods for protein kinase assay

A number of other techniques have been employed, with different advantages and disadvantages according to the type of substrate and the phosphorylation conditions used:

- SDS–PAGE (27);
- isoelectric focusing (28);
- PEI–cellulose column chromatography (29);
- binding to ferric adsorbent paper of tritiated phosphopeptide (30) (the method is radioactive, employing ^{3}H, but avoids the use of the higher energy ^{32}P radioisotope);
- assay on streptavidin-linked disks, which can capture previously biotinyl-ated peptides (31) (this method has the advantage of a low background in

crude extracts, since the binding of other phosphorylated, but not bio-tinylated, proteins does not occur);
- a thin-layer chromatography technique suitable for both acidic and basic peptides (32);
- SDS–PAGE of peptides which have been linked to amino acid polymers (33).

Moreover, methods are being developed which avoid the use of radio-isotopes, exploiting different properties such as the fluorescence of peptides which have been labelled with fluorescamine (34, 35), or separation of the phospho- and dephospho-peptides by capillary zone electrophoresis (36).

5. Assay of protein phosphatases using phosphorylated peptide substrates

Compared with protein kinases (especially PSKs), protein phosphatases do not appear to display a marked specificity for the primary sequence around their phosphoamino acid substrates (37). It is possible, however, to obtain a number of phosphorylated peptides which, irrespective of their relatedness to natural phosphorylation sites, are nevertheless readily and efficiently dephos-phorylated by protein phosphatases, either serine/threonine or tyrosine specific. These peptides provide a valuable tool for monitoring these enzymes and are frequently exploited, especially in the case of protein–tyrosine phos-phatases. A number of such peptides are listed in *Table 4*, with indications about the phosphatases that can be assayed with each of them. It should be noted once again that protein phosphatases are in general much more promiscuous than protein kinases with respect to their peptide substrates.

Table 4. Phosphopeptides used for the assay of protein phosphatases[a]

Phosphopeptide	Phosphorylating kinase[b]	Phosphatase(s) that can be assayed[c]
RRATpVA	PKA	PP2A and PP2C
RRREEETpEEE	CK2	PP2A
DLDVPIPGRFDRRVSpVAAE	PKA	PP2B > PP2A >PP2C
INGSpPRTpPRRGQNR	Cyclin-dep.PK	PP2A (trimeric)
SpEEEEE	CK2	Acid/alkaline phosphatases
KKKKKRFSpFKKSpFKLSSFSpFKKNKK	PKC	PP1, PP2A, PP2B, PP2C
EDNEYpTA	Syk	PTPases
NIDGEVNYpEE	c-Fgr	PTPases
AFLEDFFTSTEPQYpQPGENL	Csk	PTPases
TAEPDYpGALYE		H PTPases β, LAR, CD45

[a] Data from ref. 37.
[b] Abbreviations: CK2, protein kinase CK2 or casein kinase 2; Csk, C-terminal Src protein kinase.
[c] Abbreviations: PP, protein phosphatase; PTPase, protein tyrosine phosphatase.

A complication in the case of phosphopeptides as substrates for protein phosphatases is the fact that they must be phosphorylated prior to use. This is normally carried out using $[\gamma\text{-}^{32}P]$ATP as phosphate donor and a suitable protein kinase as the catalyst. The latter will be chosen for its ability to phosphorylate the desired peptide, which may give rise to some problems. It should be noted in this respect that the best peptide substrates for protein phosphatases-2A and -2C contain phosphothreonine rather than phosphoserine. Since threonine is phosphorylated much less readily than serine by most kinases, the preparation of large amounts of phosphothreonyl peptides is sometimes troublesome. On the other hand, it should be remembered that it may not be necessary to achieve exhaustive phosphorylation of the peptide, nor to free the phosphopeptide of its non-phosphorylated form. It has been shown that even the presence of a large excess of the dephosphorylated peptide does not appreciably affect the kinetic parameters of most protein phosphatases, either serine/threonine or tyrosine specific. It is therefore common to use the mixture of the two forms without separation, ignoring the presence of the dephosphorylated peptide, and assuming that the phospho-peptide concentration corresponds to that of $[^{32}P]$phosphate incorporated into it.

5.1 Use of radiolabelled peptides for protein phosphatase assays

5.1.1 Preparation

$[^{32}P]$Phosphopeptides to be used as protein phosphatase substrates can be obtained by incubating the peptides with the suitable protein kinases, as described in Section 4.1. When very pure kinases are used, the incubation time can be prolonged up to some hours, in order to obtain as high a phosphorylation degree as possible. In some cases, the addition of a protease inhibitor cocktail to the reaction mixture can help in preventing peptide degradation during very long incubations. Particular care should be taken in the planning of the phosphopeptide preparation, avoiding the presence of any possible inhibitors of the protein phosphatase, since the phosphopeptide purification method might not allow their elimination.

Phosphorylation reactions are stopped by addition of 30% acetic acid, and separation of phosphopeptides from $[\gamma\text{-}^{32}P]$ATP is achieved, whenever applicable, by using the AG 1-X8 chromatography technique. The specific radioactivity of $[\gamma\text{-}^{32}P]$ATP needs to be quite high (2000–5000 cpm/pmol) and must be exactly known, since the concentration of the phosphopeptide will be assumed to be equal to that of the $[^{32}P]$phosphate in the peptide. After AG 1-X8 anion-exchange purification, the phosphopeptide is lyophilized and resuspended in the desired amount of water, adjusting the pH if necessary.

Protocol 7. Radioactive phosphopeptide preparation for protein phosphatase assay

Equipment and materials
- disposable columns (e.g. Pasteur pipettes plugged with glass wool, and a suitable rack to hold them)
- AG 1-X8 anion-exchange resin [equilibrated in 30% (v/v) acetic acid]
- 30% (v/v) acetic acid

Method

1. Carry out the phosphorylation reaction as described under *Protocol 1*, in a total volume of 200 μl.[a]

2. Stop the reaction by addition of 85 μl of 100% (v/v) acetic acid plus 215 μl 30% (v/v) acetic acid, to reach a total volume of 0.5 ml, and a final concentration of 30% (v/v) acetic acid.

3. Prepare a number of columns, corresponding to the number of samples, by pouring about 0.5 ml of AG 1-X8 resin into each column. Wash the resin extensively with 30% (v/v) acetic acid.

4. Load the samples on to the columns and collect the flow through fraction.

5. Wash three times with 1 ml of 30% (v/v) acetic acid and collect the eluate in the same tubes as the flow through, to a total volume of 3.5 ml.

6. Count a 20 μl aliquot out of the 3.5 ml by scintillation counting.[b]

7. Lyophilize the phosphopeptides eluted from the AG 1-X8 columns.[c]

8. Resuspend the lyophilized phosphopeptide in water, or buffer at the desired pH.

9. Check the concentration of the [^{32}P]phosphopeptide by counting 2 μl in a scintillation counter.[d]

[a] The incubation volume should be chosen dependending of the amount of phosphopeptide required. The amount of 100% and 30% acetic acid to be added at step 2 will change as a consequence.
[b] The result will give an estimate of the degree of phosphorylation, and will indicate the appropriate volume in which the phosphopeptide needs to be resuspended after lyophilization in order to obtain the desired concentration.
[c] Add an equal volume of water to the sample in 30% acetic acid, to allow it to freeze at –80 °C.
[d] Divide the cpm in the phosphopeptide by the specific radioactivity of the ATP (cpm/pmol) to obtain the pmol of phosphopeptide contained in 2 μl, and hence calculate the phosphopeptide concentration.

5.1.2 Dephosphorylation reaction and assay

Phosphopeptide dephosphorylation reactions are performed under the optimal conditions for the particular protein phosphatase to be tested. Phosphopeptides are usually used at concentrations in the micromolar and submicromolar range for serine/threonine phosphatases and tyrosine phosphatase, respectively. Incubations (20–50 μl) are stopped after the appropriate time (5–30 min) by addition of trichloroacetic acid. The degree of dephosphorylation is assessed by quantification of the amount of [^{32}P]phosphate released, after its conversion into a phosphomolybdate complex. The complex can be extracted into an organic phase that is counted for ^{32}P radioactivity, while the radio-labelled phosphopeptide remains in the aqueous phase.

Protocol 8. Peptide phosphatase assay

Materials

- 5% (w/v) ammonium molybdate (prepared as in *Protocol 3*)
- TCA [trichloroacetic acid, 10% (w/v)]
- vortex mixer or rotational shaker
- water-saturated isobutanol/toluene (1:1) (prepared as in *Protocol 3*)
- liquid scintillation counter

Method

1. Stop the dephosphorylation reaction by adding 1.5 ml of 10% TCA plus 0.5 ml of 5% ammonium molybdate.

2. Add 2.5 ml of isobutanol/toluene (1:1) solution. Cap the tubes and shake for about 30 sec.[a]

3. Wait for the physical separation of the two phases; then withdraw 2 ml from the upper organic phase and count the radioactivity in 2 ml of scintillant.

4. Calculate the total amount of phosphate released by multiplying the cpm detected in the 2 ml by the factor 1.25.

[a] Mixing of the aqueous and organic phases can be obtained by vortexing, or on a rotational shaker.

5.2 Alternative methods for assaying protein phosphatases with phosphopeptides

Phosphopeptides can also be made by chemical synthesis (38, 39), the main advantage being the very high yield (approaching 100% phosphorylation) and the possibility to overcome the severe limitations imposed by the specificity and efficiency of the available protein kinases. The use of synthetic phospho-peptides may allow the assay of protein phosphatase activity without employing ^{32}P, whose short half-life discourages the expensive synthesis of radioactive phosphopeptides that can be used for only a limited period of time. This requires the development of reliable and sensitive techniques for the non-radioactive assay of phosphate release. Many attempts to achieve this have been described, exploiting different detection methods. It is worth mention the following possibilities:

- reaction of phosphate with the reagent malachite green (40, 41), a colorimetric assay which allows the detection of nanomole amounts of phosphate;

- changes in ultraviolet absorption and fluorescence resulting from the dephosphorylation of a phosphotyrosyl peptide, which can be followed continuously (42);

- changes in fluorescence intensity in response to dephosphorylation of a properly designed phosphoseryl peptide, with a tryptophan residue adjacent to the phosphorylated site (43);

- decrease in fluorescent signal from an anti-phosphotyrosine antibody, due to the release of phosphate from the tyrosyl-phosphorylated substrate (44);

- the effect of unlabelled phosphopeptides as competitive inhibitors for the dephosphorylation of a standard radiolabelled substrate (39).

Acknowledgements

The authors are indebted to Giuseppe Tasinato for his invaluable help in designing and performing technical devices to improve the assay of peptide phosphorylation, and to our colleagues Arianna Donella Deana, Flavio Meggio, and Paolo Ruzza for providing unpublished data. The work done in the laboratory of the authors was supported by grants from the European Commission (BioMed-2), Harmenise Harvard Foundation, C.N.R. (Target Project on Biotechnology), Associazione Italiana Ricerca sul Cancro, M.U.R.S.T., and Italian Ministry of Health (Project AIDS).

References

1. Pinna, L.A., Ruzzene, M. (1996). *Biochim. Biophys. Acta*, **1314**, 19.
2. Ruzzene, M., Songyang, Z., Marin, O., Donella-Deana, A., Brunati, A.M., Guerra, B., Agostinis, P., Cantley, L.C., and Pinna, L.A. (1997). *Eur. J. Biochem.*, **246**, 433.
3. Gibbs, C.S. and Zoller, M.J. (1991). *Biochemistry*, **30**, 5329.
4. Gibbs, C.S. and Zoller, M.J. (1991). *J. Biol. Chem.*, **266**, 8923.
5. Sarno, S., Vaglio, P., Issinger, O.G., and Pinna, L.A. (1996). *J. Biol. Chem.*, **271**, 10595.
6. Sarno, S., Vaglio, P., Marin, O., Issinger, O.-G., Ruffato, K., and Pinna, L.A. (1997). *Biochemistry*, **36**, 11717.
7. Taylor, S.S., Radio-Andzelm, E., and Hunter, T. (1995). *FASEB J.*, **9**, 1255.
8. Hubbard, S.R. (1997). *EMBO J.*, **16**, 5572.
9. Leighton, I.A., Dalby, L.N., Caudwell, F.B., Cohen, P.T.W., and Cohen, P. (1995). *FEBS Lett.*, **375**, 289.

10. Songyang, Z., Blechner, S., Hoagland, N., Hoeckstra, M.F., Piwnica-Worms, H., and Cantley, L.C. (1994).*Curr. Biol.*, **4**, 973.
11. Songyang, Z., Carraway, K.L., III, Eck, M.J., Harrison, S.C., Feldman, R.A., Mohammadi, M., Schessinger, J., Hubbard, S.R., Smith, D.P., Eng, C., Lorenzo, M.J., Ponder, B.A.J., Mayer, B.J., and Cantley, L.C. (1995). *Nature*, **373**, 536.
12. Brunati, A.M., Donella-Deana, A., Ruzzene, M., Marin, O., and Pinna, L.A. (1995). *FEBS Lett.*, **367**, 149.
13. Lee, T.R., Niu, J., and Lawrence, D.S. (1995). *J. Biol. Chem.*, **270**, 27022.
14. Yan, X., Corbin, J.D., Francis, S.H., and Lawrence, D.S. (1996). *J. Biol. Chem.*, **271**, 1845.
15. Glass, D.B., Masaracchia, R.A., Feramisco, J.R., and Kemp, B.E. (1978). *Anal. Biochem.*, **87**, 566.
16. Marin, O., Meggio, F., and Pinna, L.A. (1994). *Biochem. Biophys. Res. Commun.*, **198**, 898.
17. Lasa-Benito, M., Marin, O., Meggio, F., and Pinna, L.A. (1996). *FEBS Lett.*, **382**, 149.
18. Marin, O., Meggio, F., Sarno, S., Andretta, M., and Pinna, L.A. (1994). *Eur. J. Biochem.*, **223**, 647.
19. Wilson, L.K., Dhillon, N., Thorner, J., and Martin, G.S. (1997). *J. Biol. Chem.*, **272**, 12961.
20. Kemp, B.E., Bylund, D.B., Huang, T.S., and Krebs, E.G. (1975). *Proc. Natl. Acad. Sci. USA*, **72**, 3448.
21. Meggio, F., Donella, A., and Pinna, L.A. (1976). *Anal. Biochem.*, **71**, 583.
22. Martin, J.B., and Doty, D.M. (1949). Anal. Chem., 21, 965.
23. Donella Deana, A., Meggio, F., and Pinna, L.A. (1979). *Biochem. J.*, **179**, 693.
24. Bylund, D.B., and Huang, T.S. (1976). *Anal. Biochem.*, **73**, 477.
25. Perich, J.W., Meggio, F., Reynolds, E.C., Marin, O., and Pinna, L.A. (1992). *Biochemistry*, **31**, 5893.
26. Boyle, W.J., Van der Geer, P., and Hunter, T. (1991). In *Methods in enzymology* (ed. T. Hunter and B.M. Sefton), Vol. 201, p. 111. Academic Press, London.
27. Hunter, T., Ling, N., and Cooper, J.A. (1984). *Nature*, **311**, 480.
28. Fiol, C.J., Wang, A., Roeske, R.W., and Roach, P.J. (1990). *J. Biol. Chem.*, **265**, 6061.
29. Budde, R.J.A., McMurray, J.S., and Tinker, D.A. (1992). *Anal. Biochem.*, **200**, 347.
30. Toomik, R., Ekman, P., Eller, M., Järv, J., Zaitsev, D., Myasoedov, N., Ragnarsson, U., and Engström, L. (1993). *Anal. Biochem.*, **209**, 348.
31. Goueli, B.S., Hsiao, K., Tereba, A., and Goueli, S.A. (1995). *Anal. Biochem.*, **225**, 10.
32. Lou, q., Wu, J., and Lam, K.S. (1996). *Anal. Biochem.*, **235**, 107.
33. Kameshita, I., and Fujisawa, H. (1996). *Anal. Biochem.*, **237**, 198.
34. Lutz, M.P., Pinon, D.I., and Miller, L.J. (1994). *Anal. Biochem.*, **220**, 268.
35. Seethala, R., and Menzel, R. (1997). *Anal. Biochem.*, **253**, 210.
36. Dawson, J.F., Boland, M.P., and Holmes, C.F.B. (1994). *Anal. Biochem.*, **220**, 340.
37. Pinna, L.A., Donella Deana, A. (1994). *Biochim. Biophys. Acta*, **1222**, 415.
38. Kitas, E.A., Knorr, R., Trzeciak, A., and bannwarth, W. (1991). *Helv. Chim. Acta*, **74**, 1315.
39. Perich, J.W., Ruzzene, M., Pinna, L.A., and Reynolds, E.C. (1994). *Int. J. Peptide Protein Res.*, **43**, 39.

40. Lanzetta, P.A., Alvarez, L.J., Reinach, P.S., and Candia, O.A. (1979). *Anal. Biochem.*, **100**, 95.
41. Martin, B., Pallen, C.J., Wang, J.H., and Graves, D.J. (1985). *J. Biol. Chem.*, **260**, 14932.
42. Zhang, Z.-Y., Maclean, D., Thieme-Sefler, A., Roeske, R.W., and Dixon, J.E. (1993). *Anal. Biochem.*, **211**, 7.
43. Wright, D.J., Noiman, E.S., Chock, P.B., and Chau, V. (1981). *Proc. Natl. Acad. Sci. USA*, **78**, 6048.
44. Babcook, J., Watts, J., Aebersold, R., and Ziltener, H.J. (1991). *Anal. Biochem.*, **196**, 245.
45. Kemp, B.E., Graves, D.J., Benjamini, E., and Krebs, E.G. (1977). *J. Biol. Chem.*, **252**, 4888.
46. Feramisco, J.R., Glass, D.B., and Krebs, E.G. (1980). *J. Biol. Chem.*, **255**, 4240.
47. Mitchell, R.D., Glass, D.B., Wong, C-W, Angelos, K.L., and Walsh, D.A. (1995). *Biochemistry*, **34**, 528.
48. Leader, D.P., Donella Deana, A. Marchiori, F., Purves, F.C., and Pinna, L.A. (1991). *Biochim. Biophys. Acta*, **1091**, 426.
49. Ferrari, S., Marchiori, F., Borin, G., and Pinna, L.A. (1985). *FEBS Lett.*, **184**, 72.
50. Proud, C.G., Colthrust, D.R., Ferrari, S., and Pinna, L.A. (1991). *Eur. J. Biochem.*, **195**, 771.
51. Yasuda, I., Kishimoto, A., Tanaka, S., Tominaga, N., Sakurai, A., and Nishizuka, Y. (1990). *Biochem. Biophys. Res. Commun.*, **166**, 1220.
52. Toomik, R., and Ek, P. (1997). *Biochem. J.*, **322**, 455.
53. Kielbassa, K., Müller, H-J., Meyer, H.E., Marks, F. and Gschwendt, M. (1995). *J. Biol. Chem.*, **270**, 6156.
54. Nishikawa, K., Toker, A., Johannes, F.J., Songyang, Z., and Cantley, L.C. (1997). *J. Biol. Chem.*, **272**, 952.
55. Alessi, D.R., Candwell, F.B., Andjelkovic, M., Hemmings, B.A., and Cohen, P. (1996). *FEBS Lett.*, **399**, 333.
56. Dale, S., Wilson, W.A., Edelman, A.M., and Hardie, D.G. (1995). *FEBS Lett.*, **361**, 191.
57. Mitchell, B.J., Stapleton, D., Mitchelhill, K.I., House, C.M., Katsis, F., Witters, L.A., and Kemp, B.E. (1996). *J. Biol. Chem.*, **271**, 28445.
58. Lee, J.C., Kwon, Y-G., Lawrence, D.S., and Edelman, A.M. (1994). *Proc. Natl. Acad. Sci. USA*, 91, 6413.
59. White, R.R., Kwon, Y.G., Taing, M., Lawrence, D.S., and Edelman, A.M. (1998). *J. Biol. Chem.*, **273**, 3166.
60. Pearson, R.B., Misconi, L.Y., and Kemp, B.E. (1986). *J. Biol. Chem.*, **261**, 25.
61. Heierhost, J., Tang, X., Lei, J., Probst, W.C., Weiss, K.R., Kemp, B.E., and Benian, G.M. (1996). *Eur. J. Biochem.*, **242**, 454.
62. Miyano, O., Kameshita, I., and Fujisawa, H. (1992). *J. Biol. Chem.*, **267**, 1198.
63. Stokoe, D., Caudwell, B., Cohen, P.T.W., and Cohen, P. (1993). *Biochem. J.*, **296**, 843.
64. Pearson, R.B., Mitchelhill, K.I., and Kemp, B.E. (1993). In *Protein phosphorylation: a practical approach* (ed. D.G. Hardie), p. 265. Oxford University Press, New York.
65. Zhang, J., Sanchez, R.J., Wang, S., Guarnaccia, C., Tossi, A., Zahariev, S., and Pongor, S. (1994). *Arch. Biochem. Biophys.*, **315**, 415.

66. Marin, O., Meggio, F., Daretta, G., and Pinna, L.A. (1992). *FEBS Lett.*, **301**, 111.
67. Srinivasan, J., Koszelak, M., Mendelow, M., Kwon, Y-G., and Lawrence, D.S. (1995). *Biochem. J.*, **309**, 927.
68. Garriga, J., Segura, E., Mayol, X., Grubmeyer, C., and Grana, X. (1996). *Biochem. J.*, **320**, 983.
69. Kameshita, I., Ishida, A., and Fujisawa, H. (1997). *J. Biochem.*, **122**, 168.
70. Gonzales, F.A., Raden, D.L., and Davis, R.J. (1991). *J. Biol. Chem.*, **266**, 22159.
71. Joseph, C.K., Buyn, H-S., Bittman, R., and Kolesnick, R.N. (1993). *J. Biol. Chem.*, **268**, 20002.
72. Okuno, S., Kitani, T., and Fujisawa, H. (1997). *J. Biochem.*, **122**, 337.
73. Mullinax, T.R., Stepp, L.R., Brown, J.R., and Reed, L.J. (1985). *Arch. Biochem. Biophys.*, **243**, 655.
74. Chan, K-F. J., Hurst, M.O., and Graves, D.J. (1982). *J. Biol. Chem.*, **257**, 3655.
75. Onorato, J.J., Palczewski, K., Regan, J.W., Caron, M.G., Lefkowitz, R.J., and Benovic, J.L. (1991). *Biochemistry*, **30**, 5118.
76. Purves, F.C., Donella-Deana, A., Marchiori, F., Leader, D.P., and Pinna, L.A. (1986). *Biochim. Biophys. Acta*, **889**, 208.
77. Wang, Q.M., Roach, P.J., and Fiol, C.J. (1994). *Anal. Biochem.*, **220**, 397.
78. Flotow, H., Graves, P.R., Wang, A, Fiol, C.J., Roeske, R.W., and Roach, P.J. (1990). *J. Biol. Chem.*, **265**, 14264.
79. Meggio, F., Perich, J.W., Reynolds, E.C., and Pinna, L.A. (1991). *FEBS Lett.*, **283**, 303.
80. Lam, K.S., Wu, J., and Lou, Q. (1995). *Int. J. Peptide Protein Res.*, **45**, 587.
81. Cheng, H.C., Nishio, H., Hatase, O., Ralph, S., and Wang, J.H. (1992). *J. Biol. Chem.*, **267**, 9248.
82. Marin, O., Donella-Deana, A., Brunati, A.M., Fischer, S., and Pinna, L.A. (1991). *J. Biol. Chem.*, **266**, 17798.
83. Edison, A.M., Barker, S.C., Kassel, D.B., Luther, M.A., and Knight, W.B. (1995). *J. Biol. Chem.*, **270**, 27112.
84. Ruzza, P., Calderan, A., Filippi, B., Biondi, B., Donella-Deana, A., Cesaro, L., Pinna, L.A., and Borin, G. (1995). *Int. J. Peptide Protein Res.*, **45**, 529.
85. Garcia, P., Shoelson, S.E., George, S.T., Hinds, D.A., Goldberg, A.R., and Miller, W.T. (1993). *J. Biol. Chem.*, **268**, 25146.
86. Bin, X., Bird, V.G., and Miller, W.T. (1995). *J. Biol. Chem.*, **270**, 29825.

11

Cloning and expression of cDNAs encoding protein kinase subunits

DAVID CARLING

1. Introduction

Protein kinases form a large family of enzymes that play a key role in the control of a diverse variety of cellular processes, including the regulation of metabolic pathways, cell differentiation, and cell proliferation. Changes in the normal activity of protein kinases can lead to the development of a large number of different disease states. A critical step in our understanding of the precise regulatory role of protein phosphorylation within the cell, and the lesions that occur within this regulation, is the identification and characterization of all the protein kinases present in the organism. Whilst this may seem an ambitious goal at present, the rapid progress being made in several key areas of biology suggests that it will be achievable in the not too distant future. The completion of the *Saccharomyces cerevisiae* genome sequencing project has revealed that budding yeast contain 113 protein kinase genes, over half of which have known or predicted functions (1). Sequencing of the human genome is well underway, and it has been estimated that in humans there are at least 1000 genes encoding protein kinases (2), although in this case the percentage for which there is a known function is much smaller. Advances in gene cloning and manipulation have made analysis of mammalian genes, both *in vitro* and *in vivo*, a routine procedure. In order to elucidate the exact function of a specific protein kinase, isolation of the cDNAs encoding the protein kinase subunit(s) is an important and fundamental goal for the researcher. This chapter discusses the methods available for isolation of cDNAs encoding protein kinase subunits, including the new approach of 'in silico' cloning which has been made possible by the large-scale DNA sequencing projects currently being carried out around the world. Having isolated the cDNA, methods for expression of the protein kinase in heterologous systems are also described.

2. Applications

The availability of cDNAs encoding specific protein kinase subunits provides the investigator with an invaluable reagent for studying the role of protein phosphorylation in a particular system. Inspection of the deduced amino acid sequence for known protein motifs may reveal clues as to the function and/or regulation of the protein kinase. Identification of closely related protein kinases from different species may enable further insight into the function of the enzyme, or allow different experimental approaches to be considered. For instance, identification of a closely related protein kinase in yeast would open up the possibility of using reverse genetics to study protein function. The determination of mRNA levels may reveal differences in spatial and/or temporal expression that may indicate different roles for the protein kinase and point the investigator in a particular direction. Knowledge of the deduced amino acid sequence will allow specific antibodies to be generated against synthetic peptides. These antibodies can then be used to determine protein expression by Western blotting, whilst immunoprecipitation of the protein kinase subunit may reveal interactions with other proteins. Overexpression of the protein kinase subunit, or of a dominant negative mutant, in a hetero-logous system may be useful for studying downstream function (see Chapter 4). Finally, the rapid development of techniques and associated vectors for gene targetting means that it is now possible to consider a gene targetting approach as a realistic method for studying the role of a specific protein kinase *in vivo*.

3. Classification of protein kinases

Broadly speaking, protein kinases fall into two categories: protein–serine/threonine kinases and protein–tyrosine kinases. All known eukaryotic protein kinases contain a highly conserved catalytic domain of about 250–300 amino acids (3, 4). Multiple amino acid sequence alignments of these domains has revealed the presence of 12 highly conserved regions, or subdomains, that contain essentially invariant residues (*Figure 1*). Outside the catalytic domain, protein kinases exhibit considerable variability and may contain additional regulatory domains, additional subunits, or both. In addition to the division between serine/threonine kinases and tyrosine kinases, protein kinases can be classified into smaller subfamilies, the major criterion used in this scheme being the similarity in primary amino acid sequence of the catalytic domain. *Table 1* lists some of the major groupings using this classification system based on the scheme of Hanks *et al.* (4, 5). In addition to their sequence similarity, protein kinases within groups tend to be similar in overall struc-tural topology, have similar modes of regulation, and have similar substrate specificities (6).

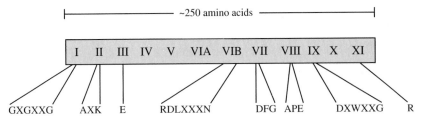

Figure 1. Schematic representation of the catalytic domain of eukaryotic protein kinases. Subdomains identified from multiple amino acid sequence alignments of protein kinases are indicated by roman numerals. Amino acid residues or motifs that are highly conserved or almost invariant throughout the entire family are shown (single-letter code, where X indicates any amino acid at non-conserved positions).

Table 1. Some of the major protein kinase classes grouped according to the amino acid sequence of their catalytic domains[a]

A. AGC Group
   (i) Cyclic nucleotide-regulated protein kinases, e.g. cyclic AMP-dependent protein kinase catalytic subunit (PKA)
   (ii) Diacylglycerol-regulated protein kinases, e.g. protein kinase C (PKC)
   (iii) Other AGC related protein kinases, e.g. ribosomal S6 kinase (p70 S6 kinase), β-adrenergic receptor kinase, type 1 (β-ARK1)

B. CaMK Group
   (i) Calcium/calmodulin-regulated protein kinases, e.g. Ca^{2+}/calmodulin-dependent protein kinase II (CaMKII), phosphorylase kinase catalytic subunit (PhK-γ)
   (ii) Snf1 subfamily, e.g. Snf1, AMP-activated protein kinase (AMPK)
   (iii) Other CaMK related protein kinases, e.g. mitogen-activated protein kinase-activated kinase 2 (MAPKAPK2)

C. CMGC Group
   (i) Cyclin-dependent protein kinases, e.g. *Schizosaccharomyces pombe* cell-division-cycle gene product (Cdc2)
   (ii) Extracellular signal-regulated/mitogen-activated protein kinases, e.g. ERK1
   (iii) Glycogen synthase kinase 3 subfamily, e.g. glycogen synthase kinase 3
   (iv) Casein kinase II subfamily, e.g. casein kinase II (CK2)
   (v) Other CMGC-related protein kinases, e.g. *S. cerevisiae* protein kinase, Yak1

D. PTK Group
   (i) Non-membrane spanning protein-tyrosine kinases, e.g. human homologue of retroviral oncoprotein Src (c-Src)
   (ii) Membrane-spanning protein–tyrosine kinases, e.g. epidermal growth factor receptor

E. OPK Group
   This final group includes protein kinases which do not fall into the other major groups, e.g. the MEK/STE7, Raf subfamily, casein kinase I subfamilies

[a] I have not attempted to give a complete list of all eukaryotic protein kinases in the above classification scheme. Instead, protein kinases are organized into major groups that share basic structural and functional properties, using alignments generated by Hanks *et al.* (3, 4).

4. Isolation of cDNAs encoding protein kinase subunits

4.1 The traditional approach

The methods available for isolating cDNAs encoding protein kinase subunits are the same as those for any other cDNA. Previously, the most popular route involved library screening using degenerate oligonucleotide probes, based on amino acid sequence information derived from the purified subunit poly-peptides. With the advent of the polymerase chain reaction (PCR) it became possible to amplify a region of the cDNA, again using degenerate oligo-nucleotides, and use this as a probe for library screening. The primary struc-ture of the catalytic domain of the protein kinase family allowed a number of cloning strategies to be developed for isolation of their cDNAs. By designing degenerate oligonucleotides, coding for stretches of highly conserved amino acid sequences within the subdomains, it was possible to use these either as probes for library screening or as primers for PCR (5). This homology-based cloning approach has been used successfully to identify novel protein kinases from a number of species (e.g. refs. 7–10). In addition to these frequently used methods, a number of more specialized approaches have been utilized for isolating protein kinase subunit cDNAs, e.g. expression cloning and library screening using the two-hybrid system.

4.2 The 'modern' approach

4.2.1 Expressed sequence tags (ESTs)

A new and exciting approach for identifying and isolating cDNAs encoding protein kinase subunits has been made possible by the rapid advances being made by the large-scale DNA sequencing projects, particularly in the expressed sequence tag (EST) project (11). The aim of this project (carried out at the University of Washington, St. Louis, and sponsored by Merck and Co.) is to produce short (300–600 bp) 5′ and 3′ terminal sequence reads of cDNA clones from normalized cDNA libraries. Since the clones are from cDNA libraries it is assumed that they represent expressed genes. The clones are arrayed in multiwell plates and are available to researchers (royalty free) from the Integrated Molecular Analysis of Genomes and their Expression (I.M.A.G.E.) Consortium. At the time of writing (July 1998) just over 1 million sequences from human cDNA libraries had been entered into the database (dbEST), and it is estimated that this represents at least 60 000 of the 80 000–100 000 genes that comprise the human genome (i.e. 60–75%). It is therefore quite likely that the cDNA encoding an investigator's favourite protein kinase subunit has already been cloned and partially sequenced! Furthermore, the large number of EST sequences mean that in many cases there will be a number of overlapping sequences for the same gene, and so it may be possible to deduce the entire sequence of a gene simply from the database. In addition to the human EST project, a number of related projects are characterizing ESTs from other species, such as mouse, rat, and the higher plant *Arabidopsis thaliana.*

4.2.2 *In silico* cloning

Public access to the sequences within the EST database, combined with the ready availability of the cDNA clones from which the sequences are derived, makes it possible to use computational searching methods to isolate specific cDNAs. The application of database searching to identify and isolate cDNA sequences has been dubbed '*in silico* cloning', although perhaps an equally apt term would be cloning by phoning! This approach bypasses much of the traditional route of library screening and gene amplification, promising to dramatically short cut the process of cloning. It should be possible to adapt *in silico* cloning for virtually any cDNA cloning exercise. By searching the database with the sequence of a known protein kinase subunit, cDNAs encoding closely related subunits from the same species (paralogues) can be identified. Similarly, by searching with a sequence from a different species the equivalent (orthologous) gene can be identified. Perhaps the most labour-saving application of *in silico* cloning, however, will be in the isolation of genes coding for previously unidentified proteins, including protein kinase subunits.

In order to isolate cDNAs encoding novel protein kinase subunits, it remains necessary to obtain some information regarding the primary structure of the protein kinase subunits of interest. Until recently, this meant obtaining amino acid sequence information by Edman degradation and using this information to search the EST database. During the last few years, however, improvements in the techniques for analysing peptides by mass spectroscopy have enabled alternative approaches to be employed for obtaining primary structural information. These methods are much more sensitive than sequencing using Edman degradation and can yield sufficient information with which to search the databases from femtomole (10^{-15}) quantities of protein extracted from SDS–gels (12) (see Chapter 6). By combining the accurate masses of proteolytically derived peptides with short stretches of primary amino acid sequence, it is possible to identify a protein unambiguously from a search of the databases (13, and references therein). Although the instruments and methodology required for this type of analysis are specialized and are not widely available at present, it seems likely that these techniques will play a key role in protein identification in the near future.

For the remainder of this chapter it is assumed that some primary amino acid sequence information is available. Where novel protein kinase subunits are being pursued this will mean that the subunits have been purified to a stage sufficient for the polypeptides to be identified, e.g. by SDS–PAGE, allowing amino acid sequence information (or accurate peptide masses) to be obtained. This procedure will almost certainly involve a significant amount of time and effort, but unfortunately there is no way around this aspect of the work at present!

4.2.3 Searching the EST database

The most convenient way to search the EST database for most investigators is to use the programmes available on the world-wide web (WWW), such as those found on the BLAST (Basic Local Alignment Search Tool) pages (http://www.ncbi.nlm.nih.gov/BLAST/), or the EST BLAST Web Tool on the Human Genome Mapping Group's site (http://www.hgmp.mrc.ac.uk/ Registered/Webapp/estblast/index.html). These facilities allow the investigator to search directly the EST division of GenBank, which contains ESTs from a variety of organisms, but mainly human and mouse. A list of sites that provide information and access to the EST databases and access to the programmes required for searching the databases is provided in *Table 2*.

Table 2. List of sites providing access and information for various databases and associated search tools

Name	URL
UK Human Genome Mapping Project Resource Centre[a]	http://www.hgmp.mrc.ac.uk/
National Centre for Biotechnology Information	http://www.ncbi.nlm.nih.gov/
I.M.A.G.E. Consortium	http://www-bio.llnl.gov/bbrp/image/image.html
Washington University Merck & Co. Human EST project	http://genome.wustl.edu/est/esthmpg.html
Stanford Genomic Resources[b]	http://genome-www.stanford.edu
Joint Genome Institute	http://www.jgi.doe.gov/

[a] Registration is required for access to this site.
[b] Includes *Saccharomyces* Genome Database and *Arabidopsis thaliana* Database.

Protocol 1. Searching the EST database

Method

1. Open database-searching programme on the WWW, e.g. BLAST.

2. Enter query sequence (amino acid or DNA). The simplest option is to copy the sequence from another file, e.g. a Word file, and paste it directly into the box provided.

3. Select the appropriate programme. Use TBLASTN if the query sequence is protein, or BLASTN if the query sequence is DNA.

4. Select the sequence databases to be searched, e.g. human ESTs only, all EST sequences.

5. Run program.

6. Check alignments visually to identify relevant hits.

7. Run program to construct a composite sequence from overlapping sequences.

8. Perform BLAST on composite sequence.

9. Repeat steps 6 to 8 until no further sequence information is retrieved.

4.2.4 Output from BLAST search

Two examples of parts of the output from BLAST searches of the EST database, which highlight some of the features of the programmes, are shown in *Figures 2* and *3*. The first search (*Figure 2*) was performed using the entire coding sequence of rat AMP-activated protein kinase (AMPK) β1 subunit (14) as the query sequence and searched the human EST database. In the second example (*Figure 3*) a short peptide sequence derived by cyanogen bromide cleavage of AMPKα2 (15) was used to search all EST sequences. In both examples a number of matches, or 'hits', were identified and these are arranged in order of significance, based on sequence identity. The details of the database accession (e.g. EM:HSAA97770) followed by a brief description (e.g. AA191558 zp81a02.s1 Stratagene HeLa cell) of each clone is shown first. An example of an EST entry is shown in *Figure 4*. The number of sequences producing an alignment is shown in the far right-hand column. The Poisson significance (P) value is in the preceding column and generally the lower the value the more likely the alignment is to represent a meaningful match. For a more comprehensive discussion of the parameters used in similarity searching of databases see Altschul *et al.* (16).

The programme automatically aligns the query sequence with each individual hit, and this is shown further down the figure. A quick visual inspection of the alignments gives the investigator an indication of the likely relevance of each individual hit. In the examples shown it is immediately apparent that the hits are highly similar to the query sequences. It is important, however, to analyse the output of the searches thoroughly. For instance, a complete analysis of the output data from the first search reveals that, in addition to identifying clones encoding the human orthologue of rat AMPKβ1, a cDNA clone encoding a second human isoform (AMPKβ2) has been identified (accession number EM:HS171329 in *Figure 2*). In the second search using peptide sequence derived from AMPKα2, cDNAs encoding human AMPKα2 and a second human isoform (AMPKα1), as well as cDNAs encoding AMPKα isoforms from different species, are identified (*Figure 3*). Routines within the BLAST programme can be used to create a multiple alignment of all of the

```
Blast output for AMPKbeta1 (rat)
                                                            Smallest
                                                              Sum
                                                    High    Probability
Sequences producing High-scoring Segment Pairs:     Score   P(N)          N

EM:HSAA97770 AA191558 zp81a02.s1 Stratagene HeLa cell s3 ...   1343   1.9e-108   2
EM:HSZZ77079 AA371954 EST83796 Pituitary gland, subtracte...   1262   1.3e-97    1
EM:HSZZ77080 AA371955 EST83797 Pituitary gland, subtracte...   1245   3.7e-96    1
EM:HSAA31039 AA131038 zo16h03.r1 Stratagene colon (#93720...    846   2.3e-90    2
EM:HSAA97878 AA191658 zq43b01.s1 Stratagene hNT neuron (#...   1134   3.1e-86    1
EM:HS171329  W07171    za93d09.r1 Soares fetal lung NbHL19W .    620   5.7e-85    3
EM:HSC2TB121 F11147    H. sapiens partial cDNA sequence; clo...  1096   1.7e-83    1
EM:HS1256032 AA461052 zx61f08.s1 Soares total fetus Nb2HF...   1055   3.7e-80    1
EM:HS1255184 AA460485 zx61f08.r1 Soares total fetus Nb2HF...    637   2.2e-76    4
EM:HSZZ36971 AA331900 EST35798 Embryo, 8 week I Homo sapi...    855   4.1e-63    1


Hit 1

>EM:HSAA97770 AA191558 zp81a02.s1 Stratagene HeLa cell s3 937216 Homo sapiens cDNA
clone 626570 3' similar to SW:SIP2_YEAST P34164 SIP2 PROTEIN.
            Length = 472

   Minus Strand HSPs:
HSP 1
Score = 1343 (371.1 bits), Expect = 1.9e-108, Sum P(2) = 1.9e-108
Identities = 303/347 (87%), Positives = 303/347 (87%), Strand = Minus / Plus

Query:   813 TCATATGGGCTTGTAGAGGAGGGTGGTGACGTACTTTTTCTTGTACCGATGGGTCGCACT 754
             ||||||||||||||| |  | ||||||||| |||| ||||||||||| ||||| || ||
Sbjct:   110 TCATATGGGCTTGTATAACAAGGTGGTGACGNACTTCTTCTTGTACCGGTGGGTTGCGCT 169

Query:   753 GAGCACCATCACTCCATCCTTGATAGAGAGTGCATAGAGGTGGTTCAGCATGACGTGGTT 694
             |||||||||||||||||||||||||||| || || || ||||||||||||||||||| ||
Sbjct:   170 GAGCACCATCACTCCATCCTTGATAGACAGCGCGTATAGGTGGTTCAGCATGACGTGATT 229

Query:   693 GGGCTCCGGAAGCAGCGCTGGATCACAAGAGATGCCCGTGTCCTTGTTCAAGATGACCTG 634
             |||||| ||||||| |||||||||||| || || ||||||||||||||||| ||||||||
Sbjct:   230 GGGCTCAGGAAGCAAAGCTGGATCACAGGAAATCCCCGTGTCCTTGTTCAGGATGACCTG 289

Query:   633 CAGCAGGTGAGGCGGGAGGATGGGCGGGGCCTTGAACCGCTCCTCTGGTTTAGAGATGTA 574
             || || || || ||||| || || || |  | |||||| ||| || ||||| ||| |||
Sbjct:   290 GAGGAGATGTGGGGGGAGAATAGGGGGTGCCCGAAAGCGCTCTTCGGGTTTGCAGACGTA 349

Query:   573 AGGCTCCTGGTGGTAGGGTCCTGGGGGGGGAACTGGACAGCTCAGATACATCGGAGCACTT 514
             |||||||| ||||||||||||||||| |||||||||||||||||||| |||||||||||||
Sbjct:   350 GGGCTCCTGATGGTAGGGTCCTGGGGGAGAACTGGACAGCTCAGACACATCGGAGCACTT 409

Query:   513 TTGGGAATCCACCATTAAAGCATCAAATACTTCAAAGTCAGTTTTCT 467
             ||||||||||||||||||||||||||||||||||| |||||||||||
Sbjct:   410 TTGGGAATCCACCATTAAAGCATCAAATACTTCANAGTCAGTTTTCT 456


HSP 2
Score = 92 (25.4 bits), Expect = 1.9e-108, Sum P(2) = 1.9e-108
Identities = 20/22 (90%), Positives = 20/22 (90%), Strand = Minus / Plus

Query:   471 TTTCTTCACTTGAATGATGTTG 450
             |||  ||||||||||||||||||
Sbjct:   451 TTTTCTCACTTGAATGATGTTG 472
```

Figure 2. Part of the output from a BLAST search using the nucleotide sequence encoding rat AMP-activated protein kinase (AMPK) β1 subunit is shown. The output consists of an identifier for database accession (e.g. EM:HSAA97770) followed by a brief one-line description of the database sequence (e.g. AA191558 zp81a02.s1 Stratagene HeLa cell). The number of sequences producing an alignment is shown in the far right-hand column (N). The probability [listed under P(N)] expresses the likelihood of the match occurring by chance: the lower the value the more likely the alignment is to represent a significant match. Following on from the one-line descriptions the local alignments are displayed (the alignments from the first hit only are shown here). The complete definition line of the database entry is listed, as well as some statistics about the match.

```
TBLASTN 1.4.11 [24-Nov-97] [Build 24-Nov-97]

Reference:  Altschul, Stephen F., Warren Gish, Webb Miller, Eugene W.
Myers, and David J. Lipman (1990).  Basic local alignment search tool.  J. Mol.
Biol. 215:403-10.

Notice:  statistical significance is estimated under the assumption that
the equivalent of one complete reading frame of the database codes for protein
and that significant alignments will involve only coding reading frames.

Query= tmpseq_1
        (18 letters)

Database:  Non-redundant Database of GenBank EST Division
           1,690,584 sequences; 632,265,608 total letters.
Searching.................................................done

                                                           Smallest
                                                           Sum
                                             Reading  High  Probability
Sequences producing High-scoring Segment Pairs:  Frame Score  P(N)      N

gb|AA826292|AA826292  od71e01.s1 NCI_CGAP_Ov2 Homo sap... +1   98  3.3e-06  1
gb|AA861450|AA861450  ak24h10.s1 Soares testis NHT Hom... -1   80  0.0013   1
gb|AA400607|AA400607  zt70g09.s1 Soares testis NHT Hom... -1   80  0.0013   1
dbj|D68582|CELK133B5F C.elegans cDNA clone yk133b5 : 5... +1   73  0.013    1
gb|AA391609|AA391609  LD10636.5prime LD Drosophila mel... +1   73  0.013    1
dbj|C66495|C66495     C.elegans cDNA clone yk222d7 : 5... +1   73  0.013    1


gb|AA826292|AA826292 od71e01.s1 NCI_CGAP_Ov2 Homo sapiens cDNA clone
IMAGE:1373400 similar to SW:AAK2_RAT Q09137 5'-AMP-ACTIVATED PROTEIN KINASE,
CATALYTIC ALPHA-2 CHAIN ;
          Length = 430

Plus Strand HSPs:
Score = 98 (46.7 bits), Expect = 3.3e-06, P = 3.3e-06
Identities = 18/18 (100%), Positives = 18/18 (100%), Frame = +1

Query:      1 MKQLDFEWKVVNAYHLRV 18
              MKQLDFEWKVVNAYHLRV
Sbjct:    223 MKQLDFEWKVVNAYHLRV 276


gb|AA861450|AA861450 ak24h10.s1 Soares testis NHT Homo sapiens cDNA
clone IMAGE:1406947 3' similar to TR:O00286 O00286 AMP-ACTIVATED
PROTEIN KINASE ALPHA-1 ;
          Length = 480

Minus Strand HSPs:
Score = 80 (38.1 bits), Expect = 0.0013, P = 0.0013
Identities = 14/18 (77%), Positives = 17/18 (94%), Frame = -1

Query:      1 MKQLDFEWKVVNAYHLRV 18
              +KQLD+EWKVVN Y+LRV
Sbjct:    399 IKQLDYEWKVVNPYYLRV 346
```

Figure 3. Part of the output from a BLAST search using a short peptide sequence (18 amino acids) derived from rat AMPKα2 is shown. The output is similar to that described in the legend to *Figure 2* and the first two alignments are shown. Identities are indicated by printing the amino acid symbol (single-letter code) between the sequences. Similarity between two different residues is indicated by a '+' between the sequences. Note that the second hit (accession number AA861450) identifies a different isoform (AMPKα1) compared to the query sequence. Also note from the one-line descriptions that related cDNAs from *Caenorhabditis elegans* and *Drosophila melanogaster* are identified.

David Carling

LOCUS   AA826292   430 bp   mRNA   EST   07-APR-1998
DEFINITION od71e01.s1 NCI_CGAP_Ov2 Homo sapiens cDNA clone IMAGE:1373400
similar to SW:AAK2_RAT Q09137 5'-AMP-ACTIVATED PROTEIN KINASE, CATALYTIC
ALPHA-2 CHAIN; mRNA sequence.
ACCESSION AA826292
NID   g2899604
KEYWORDS   EST.
SOURCE   human.
ORGANISM Homo sapiens; Eukaryota; Metazoa; Chordata; Vertebrata; Mammalia; Eutheria;
Primates; Catarrhini; Hominidae; Homo.
REFERENCE 1 (bases 1 to 430)
AUTHORS   NCI-CGAP http://www.ncbi.nlm.nih.gov/ncicgap.
TITLE   National Cancer Institute, Cancer Genome Anatomy Project (CGAP), Tumor Gene
Index
JOURNAL Unpublished (1997)
COMMENT
        Contact: Robert Strausberg, Ph.D.
        Tel: (301) 496-1550
        Email: Robert_Strausberg@nih.gov
        Tissue Procurement: Christopher A. Moskaluk, M.D., Michael R.
        Emmert-Buck, M.D., Ph.D.
        cDNA Library Preparation: David B. Krizman, Ph.D.
        cDNA Library Arrayed by: Greg Lennon, Ph.D.
        DNA Sequencing by: Washington University Genome Sequencing Center
        Clone distribution: NCI-CGAP clone distribution information can be
        found through the I.M.A.G.E. Consortium/LLNL at:
        www-bio.llnl.gov/bbrp/image/image.html

        Insert Length: 579   Std Error: 0.00
        Seq primer: -40m13 fwd. ET from Amersham
        High quality sequence stop: 427.
FEATURES   Location/Qualifiers
     source   1..430
                 /organism="Homo sapiens"
                 /note="Vector: pAMP10; mRNA made from invasive ovarian
                 tumor, cDNA made by oligo-dT priming. Non-directionally
                 cloned. Size-selected on agarose gel, average insert size
                 600 bp. Reference: Krizman et al. (1996) Cancer Research
                 56:5380-5383."
                 /db_xref="taxon:9606"
                 /clone="IMAGE:1373400"
                 /clone_lib="NCI_CGAP_Ov2"
                 /sex="female"
                 /tissue_type="ovary"
                 /lab_host="DH10B"
BASE COUNT   136 a   88 c   99 g   107 t
ORIGIN
        1 gctcgactta tggatgatag tgccatgcat attcacgcag gcctgaaacc tcatccagaa
       61 aggatgccac ctcttatagc agacagcccc aaagcaagat gtccattgga tgcactgaat
      121 acgactaagc ccaaatcttt agctgtgaaa aaagccaagt ggcatcttgg aatccgaagt
      181 cagagcaaac cgtatgacat tatggctgaa gtttaccgag ctatgaagca gctggatttt
      241 gaatggaagg tagtgaatgc ataccatctt cgtgtaagaa gaaaaaatcc agtgactggc
      301 aattacgtga aaatgagctt acaactttac ctggttgata acaggagcta tcttttggac
      361 tttaaaagca ttgatggtaa ggaagctatg catgcaggac ctacccaccg gatgtcttcg
      421 agatgatttt

Figure 4. An example of an entry from the EST database is shown. Hit 1 from the search using AMPKα2 peptide sequence (see *Figure 3*) was retrieved from the database by following the link on the WWW from the BLAST output. The entry provides a description of the clone, including the clone identification number (1373400 in this example) which is required for obtaining the clone from the I.M.A.G.E. Consortium.

264

hits. This is very useful for identifying hits with overlapping sequences (the large number of sequences in the EST database, combined with the way in which the database is constructed, often leads to multiple overlapping entries). Clones with overlapping sequence can be used to build up a composite sequence, or *contig*, which can then be used as the input sequence for a new search. It may be possible, using this iterative procedure, to obtain the entire sequence of a given cDNA without ever doing any benchwork.

4.2.5 Limitations of cloning *in silico*

The most basic limitation of this approach is that in order for it to succeed, part of the input sequence must be represented in the EST database (or at least a closely related sequence). Since most of the entries in the database are derived from human or mouse cDNAs, searching with a sequence derived from these species will have the greatest chance of success. It is likely, however, that input sequences derived from other organisms will enable the equivalent human or mouse cDNAs to be identified. If this is the case, then the human or mouse sequences can be used to design probes in order to isolate the cDNA from the original source organism. A second problem facing *in silico* cloning stems from the way in which the entry sequences are generated. The sequences are obtained from a single read and entered into the database using an automated process. Manual editing is not carried out on the data, and this inevitably leads to some errors occurring within the sequences submitted to the database. It is important, therefore, when checking alignments to take into consideration the possibility of errors in the database entries. Sequencing errors within the ESTs will be most problematic when searching the database with short peptide sequences, since the errors may lead to frameshifts within the predicted amino acid sequence. The original fluorescent sequencing data are available on the WWW, and these can be viewed in order to help confirm the reading frame. It must be stressed, however, that once a candidate clone has been identified from a database search it is important that the investigator determines the sequence of the insert unambiguously. Although *in silico* cloning can drastically reduce the time spent in identifying a cDNA it does not replace all of the experimental work!

4.2.6 Isolating cDNA containing the entire coding sequence

In most cases the eventual aim of cloning *in silico* is to isolate a cDNA clone encoding the entire protein of interest, which can then be used as a reagent to help study functional and regulatory properties. Inspection of the deduced amino acid sequences of the clones identified from the EST database will allow the investigator to determine whether any of the clones contain the predicted start and stop codons. Software within the BLAST programme can be used to identify open reading frames within the sequence, which can be confirmed by viewing the original fluorescent sequencing data. Identification of the start and stop codons may not be straightforward, especially when cDNAs encoding previously uncharacterized proteins are being analysed. Some indication of the expected size of the coding region will probably be available (from the observed molecular mass of the protein) although this can be misleading, e.g. if post-translational processing of the protein occurs. The problems associated with the identification of the start and stop codons are not restricted to the *in silico* cloning approach and occur whatever method is used for isolating cDNA. It is important, therefore, that the investigator carries out sufficient checks to ensure that the entire coding sequence has been identified (see below). It is possible that a clone will be identified from the database searching that encodes the entire protein, in which case the clone, after having being obtained from the I.M.A.G.E Consortium and its sequence confirmed, can be used directly for subsequent studies.

Protocol 2. Isolation of 5' and 3' cDNA sequences by PCR

Materials
- appropriate DNA template, e.g. Marathon-Ready cDNA (Clontech)
- 10 × PCR buffer (150 mM Tris–HCl pH 8.8, 500 mM KCl, 25 mM MgCl$_2$)
- thermostable DNA polymerase[b]
- 2 mM dNTPs (Amersham Pharmacia Biotech)
- PCR primers (gene specific and adaptor specific)[a]

Method

1. Mix together in a 0.5 ml microfuge tube the following (in the order listed):[c]

Component[d]	Volume (μl)
10 × PCR buffer	5
2 mM dNTPs	5
H$_2$O	to 50
Gene-specific primer (10 pmol/μl)	1
Adaptor-specific primer (10 pmol/μl)	1
Thermostable DNA polymerase	2.5 units
DNA template (0.1 ng/μl)	5

2. Overlay the reactions with two drops of light mineral oil to prevent evaporation.

3. Amplify for 30 cycles. The exact conditions for amplification must be determined experimentally, but the following conditions should provide a suitable starting point:
    (a) denaturation   1 min at 94°C
    (b) annealing   1 min at 55°C
    (c) extension   1 min at 72°C

The number of cycles can be increased if the amount of product is low. Following amplification the reaction is incubated for 10 min at 72°C to complete extension of all products.

4. Analyse 5 µl of the reaction products by agarose gel electrophoresis. If little or no product is observed, or if there are multiple products, amplify an aliquot of the reaction with internal primers (nested PCR).[e]

5. Resolve reaction products by agarose gel electrophoresis. Excise the gel slice containing the amplified product and purify DNA using a commercially available kit, e.g. QIAGEN gel extraction kit.

6. Ligate purified DNA into a vector suitable for cloning PCR products, e.g. pGEM-T (Promega), using the manufacturers protocol.[f]

7. Transform ligation mixture into competent *E. coli* cells, e.g. JM109, XL-1 Blue, and plate on LB-agar plates containing the appropriate supplements, e.g. ampicillin (100 µg/ml).

8. Pick individual bacterial colonies, prepare plasmid DNA, and analyse insert by DNA sequencing.

[a] Choose gene-specific primers (21–25 bp) corresponding to a region of high quality sequence read from the EST clone (viewing the original sequencing traces will help determine the choice of primers). The sequence at the beginning and end of the reads is sometimes of relatively poor quality and so primers in this region should be avoided.
[b] A DNA polymerase with proof-reading activity is recommended in order to minimize mutational errors introduced during PCR.
[c] The order of addition is important in order to avoid cross-contamination of the template DNA.
[d] Control reactions containing only a single primer should also be performed. These controls can be useful if additional non-specific products are observed in the reaction with both primers.
[e] Primers corresponding to sequence internal to the primers used in the original PCR can be used to perform a second round of amplification ('nested' amplification). In addition to further amplification of the product, this also increase the specificity of the reaction, which may be helpful if a number of products are generated in the first round.
[f] A number of vectors for cloning PCR products are available commercially. A popular type of these vectors relies on the intrinsic activity of *Taq* DNA polymerase in adding a single base (usually adenosine) to the 3'-end of the DNA (19), although vectors employing a variety of other strategies are also available.

If a clone encoding the entire coding sequence is not identified from the database search, the information gathered from partial cDNA sequences can be used to rapidly isolate and clone the cDNA encoding the entire protein. The partial cDNA sequence can be used as a probe for conventional library screening, details of which can be found in a number of cloning manuals (e.g. ref. 17) and will not be described here. An alternative and much quicker approach is to use RACE (rapid amplification of cDNA ends) PCR (18). In this method cDNA, to which short adaptors of known sequence have been added at the 5'- and 3'-ends, is used as a template for PCR. The target cDNA is amplified using a gene-specific primer based on the partial cDNA sequence, which can be either sense (for amplification of 3' sequence) or antisense (for amplification of 5' sequence), in combination with an adaptor-specific primer.

In some cases it may be necessary to carry out a second round of amplification, e.g. if the target cDNA is poorly represented within the template DNA. If a second round of amplification is required it is preferable to use a different gene-specific primer (internal to the original primer). This nested approach greatly reduces the possibility of amplifying non-specific cDNAs. The details of this method are given in *Protocol 2*.

Once the cDNA ends have been amplified and their sequences confirmed, it is a relatively straightforward procedure to isolate cDNA encoding the entire protein. Primers spanning the initiation and stop codons can be designed and used to amplify the cDNA by PCR. If the cDNA is of low abundance, it may be necessary to design overlapping sets of primers and carry out two rounds of amplification. If using this approach, ensure that the second set of primers still span the initiation and stop codons! It is often advantageous to include unique restriction sites within the primers that will facilitate subsequent cloning steps. It is worthwhile spending a few minutes deciding which restriction sites to include as this can save a lot of time later. The product can be cloned using steps 5–8 of *Protocol 2*. The next step is to verify the integrity of the cloned cDNA. This is critical since the cDNA has been generated by PCR and the error rate of *Taq* polymerase is significant. The use of thermostable DNA polymerases which have proof-reading activity will reduce the error rate, but it is still important to check carefully the sequence of the cloned cDNA. The inserts from several independent clones (preferably from independent PCR amplifications) should be sequenced. Comparison of these sequences, together with the sequence obtained from the EST database, will allow the investigator to determine any potential errors. Where it is not possible to identify a clone without errors which affect the coding sequence, a 'correct' cDNA can be assembled by subcloning regions of the insert (which do not contain any errors) into the vector. A final check of the sequence should be carried out to ensure the subcloning has not altered the reading frame.

5. Confirming the identity of the protein product

Isolating the cDNA is the easy part. The problem ahead lies in uncovering the function of the expressed protein. Initial clues may be revealed from inspection of the deduced sequence of the protein. Certainly, in the case of the catalytic subunit, the amino acid sequence will confirm its place within the protein kinase family. Similarly, identification of recognizable sequence motifs within non-catalytic protein kinase subunits may suggest a particular function, binding site, or subcellular localization for the protein. Despite these predictions, no amount of sequence comparisons will yield conclusive evidence regarding the function of the protein. One of the approaches immediately open to the researcher armed with a newly cloned cDNA is to express the recombinant protein, and use this as a tool to facilitate the study of its function. A number of expression systems are available, and the choice of a particular system will depend on the intended downstream application of the expressed protein.

5.1 Expression in bacteria

Protein expression in bacteria (usually *E. coli*) is relatively cheap, does not require expensive or specialized materials or equipment, and can yield high levels of recombinant protein in a short period of time. However, eukaryotic proteins produced in bacteria may not be correctly folded, and will lack any post-translational processing required for activity, e.g. acylation, glycosylation, or phosphorylation. In the case of protein kinase subunits, this may be particularly relevant if phosphorylation is required for regulation of activity. None the less, bacterial expression still remains a very attractive starting point for the generation of recombinant protein for antibody production, structural studies, and as a substrate for *in vitro* studies, e.g. identification of phosphorylation sites. Although the number of different systems available for expression in *E. coli* has grown enormously over the past few years, the strategies involved can be divided into two basic approaches:

- expression of native protein
- expression of a 'fusion' protein

In order to express native protein, the cDNA is inserted into a bacterial expression vector downstream of an efficient ribosomal binding site. This allows translation to occur from the initiation codon within the cDNA, generating the native recombinant protein. The second approach is to insert the cDNA so that it is in the same reading frame as vector DNA encoding a specific polypeptide sequence. Expression from this type of vector results in the generation of a fusion protein between the protein of interest and the polypeptide 'tag'. This approach ensures good initiation of translation, and has the advantage that the polypeptide tag encoded by the vector allows methods to be utilized for the rapid and convenient purification of the recombinant protein. Examples of some commonly used tags for bacterial expression of protein kinase subunits are:

(a) *Glutathione-S-transferase* (*GST*). A series of vectors are available (pGEX series from Amersham Pharmacia Biotech) which allow expression of GST fusion proteins. The fusion proteins can be readily purified by affinity chromatography using glutathione–Sepharose, eluting with buffers containing glutathione.

(b) *Maltose-binding protein* (*MBP*). The pMal range of vectors (New England Biolabs) generate fusion proteins with MBP which can be purified by chromatography on amylose resin, eluting with buffers containing maltose.

(c) *Poly-histidine [poly(His)]*. A number of vectors, e.g. pET series (Novagen) and the pQE series (QIAGEN), generate fusion proteins with a poly(His) tag (usually consisting of six consecutive histidine residues). Poly(His)-containing fusion proteins can be purified using metal (Ni^{2+})-chelating resins and eluted with buffers containing imidazole.

The wide applicability and success of these fusion protein vectors have made this type of approach the most popular method for expressing recombinant proteins in bacteria. Most vectors for expressing fusion proteins also include a sequence, immediately downstream of the tag sequence, which encodes a recognition site for a specific protease. This allows the polypeptide tag to be cleaved from the protein of interest, although the efficiency of cleavage can be very poor for some fusion proteins, and must be determined experimentally. A typical method for expressing recombinant protein in *E. coli* is shown in *Protocol 3*.

Protocol 3. Expression of recombinant protein in bacteria

Materials

- LB broth
- isopropyl β-D-thiogalactopyranoside (IPTG, 100 mM stock solution in water, store at −20°C)
- phosphate buffered saline (PBS; 140 mM NaCl, 2.7 mM KCl, 10 mM Na_2HPO_4, 1.8 mM KH_2PO_4, pH 7.3)

Methods

1. Subclone cDNA into a suitable expression vector and transform competent *E. coli* cells.[a]

2. Pick an individual colony[b] from the transformation plate and use to inoculate 100 ml LB broth containing appropriate antibiotic, e.g. ampicillin. Incubate at 37°C, with shaking, until the optical density of the cell culture measured at 600 nm is between 0.6 and 0.8.

3. Induce protein expression by addition of 100 mM stock solution of IPTG to yield a final concentration between 0.1 and 1 mM, and continue to incubate for 1–2 h.[c]

4. Harvest cells by centrifugation at 5000 *g* for 5 min at room temperature.

5. Resuspend cells in 5 ml ice-cold PBS.[d]

6. Lyse cells by sonication (6 × 15 sec bursts on high-power) on ice.

7. Remove insoluble material by centrifugation at 10000 *g* for 15 min at 4°C.[e]

8. Purify soluble recombinant protein from supernatant by affinity chromatography using the properties of the fusion tag.

9. Cleave the fusion protein with appropriate protease if required, and purify the protein of interest from polypeptide tag.[f]

[a] A wide range of *E. coli* strains are now available for protein expression and pilot studies should be performed in order to determine the best choice of host cell for a particular protein. When using a vector in which expression is driven from the T7 promoter a strain of *E. coli* containing a copy of the gene for T7 RNA polymerase is required, such as BL21 (DE3).

[b] Since there may be some variation in the level of expression of the fusion protein between different clones it is advisable to pick several colonies, grow individual cultures, and determine the expression for each of the clones.

[c] Altering the growth conditions, e.g. temperature and incubation time, can have a marked effect on the level of protein expression, as can the concentration of IPTG used for induction. The optimal conditions for protein expression, therefore, must be determined experimentally.

[d] Different buffers can be used in place of PBS, or additions, such as glycerol, dithiothreitol, etc., can be made if required.

[e] In many cases a proportion of the recombinant protein will be present in the insoluble fraction. It may be possible to alter the growth conditions (see footnote *c*) in order to increase the amount of soluble protein and this should be determined experimentally. Even in cases where no soluble protein can be recovered it may still be possible to use the expressed protein as an antigen for antibody production, e.g. following purification by SDS–PAGE.

[f] The methods for purifying the protein away from the tag depend on the particular tag being used. For short tag sequences, e.g. poly(His), the tag can be removed by dialysis or ultrafiltration. Protein tags, e.g. glutathione-*S*-transferase, can be removed by affinity chromatography using the same affinity resin used for the purification of the fusion protein.

5.2 Transient expression in mammalian cells

Protein kinase subunits can be conveniently expressed in mammalian cells. This system has two major advantages over expression in bacteria. First, mammalian cells have the ability to carry out any processing or post-translational modifications that may be required for functional activity. Secondly, by choosing the appropriate cell type for expression, it is possible to study the function of the expressed gene in a background similar to conditions found *in vivo*. Transient expression of proteins in mammalian cells is a technically simple procedure with a very high success rate. These features, combined with the other advantages of expressing the protein in a mammalian cell, have made transient transfection one of the most popular approaches for studying expression of protein kinase subunits. This system is ideal for evaluating the effect of specific mutations within a gene, or for analysing large numbers of mutant forms of a particular protein. It can also be used to determine the consequences of overexpressing the protein kinase subunit, or a mutant form that interferes with the activity of the endogenous protein kinase within the cell, on downstream targets. One major drawback of transient transfection in mammalian cells is that it does not usually yield high levels of recombinant protein, and so this approach is not suitable for producing material for structural studies.

5.2.1 Choice of vector

Many vectors are commercially available for transient expression in mammalian cells, each with specific features aimed at facilitating various aspects of

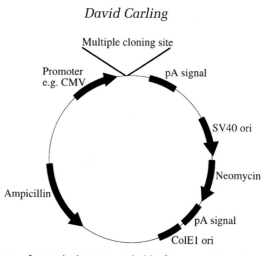

Figure 5. Key features of a typical vector suitable for expression in mammalian cells. cDNA is inserted into the vector at restriction sites present in a multiple cloning site, and RNA transcription is driven from a strong promoter, such as the human cytomegalovirus (CMV) promoter. Polyadenylation signal and transcription termination sequences (pA signal) are included downstream of the inserted gene (and downstream of the gene for selection in eukaryotic cells) in order to enhance stability of the RNA transcript. The SV40 origin (SV40 ori) and the ColE1 origin (ColE1 ori) allow replication in mammalian cells and *E. coli*, respectively. Genes encoding selectable markers allow for maintenance in *E. coli* (e.g. ampicillin resistance) and selection of stable cell lines in eukaryotic cells (e.g. the neomycin-resistance gene).

the procedure. Most of the vectors, however, share common features. The gene of interest is driven from a strong promoter, such as the immediate early gene of the human cytomegalovirus (CMV) promoter. The CMV promoter, like a number of other viral promoters, provides a high and constitutive level of expression in a wide range of cell types. Other promoters may be cell-type specific, or stimulate transcription only in the presence of an inducer. This type of promoter may be necessary if expression of the protein is toxic to the transfected cells. A linker region that contains multiple cloning sites allows insertion of the gene of interest downstream of the promoter. Transcription termination and polyadenylation signals are included within the vector in order to enhance the stability of the RNA. A bacterial resistance gene allows the vectors to be conveniently propagated in *E. coli*, and in many cases the presence of a hygromycin or neomycin resistance gene allows stable cell lines to be selected and maintained. A schematic representation of the key features of a typical vector for expression in mammalian cells is shown in *Figure 5*.

A number of mammalian expression vectors, in common with vectors for expression in *E. coli*, incorporate within them a DNA sequence encoding a polypeptide tag. cDNA is subcloned into the vector, in the same reading frame as the tag, resulting in the expression of a fusion protein. DNA sequences encoding short peptide tags (6–10 amino acids) can also be added directly to the cDNA using PCR. In this method, a PCR primer is designed which encodes the tag peptide at the 5′-end followed by gene-specific sequence at the 3′-end. The amplified product, which encodes the tag in the same reading frame as the gene product, can be cloned into any suitable mammalian expression vector. One type of tag that has proved to be particularly useful for studying the expression of protein kinase subunits in mammalian cells is the 'epitope tag'. An epitope tag is a short peptide (6–10 amino acids) which forms an antigenic determinant, or epitope, for a particular antibody. The use of antibodies directed against the tag sequence has several advantages over antibodies raised directly to the protein of interest. The properties of the anti-tag antibodies have usually been well characterized, e.g. the conditions required for immunoprecipitation, Western blotting, and immunofluorescence. This is of benefit, as antibodies generated against the protein of interest may not be well characterized and extensive characterization can be very time consuming. Additionally, control experiments can be performed using anti-tag antibodies by conducting parallel immunoprecipitations or blots on cell extracts in which the tagged protein is not expressed. Similar controls are not possible with antibodies raised to the target protein itself. Anti-tag antibodies recognize only the recombinant tagged protein, and this is particularly useful when discriminating between closely related proteins, e.g. protein isoforms. *Table 3* lists four epitope tags which have been used in a wide variety of studies investigating the expression of protein kinase subunits in mammalian cells (e.g. refs 20–23).

Another tag which has recently become popular for studying the expression of protein kinase subunits is the green fluorescent protein (GFP). GFP is a 27 kDa protein, from the jellyfish *Aequorea victoria*, which fluoresces when exposed to UV light. Expression of a fusion protein with GFP enables the subcellular localization of the fusion protein to be studied in living cells. Since the expression of the GFP fusion protein can be monitored in living cells, this system is particularly suited for investigating the localization of protein kinase

---

Table 3. Description of some commonly used epitope tags for studying the expression of protein kinase subunits in mammalian cells

Tag	Amino acid sequence of tag	Source of tag	Antibody used for detection
Poly(His)[a]	HHHHHH	Synthetic peptide	Anti-(H)4; Anti(H)5
FLAG	DYKDDDDK	Synthetic peptide	Anti-FLAG M2[b]
HA	YPYDVPDYA	Influenza haemagglutinin	12CA5[c]
myc	EQKLISEEDL	Human c-myc	9E10[d]

[a] The poly(His) tag can also be used for purification of the recombinant protein using a metal-chelating resin.
[b] See ref. 34.
[c] See ref. 35.
[d] See ref. 36.

subunits under a variety of conditions in real time (e.g. ref. 24). Furthermore, since the proteins are visualized in living cells, potential artefacts caused by the methods used for fixation of cells are eliminated.

5.2.2 Choice of cell type for expression

The main criteria for choosing which cell type to transfect are:

- suitability of cell type for functional studies
- ease and efficiency of transfection

COS cells (COS-1, COS-7) are probably the most popular cell type for carrying out transient expression studies, and provide a good starting point for initial experiments. COS cells express the SV40 large T antigen, and vectors containing an SV40 origin of replication undergo episomal replication (25, 26). This ability to amplify transfected DNA to more than 100 000 copies per cell allows very high levels of expression of mRNA and protein in COS cells. A potential drawback with COS cells is that they are derived from African green monkey kidney, and as such may not provide the most suitable background for studying the function of a particular protein. The list of alternative cell types that have been used as hosts for transient expression is almost endless, and the investigator must assess the individual merits of the different cell types with respect to the problem being addressed.

5.2.3 Methods for transfection

Several methods for introducing DNA directly into mammalian cells have been developed. Chemical methods include formation of complexes of DNA with calcium phosphate, DEAE-dextran, or cationic lipids (lipofection). In addition, DNA can be electroporated into cells using a brief high voltage electric pulse. Each of these methods has its own advantages and disadvantages, and it is advisable to compare the efficiency of transfection using the different methods. Furthermore, since the efficiency of transfection for a particular method may vary between cell types, it is worth carrying out this comparison for each cell type used. *Protocols 4* and *5* describe methods for transfecting cells by calcium phosphate co-precipitation (27) and electroporation (28). These methods are used routinely in my laboratory, and yield high and reproducible levels of transfection in a variety of cell types.

Protocol 4.   Transient transfection by calcium phosphate co-precipitation

Materials

- 2.× HBS (Hepes buffered saline: 270 mM NaCl, 10 mM KCl, 1 mM Na_2HPO_4, 10 mM glucose, 40 mM Hepes, pH 7.1)[b]
- supercoiled plasmid DNA[a]
- 2 M $CaCl_2$

Method

1. Plate cells the day before transfection in 6 cm diameter tissue-culture dishes[c] at a density that will yield approximately 80% confluence in 3–4 days.[d]

2. Three hours prior to transfection remove the medium (including any dead cells) and replace with 4 ml fresh medium per dish.

3. Warm all reagents to room temperature and mix thoroughly.

4. For each transfection mix 10 μg DNA with 37 μl 2 M CaCl$_2$ in a final volume of 0.3 ml (made up with sterile distilled water).

5. In a tissue-culture hood add the DNA solution dropwise to 0.3 ml 2 × HBS. Gently vortex the mixture throughout the addition of DNA.

6. Incubate the DNA mixture for 30 min at room temperature to allow the precipitate to form.[e]

7. Add the DNA suspension dropwise to the dish and swirl the dish to ensure an equal covering of the suspension over the cells.

8. Incubate the cells at 37°C for 12 h.

9. Remove the medium and wash the cells with 3 × 5 ml serum-free medium, in order to remove the co-precipitate.

10. Add 4 ml fresh growth medium per dish and continue to incubate for 48–72 h.[f]

11. Harvest the cells and analyse for protein expression.

[a] High purity plasmid DNA, free from contaminating protein and RNA, must be used.
[b] All materials should be prepared using tissue-culture grade reagents. The pH of 2 × HBS is critical and should be adjusted by the addition with 0.5 M sodium hydroxide to pH 7.1 (± 0.05). Filter sterilize and store 5 ml aliquots at −20°C.
[c] If necessary larger dishes can be used and the volumes listed adjusted accordingly.
[d] The plating density depends on the growth rate of the particular cell type being used. As a general guideline about 1–2 × 10^5 cells per 6 cm dish should be seeded.
[e] The mixture should appear slightly cloudy or opaque due to the formation of the calcium phosphate–DNA co-precipitate.
[f] The time required for optimum expression may vary and should be determined experimentally.

Protocol 5. Transient transfection by electroporation

Materials

- electroporation buffer (140 mM NaCl, 0.75 mM Na$_2$HPO$_4$, 20 mM Hepes, pH 7.15)
- supercoiled plasmid DNA

Method

1. Grow cells to approximately 70% confluence. Harvest by scraping and resuspend in electroporation buffer at a concentration of 2–20 × 10^6 cells/ml.

Protocol 5. *Continued*

2. Transfer 0.8 ml of the cell suspension into a BioRad Gene Pulsar Cuvette.

3. Add 20 μg plasmid DNA and incubate on ice for 10 min.

4. Electroporate at 450 V and 250 μF using a BioRad Gene Pulsar apparatus.[a]

5. Place the cuvette on ice for 10 min.

6. Plate the cells on 6 cm diameter dishes and incubate at 37°C for 48–72 hours.

7. Harvest the cells and analyse for protein expression.

[a] The strength of the electric field required for efficient transfection varies with cell type and must be determined empirically.

5.3 Other expression systems

The same methods employed for transient transfection can be used to generate cell lines in which the protein of interest is stably expressed. Indeed, most commercial vectors used for transient transfection include selectable markers, e.g. resistance to neomycin, suitable for the production of stable cell lines. Another source of recombinant protein that has proved successful in a large number of cases is overexpression in insect cells using the baculovirus system (29). This approach has the advantage that it can yield high levels of protein, suitable for biochemical and structural studies. Since expression is carried out in a eukaryotic cell there is a reasonable likelihood that the protein has undergone the correct processing. Until recently, however, expression using the baculovirus system was relatively time consuming compared with bacterial or mammalian expression systems. Improvements in the methods for generating high titre recombinant viral stocks, coupled with the use of vectors encoding specific peptide tag sequences to facilitate purification of the recombinant protein, have greatly reduced the time and effort required for protein expression using the baculovirus system. Consequently, this approach may provide the investigator with a valuable alternative to expression in bacteria for the production of large quantities of protein.

Expression in yeast provides another alternative for studying the properties of a particular protein. High level expression may be possible using the *Pichia pastoris* expression system (30), although at present there are few examples of protein kinase subunits that have been expressed successfully using this system. Protein–protein interactions can be studied in yeast using the two-hybrid system (31) (see Chapter 14). This system is particularly amenable for identifying regions within proteins that interact, although it is important to confirm the results obtained using the two-hybrid system with data from independent experiments using a different technique. Yeast also provides the possibility of studying the function of a protein kinase subunit by complementation analysis. If the protein kinase subunit under investigation has a closely related counterpart in yeast, then the gene encoding this protein in yeast can be deleted, or, alternatively, mutations can be introduced into the gene, in order to disrupt the endogenous activity of the protein kinase in yeast. The effect of expressing the related gene product in the mutant strain on protein kinase activity and downstream pathways can then be determined. Access to the complete sequence of the *S. cerevisiae* genome makes this 'reverse genetics' approach an attractive and potentially powerful means for studying protein kinase function (32, 33). In order to carry out these studies, however, sufficient preliminary information regarding the function of the yeast protein kinase subunit is required, e.g. an *in vitro* assay, phenotype of the mutant yeast strains, etc.

6. Concluding remarks

In this chapter I have described the methods for isolating cDNAs encoding protein kinase subunits, and some of the approaches that can be used to express the protein products in order to study their function. It is already apparent that the phenomenal progress being made within the large-scale DNA sequencing projects has changed the way in which cDNA cloning projects are undertaken. It seems likely that from now on the isolation of most, if not all, cDNAs will take advantage of *in silico* cloning. Major improvements in the techniques for analysing proteins by mass spectrometry have made it possible to obtain sufficient primary structural information with which to identify a protein unambiguously from the database, from sub-picomole quantities of the protein. Combining the developments in protein sequencing with the explosion in the amount of information available in the databases should serve to reduce significantly the time required to isolate cDNAs encoding a protein of interest, including protein kinase subunits. Having isolated the cDNA, the next step is to confirm the identity and function of the protein kinase subunit. One approach, which I have described here, is to use bacterial and mammalian expression systems in order to study structural, regulatory, and functional properties of the protein. High level expression in bacteria may allow sufficient material to be generated for structural studies using X-ray crystallography, and will almost certainly be useful in producing antibodies against the protein. Overexpression in mammalian cells can be used to address the role of the protein kinase subunit in the regulation of cellular processes. By expressing mutated forms of the protein it may be possible to interfere with endogenous activity and use this as a system to study the function of the protein kinase (see Chapter 4). Whatever approach is used, it seems clear that the next few years will witness a huge expansion in the isolation and characterization of cDNAs encoding protein kinase subunits.

Acknowledgements

Work in the author's laboratory is funded by the Medical Research Council (UK). I would like to thank Nicky Gumpel for advice and help with writing this chapter.

References

1. Hunter, T. and Plowman, G. D. (1997). *Trends Biochem. Sci.*, **22**, 18.
2. Hunter, T. (1987). *Cell*, **50**, 823.
3. Hanks, S. K., Quinn, A. M., and Hunter, T. (1988). *Science*, **241**, 42.
4. Hanks, S. K. and Quinn, A. M. (1991). In *Methods in enzymology* (ed. T. Hunter and B. M. Sefton), Vol. 200, p. 38. Academic Press, London.
5. Hanks, S. K. and Lawton, M. A. (1993). In *Protein phosphorylation: a practical approach* (ed. D. G. Hardie), 1st edn, p. 173. IRL Press, Oxford.
6. Johnson, L. N., Noble, M. E. M., and Owen, D. J. (1996). *Cell*, **85**, 149.
7. Levin, D. E., Hammond, C. I., Ralston. R. O., and Bishop, J. M. (1987). *Proc. Natl. Acad. Sci. USA*, **84**, 6035.
8. Wilks, A. F. (1989). *Proc. Natl. Acad. Sci. USA*, **86**, 1603.
9. Lindberg, R. A. and Hunter, T. (1990). *Mol. Cell. Biol.*, **10**, 6316.
10. Feng, X.-H. and Kung, S.-D. (1991). *FEBS Lett.*, **282**, 98.
11. Hillier, L., Lennon, G., Becker, M., Bonaldo, M. F., Chiapelli, B., Chissoe, S., Dietrich, N., Dubuque, T., Favello, A., Gish, W., Hawkins, M., Hultman, M., Kucaba, T., Lacy, M., Le, M., Le, N., Mardis, E., Moore, B., Morris, M., Parsons, J., Prange, C., Rifkin, L., Rohlfing, T., Schellenberg, K., Soares, M.B., Tan, F., Thierrymeg, J., Trevaskis, E., Underwood, K., Wohldman, P., Waterston, R., Wilson, R., and Marra, M. (1996). *Genome Res.*, **6**, 807.
12. Wilm, M., Shevchenko, A., Houthaeve, T., Breit, S., Schweigerer, L., Fotsis, T., and Mann, M. (1996). *Nature*, **379**, 466.
13. Mann, M. (1996). *Trends Biochem. Sci.*, **21**, 494.
14. Woods, A., Cheung, P. C. F., Smith, F. C., Davison, M. D., Scott, J., Beri, R. K., and Carling, D. (1996). *J. Biol. Chem.*, **271**, 10282.
15. Carling, D., Aguan, K., Woods, A., Verhoeven, A. J. M., Beri, R. K., Brennan, C. H., Sidebottom, C., Davison, M. D., and Scott, J. (1994). *J. Biol. Chem.*, **269**, 11442.
16. Altschul, S. F., Boguski, M. S., Gish, W., and Wootton, J. C. (1994). *Nature Genet.*, **6**, 119.
17. Sambrook, J., Fritsch, E. F., and Maniatis, T. (ed.) (1989). *Molecular cloning: a laboratory manual*, 2nd edn. Cold Spring Harbor Laboratory Press, Cold Spring Harbor, NY.
18. Frohman, M. A. (1993). In *Methods in enzymology* (ed. R. Wu), Vol. 218, p. 340 . Academic Press, London.
19. Clark, J. M. (1988). *Nucl. Acid Res.*, **16**, 9677.
20. Howe, L. R., Leevers, S. J., Gomez, N., Nakielny, S., Cohen, P., and Marshall, C. J. (1992). *Cell*, **71**, 335.
21. Her, J. -H., Lakhani, S., Zu, K., Vila, J., Dent, P., Sturgill, T. W., and Weber, M. J. (1993). *Biochem. J.*, **296**, 25.

22. Choi, K.-Y., Satterberg, B., Lyons, D. M., and Elion, E. A. (1994). *Cell*, **78**, 499.
23. Derijard, B., Hibi, M., Wu, I. -H., Barrett, T., Su, B., Deng, T., Karin, M., and Davis, R. (1994). *Cell*, **76**, 1025.
24. Pines, J. (1995). *Trends Genet.*, **11**, 326.
25. Gluzman, Y. (1981). *Cell*, **23**, 175.
26. Mellon, P., Parker, V., Gluzman, Y., and Maniatis, T. (1981). *Cell*, **27**, 279.
27. Chen, C. and Okayama, H. (1987). *Mol. Cell. Biol.*, **7**, 2745.
28. Chu, G., Hayakawa, H., and Berg, P. (1987). *Nucl. Acid Res.* **15**, 1311.
29. Luckow, V. A. and Summers, M. D. (1989). *Virology*, **170**, 31.
30. Tschopp, J. F., Brust, P. F., Cregg, J. M., Stillman, C., and Gingeras, T. R. (1987). *Nucl. Acid Res.*, **15**, 3859.
31. Fields, S. and Song, O.-K. (1989). *Nature*, **340**, 245.
32. Hieter, P., Bassett, D. E., Jr, and Valle, D. (1996). *Nature Genet.*, **13**, 253.
33. Johnston, M. (1996). *Curr. Biol.*, **6**, 500.
34. Hopp, T.P., Prickett, K.S., Price, V., Libby, R.T., March, C.J., Cerretti, P., Urdal, D.L., and Conlon, P.J. (1988). *Biotechnology*, **6**, 1205.
35. Wilson, I.A., Niman, H.L., Houghten, R.A., Cherenson, A.R., Connolly, M.L., and Lerner, R.A. (1984). *Cell*, **37**, 767.
36. Evans, G.I., Lewis, G.K., Ramsay, G., and Bishop, J.M. (1985). *Mol. Cell. Biol.*, **5**, 3610.

12

Interaction cloning of protein kinase partners

JULIE M. STONE and JOHN C. WALKER

1. Introduction

Historically, protein kinase partner proteins have been isolated by biochemical means, which requires purification of the interacting protein for antibody production or microsequencing before a cDNA clone can be obtained. However, two molecular biological approaches, the yeast two-hybrid system (Chapter 14) and interaction (or expression) cloning (1), directly yield a cDNA clone encoding the interacting protein. In this chapter we describe some of the recent advances in interaction cloning as a simple, rapid, and powerful alternative to biochemical approaches to identify protein kinase partners. The recombinant protein kinase of interest is enzymatically labelled with ^{32}P by virtue of a five amino recognition site for cAMP-dependent protein kinase (PKA), to facilitate the identification of cDNA clones encoding interacting partners. Briefly, expression of polypeptides encoded by cDNA inserts from a bacteriophage λ expression library is induced, and the polypeptides adsorbed on to nitrocellulose filters. The library is then screened with the radiolabelled protein probe to identify cDNA clones expressing interacting protein kinase partners (see *Figure 1*).

Interaction cloning is relatively easy to perform, and we will discuss some of the advantages of interaction cloning compared with the yeast two-hybrid system, as well as some potential pitfalls. The advantages include:

- no specialized cDNA libraries are required; bacteriophage λ cDNA expression libraries generated from the organism of interest are usually readily available;

- no specialized equipment or chemicals are required that are not routinely available in most laboratories practising basic molecular biology;

- the interaction cloning technique is rapid and simple, requiring considerably less time and effort than a two-hybrid screen.

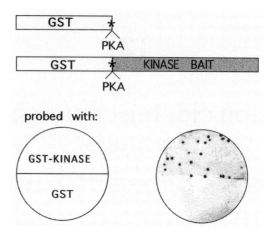

Figure 1. Representative results of a successful interaction cloning screen to identify a protein kinase partner. Recombinant fusion protein containing the protein kinase bait was labelled with ^{32}P (*) at a five amino acid protein kinase A (PKA) recognition site. This radiolabelled protein probe was used to screen a bacteriophage λ cDNA expression library to identify protein kinase partners. To ensure that the observed interaction is specific to the bait portion of the protein probe, a control experiment is performed prior to purification of the bacteriophage λ interacting clone. A representative autoradiogram of a specific interacting partner is shown.

The possible disadvantages include:

• some biologically relevant interactions that might be observed in the yeast two-hybrid system may not be detected in a filter binding assay;

• enzyme/substrate interactions are often transient; these might be detected in the two-hybrid system, but overlooked in the interaction cloning technique.

Neither interaction cloning nor the two-hybrid system will detect protein–protein interactions which are dependent on post-translational modifications that would not occur in bacteria or yeast. However, the proven success of the technique warrants its application.

2. Applications of interaction cloning to identify protein kinase partners

The screening of bacteriophage λ cDNA expression libraries with radio-labelled DNA probes to identify DNA-binding proteins (2), or with anti-bodies to identify any protein antigen (3), are well-established techniques. These approaches can be modified to identify interacting proteins by probing the filters with the protein of interest (the 'bait') and detecting the bait using antibodies (4). However, by screening with a ^{32}P-labelled bait protein as described here, the additional incubations and washes necessary for immuno-detection, which increase the likelihood of disrupting weak protein–protein interactions, can be avoided. ^{125}I-labelled protein probes have also been used successfully to screen for interacting proteins (5, 6), but the problems associated with handling ^{125}I can be considerable.

Protein kinases autophosphorylated with ^{32}P were originally used to screen cDNA expression libraries for the isolation of proteins which interact with receptor protein kinases, a technique referred to as CORT (cloning of receptor targets) (7–9). However, by introducing a PKA recognition site into the protein bait as a target for ^{32}P-labelling, the technique can be extended to proteins that do not have inherent autophosphorylation capabilities (10, 11).

The utility of interaction cloning has been well demonstrated, and several protein kinase substrates have been identified using this technique, or modifications of it (4, 7, 12, 13). However, because the success of the interaction cloning screen is dependent on the strength of the interaction, and many protein phosphorylation reactions occur via transient enzyme–substrate intermediates, interaction cloning is most likely to yield regulatory proteins or subunits rather than substrates. Indeed, these types of interactions are commonly observed (7, 8, 14). Another recently described modification of the interaction cloning technique has been used to identify protein kinase substrates (see Chapter 13). In this approach, solid phase phosphorylation of proteins encoded by a λ phage library has been reported using a purified protein kinase and [γ-^{32}P]ATP (15, Chapter 13). This technique may permit isolation of substrates that form a transient intermediate, where binding affinity may be low.

3. Design of the protein probe and expression library

Two important choices are required in interaction cloning:

- the design of the 'bait' protein containing the phosphorylation site for labelling
- the construction or acquisition of an appropriate cDNA expression library

3.1 Design of the 'bait' protein

Although radiolabelling of protein kinases by autophosphorylation has been used successfully to identify binding partners, inclusion of an artificial PKA phosphorylation site can lead to labelling at high specific activity, and facilitates interaction cloning with protein kinases that have low intrinsic kinase activity or an inability to autophosphorylate. To facilitate one-step purification of the recombinant protein kinase bait, an affinity tag is included in the design of the protein probe. Furthermore, the bait portion of the probe should be carefully considered. For example, inclusion of different portions of the protein kinase (e.g. catalytic domain, regulatory domains) in the bait might yield different groups of interacting partners. This approach could be informative in the later stages of determining the relevance of the observed interaction.

Vectors for recombinant fusion protein expression that contain a PKA recognition site can be obtained commercially. Vectors yielding glutathione-*S*-transferase (GST) fusions are the most commonly used, but the method can be adapted to vectors yielding other affinity tags. Several companies market PKA site vectors with affinity tags such as GST (Pharmacia), poly-histidine (Novagen), or calmodulin-binding protein (Stratagene). Alternatively, one can engineer a PKA site into existing vectors by introducing the five amino acid recognition site (RRASV) with synthetic oligonucleotides.

3.2 The cDNA expression library

Bacteriophage λ cDNA expression libraries constructed by reverse transcription of mRNA isolated from the tissue and organism of interest are generally widely available. However, if it is necessary to construct a library for a particular purpose, several options should be considered in the design of the library. For example:

- the first strand synthesis of the cDNA inserts may employ either an oligo(dT) primer or random oligonucleotide primers
- a suitable bacteriophage λ expression vector must be chosen

First-strand cDNA synthesis employing oligo(dT) primers is widely used. This permits directional cloning of the cDNA inserts, which will reduce the number of phage to be screened by twofold. However, this approach tends to select for 3'-ends of mRNA, which might bias the library. In practice because screening the library is straightforward, the complexity introduced by using random primers for first-strand synthesis is not a significant problem. Moreover, libraries generated using random primers may prove advantageous, because multiple clones representing different domains of the same protein may be identified. This could provide information delimiting the region of the protein kinase partner responsible for the observed interaction. Full-length cDNA clones can subsequently be obtained from other libraries.

The most widely used expression vector has been λgt11 (3). However, modified versions of this vector facilitate the recovery of the cDNA inserts by circumventing time-consuming λ phage DNA preparations. Vectors are available which utilize Cre-Lox recombination for *in vivo* conversion of recombinant phages into plasmid DNA in Cre recombinase-expressing host strains (e.g. λZipLox, Life Technologies), as well as those that employ helper phage for *in vivo* excision of a phagemid vector (e.g. λZAP vectors, Stratagene). Whereas most λ vectors available have cloning sites in the *lacZ* gene which produce fusions of the cDNA inserts to β-galactosidase, some λ vectors (e.g. λSCREEN-1, an improved version of λEXLOX, Novagen) utilize the T7 polymerase and may yield higher expression of the library proteins (7).

4. Preparation of protein probe

A number of excellent sources are available for descriptions of subcloning into the appropriate expression vector and subsequent purification (16), therefore these will not be discussed here. Radiolabelling of PKA site-containing recombinant protein, as described in *Protocol 1*, is adapted from the method described by Blanar and Rutter (11).

Protocol 1.   Preparation of ^{32}P-labelled protein probe

Equipment and reagents

- purified recombinant protein kinase 'bait' with a PKA recognition site
- cAMP-dependent protein kinase (Sigma P-2645; catalytic subunit of PKA from bovine heart, available in 250 Unit aliquots)
- DTT (40 mM; freshly prepared)
- 10 × PKA buffer (200 mM Tris–HCl, pH 7.5, 10 mM DTT, 1 M NaCl, 120 mM MgCl$_2$)
- Sephadex G-50 (medium grade)

- [γ-^{32}P]ATP (6000 mCi/mmol; BLU-002Z, NEN^{TM} Life Technologies)
- Z'-KCl (25 mM Hepes–KOH, pH 7.4, 12.5 mM MgCl$_2$, 20% glycerol, 100 mM KCl, 1 mg/ml BSA, 1 mM DTT; filter sterilized and stored at 4°C)
- disposable plastic columns (3 ml; or disposable syringes with a glass wool plug)
- scintillation counter

Method

1. Reconstitute 250 Units cAMP-dependent protein kinase in 25 μl of freshly prepared 40 mM DTT. Let the reconstituted enzyme stand at room temperature for approximately 10 min before use, store at 4°C.[a]

2. Phosphorylate the purified recombinant fusion protein(s), containing a PKA recognition site, by incubating for 1 h at room temperature in a reaction containing:

   3 μl 10 × PKA buffer
   5 μl [γ-^{32}P]ATP
   1–10 μl (~1 μg) of purified fusion protein containing the PKA site
   1 μl (10 Units) PKA (reconstituted as in step 1)
   H$_2$O to 30 μl total volume

3. After the 1 h incubation period, add 170 μl of ice-cold Z'-KCl and store on ice.

4. Prepare a gel filtration column to remove unincorporated ^{32}P. Pour Sephadex G-50 (medium) equilibrated in Z'-KCl into a disposable plastic column to give a bed volume of ~3 ml.

5. Allow the column to drain until the level of the buffer is at that of the top of the column bed. Load the entire reaction (from step 3) on to the G-50 column and collect the effluent in an Eppendorf tube. Place another tube under the column, add 200 μl Z'-KCl to the column, and collect the effluent. Repeat the addition and collection steps to collect a total of 12–200 μl fractions. Measure the radioactivity by Cerenkov

---

Protocol 1. *Continued*

counting in a scintillation counter, to determine the fractions with the highest specific activities and cpm/ml.

6. Optional: analyse the ^{32}P-labelled protein probe by SDS–polyacrylamide gel electrophoresis (PAGE) and autoradiography.

[a] It is important to use freshly prepared 40 mM DTT for reconstitution; the cAMP-dependent protein kinase is extremely unstable after reconstitution, and retains activity for only 2–3 days when stored at 4°C.

4.1 Comments on preparation of the ^{32}P-labelled probe

Both the affinity tag–bait fusion protein and a control containing the affinity tag alone, or the tag fused to an unrelated protein, should be radiolabelled simultaneously. The control will be used at later stages of screening to ensure that only clones that specifically interact with the bait portion of the fusion are pursued.

The Sephadex G-50 can be swelled in H$_2$O, or buffer, and stored at 4°C for months. In this case, five to ten column volumes of Z'-KCl should be used to equilibrate the column prior to loading the sample.

Two peaks of radioactivity are typically eluted from the Sephadex G-50 column. The first peak (usually fractions 5–8) corresponds to labelled protein probe, whereas the second peak (usually fractions 10–12) corresponds to unincorporated ATP and should be discarded. Fractions can be stored at 4°C for two weeks.

A small amount (1–2 μl) of the labelled protein probe can be analysed by SDS–PAGE and autoradiography. Two bands are typically evident; one of the predicted size of the GST–protein kinase fusion protein and the other of the predicted molecular weight of the affinity tag alone. This is due to the inherent susceptibility to proteolytic cleavage at the fusion junction. The presence of labelled GST protein in the probe should not interfere with the screen, since a control experiment to eliminate clones which interact with the affinity tag will be performed at the later stages of screening (see *Figure 1*).

5. Identification of interacting clones

Protocol 2. Screening the cDNA expression library

Equipment and reagents

- *E. coli* host bacteria (e.g. Y1090r⁻)
- bacteriophage λ cDNA expression library
- 150 mm or 100 mm LB-agar plates (LB containing 12 g/l Bacto agar)
- 0.7% agarose in LB

- Luria–Bertani (LB) medium (10 g/l Bacto tryptone, 5 g/l Bacto yeast extract, 5 g/l NaCl, pH 7.5; supplemented with 10 mM MgSO$_4$ and 0.2% maltose plus appropriate antibiotic for selection)

- nitrocellulose membrane filters (137 mm and 82 mm)
- IPTG (10 mM isopropyl β-D-thiogalacto-pyranoside)
- TBS–T (Tris buffered saline–Triton; 10 mM Tris–HCl, pH 8.0, 150 mM NaCl, 0.05% Triton X-100)
- HBB (Hepes blocking buffer; 20 mM Hepes–KOH, pH 7.4, 5 mM MgCl$_2$, 1 mM KCl, 5% non-fat dry milk)
- 22G needle, India ink
- binding buffer (20 mM Hepes–KOH, pH 7.4, 7.5 mM KCl, 0.1 mM EDTA, 2.5 mM MgCl$_2$, 1% non-fat dry milk)
- suspension medium (20 mM Tris–HCl, pH 7.4, 100 mM NaCl, 10 mM MgSO$_4$, 2% gelatin)
- autoradiographic film

Method A. Preparation of host strain cells and plating of cDNA library

1. Grow a 50 ml culture of Y1090r⁻ (or other appropriate *E. coli* host strain) in LB medium supplemented with 10 mM MgSO$_4$ and 0.2% maltose, and appropriate antibiotic, overnight at 37°C to saturation. Centrifuge to pellet the cells, and resuspend in 25 ml of 10 mM MgSO$_4$. The resuspended cells can be stored at 4°C and used for up to one week.

2. Determine the titre of the bacteriophage cDNA library by performing a 'spot titre'.

3. Prepare a 'spot titre' plate. Add 200 μl of the resuspended *E. coli* host cells to 3.5 ml of 0.7% agarose in LB (pre-warmed to 47°C) and pour on to a 100 mm LB-agar plate (supplemented with antibiotic, if necessary); allow to solidify.

4. Prepare a serial dilution of the bacteriophage λ cDNA expression library stock, in suspension medium. Pipette 5 μl of each dilution on to the surface of the plate and incubate at 37°C until plaques are visible (several h to overnight). Calculate the pfu/ml of the original phage stock.

5. Incubate eight 1.5 ml Eppendorf tubes, containing ∼40 000 pfu of the bacteriophage λ cDNA library, with 0.6 ml *E. coli* host (*Protocol 2A,* step 1) at 37°C for 15 min. Add the contents of each tube to 7 ml 0.7% agarose in LB at 47°C, and pour on to 150 mm LB-agar plates (supplemented with antibiotic, if necessary).

6. Incubate the plates at 42°C for ∼3 h until small plaques are visible. Formation of plaques in the absence of expression from the *lacZ* gene promoter ensures that any toxic fusion proteins present in the recombinants will not bias the library. While the plates are incubating, prepare the filters according to *Protocol 2B,* step 1.

Method B. Preparation of filters and adsorption of bacteriophage

1. Soak nitrocellulose filters in IPTG for 15 min at room temperature and air-dry. IPTG-impregnated filters can be stored in a Petri dish until use.

Protocol 2. *Continued*

2. When plaques are visible (*Protocol 2A*, step 6) overlay the plates with IPTG-impregnated filters and incubate at 37 °C an additional 6–8 h.

3. Chill the plates at 4 °C for 15 min (or overnight) and mark the filter orientation by piercing in several locations with a needle dipped in India ink. Store master plates at 4 °C.

Method C. Blocking and probing the filters

1. Remove filters and wash for 15 min in TBS–T at room temperature with shaking.

2. Reduce non-specific binding by blocking filters for 1–4 h with rocking (or shaking) at 4 °C in 100 ml HBB. Blocking can be performed overnight for convenience.

3. Incubate overnight with rocking/shaking at 4 °C in binding buffer containing 2.5–5 × 10^5 cpm/ml of radiolabelled fusion protein.

4. Wash the membrane filters three times for 10 min each in 100 ml binding buffer with shaking at room temperature, air dry, and expose to autoradiography film.

Method D. Isolation and purification of interacting clones

1. On a light-box, align the plates using the needle marks as a guide. Remove an agarose plug using the large end of a glass Pasteur pipette at the position of the positive clone(s), and transfer to a 1.5 ml Eppendorf tube. Add 1 ml suspension medium and 1 drop of $CHCl_3$, and store at 4 °C.

2. Determine the titre by serial dilution and 'spot titre' as in *Protocol 2A*, steps 2–4, and perform successive screening procedures to obtain purified clones. Subsequent screens are performed on 100 mm LB plates with ~2000 pfu/plate for a secondary screen, ~300–500 pfu/plate for a tertiary screen. Before purified clones are obtained (usually during the tertiary screen), a control to eliminate clones which might interact with the GST portion of the probe should be performed. To do this, the filters are cut in half, one half is probed with the labelled GST–bait and the other half probed with labelled GST (or other affinity tag) alone (see *Figure 1*).

5.1 Comments on expression library screening

(a) All eight filters can be placed in a 150 mm dish and probed with 30–40 ml of solution. The plate is wrapped in Parafilm and placed in a Plexiglass box for shielding. Alternatively, the filters can be incubated with solution in a heat-sealed bag. The probe solution can be stored at 4 °C and used for the subsequent secondary and tertiary screenings.

(b) Agarose plugs in suspension medium can be stored at 4°C for several months. Therefore, if many putative positive clones are identified in the primary screen, they can be stored for future analysis. For longer term storage, add dimethyl sulfoxide (DMSO) to a final concentration of 7% and store indefinitely at −70°C.

(c) The method for preparing the phage DNA will depend on the cloning vector used to construct the cDNA expression library. Instructions are provided by the suppliers.

5.2 Troubleshooting and optimization of the screen

Some of the problems that might appear during the screening process include:

- a high background
- very few or no positive interacting clones
- too many positive clones

A high signal-to-noise ratio could be due to a poor choice of nitrocellulose membrane filters. Side by side comparisons of the background on membranes from different suppliers suggests that the optimum membrane choice is rather empirical, with results dependent on the protein probe (J.M. Stone and J.C. Walker, unpublished results). The stringency of the screen can be altered by varying the salt and/or detergent concentrations of the washing and blocking buffers, or the length of time for blocking and washing. Adsorption of fusion proteins may alter the conformation of the library proteins, preventing detection of some interactions. This problem might be circumvented by denaturation/renaturation of proteins using 6 M guanidine-HCl (2, 3).

The relevance of the interaction between the bait and a putative protein kinase partner must be confirmed by other means. The PKA-containing fusion can be used to confirm interaction by what is commonly referred to as overlay assays, or far-Western analysis (17, 18). Alternatively, a solution-binding assay can be performed (19). One should assess whether the interacting partner is a substrate for the protein kinase using *in vitro* phosphorylation, and ultimately demonstrate interaction *in vivo* by co-immunoprecipitation.

References

1. Stone, J. M. (1997). In *Current protocols in molecular biology* (ed. F. M. Ausubel, R. Brent, R. E. Kingston, D. D. Moore, J. G. Seidman, J. A. Smith, and K. Struhl), Unit 20.3. John Wiley & Sons, Inc., New York.
2. Singh, H. (1991). In *Current protocols in molecular biology* (ed. F. M. Ausubel, R. Brent, R. E. Kingston, D. D. Moore, J. G. Seidman, J. A. Smith, and K. Struhl), Unit 12.7. John Wiley & Sons, Inc., New York.
3. Huynh, T. V., Young, R. A., and Davis, R. W. (1985). In *DNA cloning: a practical approach* (ed. D. M. Glover), Vol. I, p. 49. IRL Press, Oxford.

4. Chapline, C., Ramsay, K., Klauck, T., and Jaken, S. (1993). *Journal of Biological Chemistry*, **268**, 6858.
5. Mathews, L. S. and Vale, W. W. (1991). *Cell*, **65**, 973.
6. Lin, H. Y., Wang, X.-F., Ng-Eaton, E., Weinberg, R. A., and Lodish, H. F. (1992). *Cell*, **68**, 775.
7. Margolis, B., Silvennoinen, O., Comoglio, F., Roonprapunt, C., Skolnik, E., Ullrich, A., and Schlessinger, J. (1992). *Proceedings of the National Academy of Sciences USA*, **89**, 8894.
8. Lowenstein, E. J., Daly, R. J., Batzer, A. G., Li, W., Margolis, B., Lammers, R., Ullrich, A., Skolnik, E. Y., Bar-Sagi, D., and Schelssinger, J. (1992). *Cell*, **70**, 431.
9. Skolnik, E. Y., Margolis, B., Mohammadi, M., Lowenstein, E., Fischer, R., Drepps, A., Ullrich, A., and Schlessinger, J. (1991). *Cell*, **65**, 83.
10. Kaelin, W. G. J., Krek, W., Sellers, W. R., DeCaprio, J. A., Ajchenbaum, F., Fuchs, C. S., Chittenden, T., Li, Y., Farnham, P. J., Blanar, M. A., Livingston, D. M., and Flemington, E. K. (1992). *Cell*, **70**, 351.
11. Blanar, M. A. and Rutter, W. J. (1992). *Science*, **256**, 1014.
12. Chapline, C., Mousseau, B., Ramsay, K., Duddy, S., Li, Y., Kiley, S. C., and Jaken, S. (1996). *Journal of Biological Chemistry*, **271**, 6417.
13. Stone, J. M., Collinge, M. A., Smith, R. D., Horn, M. A., and Walker, J. C. (1994). *Science*, **266**, 793.
14. Mochly-Rosen, D. and Gordon, A. S. (1998). *FASEB Journal*, **12**, 35.
15. Fukunga, R. and Hunter, T. (1997). *EMBO Journal*, **16**, 1921.
16. Ausubel, F. M., Brent, R., Kingston, R. E., Moore, D. D., Seidman, J. G., Smith, J. A., and Struhl, K. (ed.) (1998). *Current protocols in molecular biology*. John Wiley & Sons, Inc., New York.
17. Hausken, Z. E., Coghlan, V. M., and Scott, J. D. (1998). In *Methods in molecular biology: protein targeting protocols* (ed. R. A. Clegg), Vol. 88, p. 47. Humana Press, Totowa.
18. Carr, D. W. and Scott, J. D. (1992). *Trends in Biochemical Sciences*, **17**, 246.
19. Swaffield, J. C. and Johnston, S. A. (1996). In *Current protocols in molecular biology* (ed. F. M. Ausubel, R. Brent, R. E. Kingston, D. D. Moore, J. G. Seidman, J. A. Smith, and K. Struhl), Unit 20.2. John Wiley & Sons, Inc., New York.

13

Identifying protein kinase substrates by expression screening with solid-phase phosphorylation

RIKIRO FUKUNAGA and TONY HUNTER

1. Introduction

The biological activities of protein kinases are evoked through phosphorylation of their substrate proteins. Phosphorylation of target proteins causes changes in their structure, stability, enzymatic activity, ability to interact with other molecules, or subcellular localization, leading to regulation of a wide variety of cellular processes. Indeed, the identification of physiological targets has been a high priority ever since the first protein kinase was purified. However, the conventional approach of purifying substrate proteins by biochemical techniques is laborious and time consuming, and is especially difficult in the case of scarce proteins. Among the approaches developed to identify protein kinase substrates in a systematic manner are various techniques for determining consensus phosphorylation site sequences using oriented peptide libraries (1) (see Chapter 16), interaction screening for protein substrates by the far-Western method (2), or the yeast two-hybrid system (3, 4) (see Chapter 14).

In this chapter, we describe an alternative screening method for identifying protein kinase substrates. This technique, termed 'phosphorylation screening', utilizes a cDNA-expressing λ phage library, with phosphorylation of expressed proteins in the solid-phase. Application of this strategy to the ERK1 MAP kinase system resulted in the isolation of several cDNAs encoding both known and novel substrates (5). We have also used phosphorylation screening to identify substrates for cyclin E/Cdk2 (6), and we expect that it will be generally applicable for direct identification of physiological targets of various protein kinases. A similar phosphorylation screening method has recently been developed for the identification of protein–tyrosine kinase substrates using a λgt11 cDNA expression library (7).

2. Phosphorylation screening of a phage expression library

In this method, a λgt11-like phage expression library is screened using *in vitro*, solid-phase phosphorylation with [γ-^{32}P]ATP and the soluble protein kinase of interest. The method may be summarized as follows:

(a) A cDNA library is prepared using λGEX5 phage vector, in which a cDNA is inserted downstream of the glutathione-*S*-transferase (GST) coding region (5, 8).

(b) Phage plaques of the cDNA library are formed on agar plates, and GST-fused recombinant proteins expressed in the plaques are transferred and immobilized on nitrocellulose filters.

(c) The plaque filters are incubated with a purified, active protein kinase in the presence of [γ-^{32}P]ATP to allow solid-phase phosphorylation. The phosphorylated plaques are visualized by autoradiography.

(d) The positive clones are characterized by cDNA sequencing, and *in vitro* and *in vivo* analyses, to identify physiological substrates for the protein kinase.

The use of a solid-phase phosphorylation screening protocol is based on findings that cellular proteins immobilized on a membrane filter can be phosphorylated by a soluble protein kinase with similar specificity to that obtained in conventional liquid-phase phosphorylation (9, 10). Phosphorylation screening has several advantages over other methods including conventional substrate purification, determination of consensus peptide sequences, or interaction screening:

(a) Even scarce proteins can be detected as long as they are good substrates.

(b) Unlike peptide library screening methods, naturally existing proteins can be directly identified.

(c) Unlike interaction screening methods, substrates that do not form a stable complex with the protein kinase can be detected.

(d) Phosphorylation screening identifies only substrate proteins, whereas interaction screening is likely to detect also other kinase-interacting molecules such as subunits or regulatory proteins.

(e) Cloned cDNAs can be easily sequenced, and because they are produced as GST fusion proteins they can be readily purified for antibody preparation.

There are, however, some potential disadvantages and practical problems in the application of this technique:

(a) If the protein kinase of interest phosphorylates endogenous protein(s) derived from *E. coli* or λ phage, it is difficult to distinguish positive plaques from negatives owing to high background. This problem, how-ever, might be minimized if an appropriate affinity system in which the recombinant products are selectively retained on a membrane filter is available [e.g. a glutathione (GSH)-derivatized cellulose filter, see Section 8].

(b) A significant amount (e.g. microgram scale) of purified protein kinase is required in an active and soluble form.

(c) The screening may isolate not only physiological targets, but also proteins that are not physiological targets but, fortuitously, are good substrates, especially in the case of a protein kinase that has relatively low specificity.

(d) Protein kinases, or other proteins that autophosphorylate, may also score as positive in the screen. This problem can be reduced by pre-incubation of the filters with unlabelled ATP and is also minimized if the cDNA inserts are not too long. Using randomly primed cDNAs of ~1 kb in size for the library preparation might minimize problems with auto-phosphorylating protein kinases (see Section 3.3).

Prior to using the phosphorylation screening technique, these advantages and disadvantages must be fully considered to judge whether this approach is suitable for the protein kinase in question.

3. Construction of λGEX5 cDNA library

3.1 λGEX5 vector

For the new screening method, we modified the phage vector λgt11 to pro-duce λGEX5 (5). The λgt11 vector has previously been successfully used for various expression screening strategies such as immunoscreening, nucleic acid–protein interaction (south-Western), and protein–protein interaction (far-Western) (11, 12). The λGEX5 vector contains a plasmid sequence between the two *Not*I sites, consisting of a ColE1 origin, the ampicillin resistance gene, and a GST gene followed by a small (0.43-kb) stuffer sequence (*Figure 1*). The nucleotide sequence of the plasmid region (pGEX-PUC-3T, see *Figure 4*) is available in the DDBJ/EMBL/Genbank database with the accession number AB014641. The *Sfi*I sites at the ends of the stuffer region are used for insertion of cDNA of up to 9 kb in size, which can be expressed as GST fusion proteins upon induction by isopropyl β-D-thiogalactopyranoside (IPTG). The λGEX5 vector system has the following advantages over the original λgt11 vector:

- Isolated clones can be rapidly converted into plasmid clones by excision rescue without purifying cDNA fragments.
- The rescued plasmids can be directly utilized not only for cDNA sequencing but also for expression of the GST fusion proteins for further character-ization.

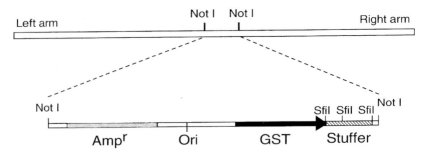

Figure 1. Structure of λGEX5 cloning vector. The plasmid region between two *Not*I sites is expanded in the lower part.

- GST, the N-terminal fusion partner of the recombinant product expressed by λGEX5, is highly soluble, easy to purify by GSH–agarose chromatography, and much smaller (27 kDa) than the β-galactosidase (114 kDa) component of fusion proteins expressed by λgt11 (12).

These advantages enable rapid isolation, subcloning, and characterization of a large number of positive clones at the same time.

3.2 Preparation of vector arms

λGEX5 phage DNA is purified from large quantities (10^{12}–10^{13} pfu) of phage lysate by established methods (13, 14). The λGEX5 phage contains an amber

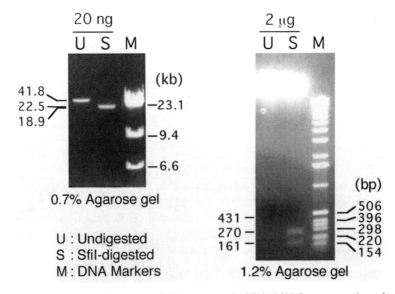

Figure 2. Confirmation of complete *Sfi*I digestion of λGEX5 DNA for preparation of vector arms. Undigested or *Sfi*I-digested λGEX5 DNA was analysed by electrophoresis on 0.7% (left) and 1.2% (right) agarose gels. The left arm (22.5 kb) and right arm (18.9 kb) are not clearly separated. Note that no DNA fragment is visible at the position of 431 bp.

mutation (*Sam*100) and should be propagated using *E. coli* BB4 (15) as the host strain. Other strains, such as Y1090 (12) and its derivatives, may not be suitable. If it is hard to obtain a high titre phage lysate by liquid culture a large-scale plate lysate (e.g. using 20 × 150 mm agarose plates) should be used instead (13, 14).

Vector arms for construction of a cDNA library are prepared by *Sfi*I digestion of the λGEX5 DNA. Dephosphorylation of the cleaved sites is not necessary because the 3′-protruding, single-stranded termini (3′-CGT) of the arms are not compatible with each other. The existence of an additional *Sfi*I site in the stuffer sequence helps to check whether or not the *Sfi*I digestion is complete (*Figure 2*). After confirming that the digestion is complete, the left arm (22.5 kb) and right arm (18.9 kb) are purified together by sucrose density gradient centrifugation or by preparative agarose gel electrophoresis (13, 14).

Protocol 1. Preparation of λGEX5 vector arms

Reagents

- λGEX5 DNA[a]
- TE buffer (10 mM Tris–HCl, pH 8.0 or 7.5, 1 mM EDTA)

- SDG buffer containing 10% or 40% sucrose [10 mM Tris–HCl, pH 8.0, 0.9 M NaCl, 5 mM EDTA, 10 or 40% (w/v) sucrose]

Method

1. Digest 80 μg of λGEX5 DNA with *Sfi*I.

2. Analyse two aliquots (20 ng and 2 μg) of the digested DNA by 0.7% and 1.2% agarose gel electrophoresis, respectively. It is important to confirm that no DNA is visible at the position of the undigested λGEX5 DNA on the 0.7% gel. On the 1.2% gel, 270 bp and 161 bp bands, but not a 431 bp band, should be visible. If the 431 bp band is detectable, it means that the *Sfi*I digestion is incomplete, which will cause a high ratio of empty phage clones in the library.

3. Extract the digested DNA once with phenol/chloroform, and once with chloroform, and recover the DNA by ethanol precipitation.

4. Dissolve the DNA in 0.4 ml of TE buffer pH 8.0, add 4 μl of 1 M MgCl$_2$, and incubate for 1 h at 42°C to allow the cohesive termini of the arms to anneal.

5. Prepare a 10–40% linear sucrose gradient (12 ml) in two centrifuge tubes [Beckman SW41.Ti rotor (or equivalent)].

6. Load 0.2 ml of the annealed DNA (40 μg) on to each gradient and centrifuge at 180 000 g (at r_{max}) for 16 h at 15°C in a Beckman SW41.Ti rotor (or equivalent).

7. Collect 10 drops (0.6–0.7 ml) of fractions through a 20G needle from the bottom of the centrifuge tube.

Protocol 1. *Continued*

8. Remove 10 μl of each fraction and analyse by 0.7% agarose gel electrophoresis.

9. Pool the fractions that contain the annealed and unannealed arms. Dilute the pooled fractions with an equal volume of TE buffer pH 7.5 and recover the DNA by ethanol precipitation.

[a] The λGEX5 vector is available on request through the DNA bank laboratory at Tsukuba Life Science Center, the Institute of Physical and Chemical Research (RIKEN), Tsukuba 305-0074, Japan. The ID number of the λGEX5 vector is RDB1911.

3.3 Preparation of *Sfi*I adaptor-ligated cDNA

Although synthesis of double-stranded cDNA is an important step for cDNA cloning, a detailed description of this step is beyond the scope of this chapter. Double-stranded cDNA can be synthesized from poly(A)$^+$RNA using an oligo(dT) primer or random hexanucleotide primers by a general method for cDNA synthesis (13, 14) or using an appropriate kit. Oligo(dT)-primed cDNA libraries are prone to be rich in clones that contains only the C-terminal part of a protein. On the other hand, randomly primed cDNA libraries evenly cover every part of a protein in principle, although they often contain a high percentage of cDNAs for ribosomal RNAs. At the final stage of cDNA synthesis, the ends of the double-stranded cDNA should be blunted for adaptor ligation. Two 5'-phosphorylated oligonucleotides are annealed to form an adaptor, which has a blunt end and a non-palindromic, single-stranded overhang (3'-ACG) compatible with the cloning sites of the *Sfi*I-digested λGEX5 arms. After adaptor ligation, the cDNA is separated by agarose gel electrophoresis, and the cDNA whose size is larger than ~0.8 kb in length is recovered by electroelution. If the cDNA is synthesized with random primers, it may be a good idea to fractionate only cDNA of ~1 kb in size, in order to minimize problems with autophosphorylating protein kinases. A large amount of excess oligonucleotide can be removed through the size fractionation step.

Protocol 2. Adaptor ligation and size fractionation of double-stranded cDNA

Reagents

- 5'-phosphorylated oligonucleotides [12 mer: 5'-pd(CCAGCACCTGCA)-3'; 9 mer: 5'-pd(AGGTGCTGG)-3']
- 10 × ligase buffer (0.5 M Tris–HCl, pH 7.5, 0.1 M MgCl$_2$, 0.1 M DTT, 10 mM ATP)

Method

1. Make up an oligonucleotide mixture containing 2.4 and 1.8 μg of the 12-mer and 9-mer oligonucleotides, respectively, in 20 μl of 10 mM MgCl$_2$.

2. Incubate the mixture at 80 °C for 2 min, allow to cool slowly to room temperature over a period of about 60 min, and then chill on ice.

3. Mix in the following order:

   55 μl of H_2O
   20 μl of the annealed oligonucleotides
   10 μl (1–2 μg) of double-stranded cDNA
   10 μl of 10 × ligase buffer
   1 μl of 10 units/μl of T4 polynucleotide kinase
   4 μl of 400 units/μl T4 DNA ligase

4. Incubate the mixture (100 μl) for 6–14 h at 16 °C. Recover the DNA by phenol/chloroform extraction and ethanol precipitation.

5. Dissolve the adaptor-ligated cDNA in 20–50 μl of TE buffer pH 8.0, and separate the DNA fragments by preparative electrophoresis on a 1% agarose gel.

6. Locate the region of DNA fragments larger than 0.8 kb in size by using appropriate DNA size markers, and recover the size-fractionated DNA from the agarose gel by electroelution.

3.4 Ligation and *in vitro* packaging

The adaptor-ligated and size-fractionated cDNA molecules can now be ligated to the *Sfi*I-digested vector arms, followed by packaging of the ligated DNA into bacteriophage λ particles using an *in vitro* packaging reaction. We usually use commercially available kits for this ligation (e.g. DNA ligation kit Version 1, PanVera/Takara) and *in vitro* packaging (Stratagene Gigapack Gold, or equivalent) according to the manufacturer's instructions. It is desirable to carry out pilot ligations to optimize the ratio of cDNA to vector arms for efficient production of recombinant phages. Also, a control ligation without cDNA is essential to check the background level of empty phage. The titre of the *in vitro*-packaged phage is determined on *E. coli* BB4. The pfu in the original, packaged phage is the number of independent clones in the library. If necessary, the cDNA library can be amplified once by propagating phages in *E. coli* BB4 on agar plates (13, 14).

3.5 Construction of a positive control phage for screening

In phosphorylation screening, conditions for screening must be determined experimentally for each protein kinase. For this purpose, ideally one needs to have a positive control phage that expresses the GST fusion protein of an appropriate substrate protein (or peptide) for the protein kinase being used for the screening (assuming that one is known). The positive control phage can be constructed by ligating the *Sfi*I-digested λGEX5 arms to an appropriate substrate cDNA produced by PCR with *Sfi*I site-containing primers. For the screening of ERK1 MAP kinase substrates, we constructed a λGEX5 recombinant encoding the C-terminal region of a transcription factor Elk-1 (λGEX-Elk-C), which contains multiple ERK phosphorylation sites (5, 16). Similarly, a GST fusion protein of the C-terminal domain of the retinoblastoma protein (GST–Rb) worked well as a positive control in a screen for cyclin E/Cdk2 substrates (6).

4. Phosphorylation screening

In this section we describe protocols for solid-phase phosphorylation, using as an example substrate screening with ERK1 MAP kinase (5). Points to be considered for application to other protein kinases will also be discussed.

4.1 Preparation of protein kinase for phosphorylation screening

The protein kinase used for phosphorylation screening must be soluble, active, and sufficiently pure. Care should be taken to avoid the presence of significant amounts of other protein kinases. Large-scale production and purification of recombinant protein kinase would be the best for this purpose. To obtain a large amount of activated ERK1 MAP kinase, we produced recombinant human ERK1 using a baculovirus expression system, in which insect Sf9 cells were co-infected with three recombinant baculoviruses encoding v-Ras, c-Raf-1, and ERK1 as described (5, 17). Similar co-expression strategies utilizing a eukaryotic expression system would also be effective for large-scale production of a protein kinase that requires a subunit protein or an activating modification such as phosphorylation. For example, active cyclin-Cdk complexes can be efficiently produced by co-expression of a hexahistidine-tagged cyclin and its partner Cdk in Sf9 cells, and easily purified using nickel-affinity column chromatography (6).

4.2 Preparation of plaque-immobilized filters

Procedures in this step are essentially the same as those used for expression screens using the λgt11 phage system (11–14). A λGEX5 cDNA library is plated on agar plates, and expression of the encoded GST fusion proteins is induced by overlaying IPTG-containing nitrocellulose filters on to the plaques. Plating density should be in the range of $1.5-3 \times 10^4$ plaques per 150 mm agar plate. High density plating ($>5 \times 10^4$ plaques per plate) may result in weak and small signals that are indistinguishable from false-positive signals.

Protocol 3. Plating out λGEX5 library and immobilization of plaques on filters

Reagents

- *E. coli* BB4 strain (Stratagene)
- SM (50 mM Tris–HCl, pH 7.5, 100 mM NaCl, 10 mM MgSO$_4$ and 2% gelatin)
- NZCYM medium (1% NZ amine, 0.1% casamino acids, 0.5% bacto-yeast extract, 1% NaCl, 10 mM MgSO$_4$; adjust the pH to 7.5 with 1 N NaOH)
- agar plates (60–80 ml of 1.5% agar in NZCYM medium per 150 mm plate)

- TB medium (1% bacto-tryptone, 0.5% NaCl)
- top agarose (0.7% agarose in NZCYM medium)
- nitrocellulose filters [137 mm in diameter, Schleicher & Schuell BA85 (0.45 μm) or equivalent]
- isopropyl β-D-thiogalactopyranoside (IPTG, 10 mM)

Method

1. Pick up a single colony of BB4 and grow cells in TB medium containing 0.2% maltose, 10 mM MgSO$_4$, and 12.5 μg/ml tetracycline[a] at 30°C overnight. Centrifuge the cells at 1500 *g* for 10 min and resuspend in 10 mM MgSO$_4$ at a density of A_{600} = 2.0. The plating bacteria can be stored at 4°C for up to a week.

2. Prepare 1.5–2 × 10^5 pfu/ml of λGEX5 library phage in SM.

3. Mix 0.1 ml (i.e. 1.5–2 × 10^4 pfu) of the phage with 0.5 ml of plating bacteria and incubate for 15 min at 37°C.

4. Add 8 ml of molten (50°C) top agarose, mix well, and pour on to 1.5% agar plates pre-warmed at 37°C. Leave the plates at room temperature for 15 min to harden the top agarose.

5. Incubate at 42°C for 3–4 h.

6. Soak nitrocellulose filters with 10 mM IPTG and remove the excess liquid by laying on Whatman 3MM paper.

7. Overlay the plates with the IPTG-impregnated nitrocellulose filters. Do not allow the plates to cool.

8. Incubate at 37°C for another 6–10 h.

[a] As BB4 contains F' factor encoding *lacI*^{q} and *tet*^{r} genes, the original strain should be maintained in the presence of tetracycline.

4.3 Solid-phase phosphorylation

The next step is incubation of the plaque-immobilized filters in a reaction buffer containing protein kinase and [γ-^{32}P]ATP. Prior to the phosphorylation step, the filters are incubated in the presence of unlabelled ATP. The purpose of this pre-incubation step is to reduce the frequency of isolating clones whose products have an autophosphorylating or ATP-binding activity, as discussed in Section 2. *Protocol 4* describes an example of this step with conditions that were optimized to carry out substrate screening for ERK1 MAP kinase (5).

Protocol 4. Solid-phase phosphorylation with ERK1 MAP kinase

Equipment and reagents

- rotating platform
- blocking solution (20 mM Tris–HCl, pH 8.0, 150 mM NaCl, 3% BSA, 1% Triton X-100)
- Triton wash buffer (20 mM Tris–HCl, pH 7.5, 150 mM NaCl, 10 mM EDTA, 1 mM EGTA, 0.5% Triton X-100, 1 mM DTT, 0.2 mM PMSF)
- MAPK reaction buffer (20 mM Hepes–NaOH, pH 7.5, 10 mM MgCl$_2$, 50 mM Na$_3$VO$_4$, 5 mM β-glycerophosphate, 5 mM NaF, 2 mM DTT, 0.1% Triton X-100)
- MAPK wash buffer (20 mM Tris–HCl, pH 7.5, 150 mM NaCl, 10 mM EDTA, 1 mM EGTA, 20 mM NaF, 0.1% Triton X-100)

Method

1. Cool the plates to room temperature. Mark the filters asymmetrically with a needle.

2. Carefully peel the filters off and immerse one by one into a large volume (at least 200 ml for 20 filters) of blocking solution. Throughout the following steps, filters should be kept wet and under gentle agitation (unless stated otherwise).

3. Agitate the filters slowly on a rotating platform for 60 min at room temperature.

4. Wash the filters three times for 20 min at room temperature in 200–300 ml of Triton wash buffer. Slightly more vigorous agitation is required to remove bacterial debris from the filters.

5. Wash the filters for 10 min at room temperature in 200 ml of MAPK reaction buffer.

6. Incubate the filters for 60 min at room temperature in 200 ml of MAPK reaction buffer containing 25–100 μM unlabelled ATP to mask proteins that have autophosphorylating and/or ATP-binding activities.

7. Wash the filters for 10 min in 200 ml of MAPK reaction buffer without ATP.

8. Incubate the filters for 60 min at room temperature (or 30°C if an air incubator is available) with gentle shaking in the MAPK reaction buffer containing 25 μM unlabelled ATP, 5 μCi/ml [γ-^{32}P]ATP, and 1 μg/ml purified human ERK1 MAP kinase. Use at least 2 ml of the solution per 137 mm filter. A tissue culture dish (e.g. Falcon 150 mm dish, #3025) is convenient for the incubation of up to 20 filters.

9. Wash the filters 6–7 times for 5–10 min at room temperature in 100–200 ml of MAPK wash buffer. Relatively vigorous agitation helps to reduce false-positive signals. Finally, wash the filters once with MAPK wash buffer without Triton X-100.

---

10. Dry the filters on paper towels, and arrange on a paper sheet for autoradiography.

4.4 Identification of cDNA clones encoding substrate candidates

The phosphorylated filters are exposed to X-ray film to identify positive plaques by autoradiography. The intensities of radioactive signals are generally quite variable from plaque to plaque, presumably because the recombinant proteins are expressed at different levels, and also because they are not all equally efficient substrates and are therefore phosphorylated to different extents. Therefore, for practical reasons, it is advisable to decide to pick up only a limited number (perhaps ~100) clones, in order of intensity, from the plates in the primary screening. It may often be difficult to discriminate truly phosphorylated plaques from false-positive spots. This problem can be overcome by making an additional autoradiogram of the same filter with much shorter exposure. Generally speaking, pinpoint signals with a sharp rim found in the short exposure are false-positives, whereas true-positive plaques give rather dull, fuzzier signals with a certain range of sizes. Therefore, most false-positive signals can be easily excluded by carefully comparing signals in the corresponding position in the two autoradiographs. If it is possible, it will also help to include a positive λGEX substrate control plate, since this will indicate the sort of signal intensity and morphology to expect for true positive plaques in the screen (see Section 4.5). Alternatively, primary screening may be carried out in duplicate (13). By comparing the duplicate filters, true plaques can be easily identified as reproducible signals. Since most plaques are in contact with one another in the primary screening, phage clones must be purified by secondary screening in which the phages are plated out more sparsely. This step also confirms whether the proteins expressed in the plaques identified in the primary screening are truly phosphorylated by the protein kinase.

Protocol 5. Identification and isolation of positive clones

Equipment and reagents
• as *Protocols 3* and *4*

Method

1. Expose the phosphorylated filters to X-ray films for 4–72 h at –70°C with an intensifying screen. Make at least two sets of autoradiograms with different exposure times, one with a long exposure and the other with a much shorter exposure.

2. Mark strong signals on the long exposure films and carefully observe

---

Protocol 5. *Continued*

their corresponding signals on the short exposure films. Identify sharp, pinpoint signals on the short exposure films, which should be false-positives. Mark them again differentially on the long exposure films for exclusion.

3. Align the long exposure films with the filter sheets. Mark the positions of the asymmetric needle spots on the filters.

4. Identify the locations of positive plaques, and remove a relatively large agar plug (2–4 mm in diameter) from each position to be certain of recovering the positive clone.

5. Transfer the agar plug to 1 ml of SM containing 2 drops of chloroform, and incubate for 1–2 h at room temperature. Measure the titre of each phage stock.

6. Replate each phage stock on to a 90 mm agar plate at a low density (200–500 plaques per plate) and transfer the plaques to 85 mm nitrocellulose filters as described in *Protocol 3*.

7. Repeat solid-phase phosphorylation of the plaque filters as described in *Protocol 4*. Use 0.7 ml of the kinase reaction buffer per filter. Expose the phosphorylated filters to X-ray films.

8. Identify a single, well-isolated positive plaque (clone) on each plate. If it is difficult to obtain a single plaque, repeat steps 3–8 to purify a phage clone.

Figure 3 shows a typical result of solid-phase phosphorylation using ERK1 MAP kinase. In the control experiment with positive (λGEX-Elk-C) and negative (λGEX5) control phages, these two phages gave a clear contrast in intensity of radioactive signals. In the actual screening of a HeLa cDNA library, however, signals of various degrees of intensity are observed as discussed above.

4.5 Optimization of screening conditions using control phages

Since optimum conditions for phosphotransfer reactions are quite variable amongst protein kinases, screening conditions should be experimentally determined for each protein kinase being used. Components of reaction buffer to be considered would be pH, species and concentrations of monovalent and divalent cations, type of detergent, protease inhibitors, phosphatase inhibitors, other additives, and so forth, depending on the enzymological properties of the protein kinase. The presence of detergent (e.g. Triton X-100, Nonidet P-40, Tween 20, etc.) in the buffers is helpful to reduce false-positive signals, most of which seem to be derived from dust or bacterial debris. Thus, it is

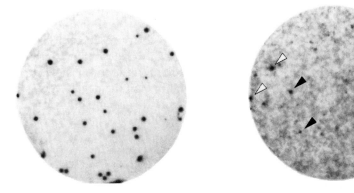

λGEX-Elk-C : λGEX5 = 1:100   HeLa cDNA library

Figure 3. Autoradiogram of a solid-phase phosphorylation screen. Phage plaques (4000 pfu per 90 mm agar plate) were transferred to 85 mm nitrocellulose filters and then subjected to phosphorylation screening with ERK1 MAP kinase according to *Protocols 3* and *4*. Left: a 1:100 mixture of the positive control phage (λGEX-Elk-C) and a negative control phage (λGEX5). Right: screening of HeLa cDNA library. Filled triangles indicate positive plaques whereas open triangles indicate false-positives. Although these signals cannot be distinguished in this figure, signals indicated by open triangles showed up as pinpoint spots on a short exposure film (not shown).

desirable to include a detergent at least in the washing buffers unless the protein kinase is highly sensitive to it. There are no general rules for the concentration and specific radioactivity of [γ-^{32}P]ATP, concentration of protein kinase, reaction temperature, and duration. Therefore, the first trial should be performed in the reaction mixture that is usually used for the in-solution kinase assay of the protein kinase. In practice, however, a simple scaling up from test-tube (10–50 μl) to culture dish (10–50 ml) may not be easy, especially for the quantities of the protein kinase and radioactive ATP needed. Fortunately, a reduction in concentration of these components may be compensated to some extent by raising the specific activity of [γ-^{32}P]ATP (i.e. reducing the concentration of unlabelled ATP) and by changing the incubation time for phosphorylation and the exposure time of autoradiography. Optimization of these three parameters is particularly important to obtain a good signal-to-noise ratio in the autoradiogram.

Ideally, determination of screening conditions should be carried out using positive and negative control phages. If a substrate for the protein kinase in question is already known, the positive control phage can be constructed as described in Section 3.5. If not, reaction conditions may need to be checked while an actual screen is being performed with a cDNA library, assuming that positive clones exist. The λGEX5 vector phage can usually be used as a negative control phage, although this phage expresses a GST fusion protein with an artificial C-terminal sequence derived from the stuffer region (GST–

stuffer). If the stuffer-derived sequence (58 amino acids; DDBJ/EMBL/ GenBank accession number AB014641) contains any potential phosphorylation sites for the protein kinase, it may be necessary to construct another control phage expressing only the GST region. It is desirable to check in advance whether the GST–fused substrate protein (or peptide) is phosphorylated by the protein kinase much more efficiently than the negative construct (GST–stuffer or GST itself) in a test-tube assay (for example, see *Figure 5*, lanes of GST and GST-Elk-C). If the fusion protein is not phosphorylated or is poorly phosphorylated, other constructs should be considered. Also, there is a possibility that GST itself may act as a good substrate for the protein kinase. In this case, other vector systems such as λgt11 (11, 12) or λZAP (18) may have to be used instead.

The control experiment can be performed using 90 mm agar plates on which either of the control phages and a series of mixtures of the two control phages are plated out at a density of 4–8 × 10³ plaques per filter. The ratios of the positive and negative control phages in the mixtures should be, for example, 1:10, 1:100, and 1:1000. The autoradiogram of solid-phase phosphorylation of these filters helps to estimate the difference in intensity between positive and negative signals, the background level in radioactivity of plaques and *E. coli* lawn, and the frequency of appearance of false-positive signals. If the model experiment works well, the number of strong signals should change amongst the filters in direct proportion to the ratio of the positive phages in the mixtures (*Figure 3*).

5. Conversion of phage clones into plasmids

If positive plaques give clearly stronger signals than the remaining clones in the secondary screening, this means that the first step of the screening is successful. For further characterization, positive clones now have to be converted into plasmids that contain their cognate cDNA. *Protocol 6* describes a rapid plate

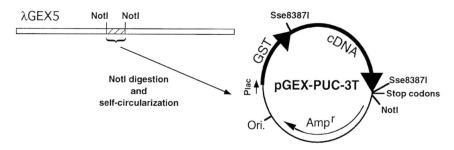

Figure 4. Schematic representation of excision rescue. After digestion of phage DNA with *Not*I, the plasmid region containing cDNA can be recovered by self-ligation. In the resulting pGEX-PUC-3T plasmid, the *Sfi*I sites used for cDNA cloning are not regenerated but *Sse*8387I sites are available to excise the cDNA fragment.

lysate method for preparing DNA from multiple (10–100) phage clones at the same time. The cDNA-containing plasmid (pGEX-PUC-3T) can be recovered by *Not*I digestion of the phage DNA followed by self-circularization, as described in *Protocol 7*. We recommend *E. coli* XL1-Blue (15) as the host strain for the transformation of this plasmid. A schematic representation of this process is shown in *Figure 4*.

Protocol 6. Rapid, small-scale preparation of λGEX5 phage DNA

Reagents

- λ diluent (10 mM Tris–HCl, pH 7.5, 10 mM MgSO₄)
- 50 × DNase/RNase mixture (0.1 mg/ml DNase I, 1 mg/ml RNase A; make up 1 × DNase/RNase mixture in λ diluent just before use)
- 35 mm agarose[a] plates in 6-well tissue culture plate (3 ml of 1.5% agarose in NZCYM medium per well)
- extraction buffer (10 mM Tris–HCl, pH 8.0, 0.1 M NaCl, 10 mM EDTA, 0.1% SDS)

Method

1. Using a pasteur pipette, pick a single plaque and place in 0.5 ml of SM containing one drop of chloroform. Incubate the suspension for 1–2 h at room temperature.

2. Mix 20 μl of the phage suspension with 30 μl of plating bacteria (BB4, A_{600} = 2.0) and incubate for 15 min at 37 °C.

3. Add 0.4 ml of molten (50 °C) top agarose[a] (0.7% agarose in NZCYM medium), and spread the bacterial suspension on the surface of 35 mm agarose plates in 6-well tissue culture plates.

4. Incubate at 37 °C overnight (10–16 h) to reach confluent lysis.

5. Directly add 0.65 ml of λ diluent containing 2 μg/ml DNase I and 20 μg/ml RNase A on to the surface of the top agarose, and incubate for 1.5–2 h at room temperature with constant, gentle shaking.

6. Transfer the phage suspension (usually 0.4–0.5 ml) to a microfuge tube, add 30 μl of chloroform, and vortex for 5–10 sec.

7. After centrifugation for 1 min, transfer 0.4 ml of the aqueous supernatant to a microfuge tube. Do not take any bacterial debris.

8. Add 0.4 ml of λ diluent containing 20% (w/v) polyethylene glycol 8000 and 2 M NaCl. Mix by vortexing, and incubate for 1 h on ice.

9. Centrifuge at 10 000 g for 10 min at 4 °C. Remove the supernatant, leaving 30–50 μl behind. Recentrifuge briefly to bring the liquid on the walls of the tube to the bottom, and remove the remaining supernatant.

10. Add 100 μl of extraction buffer, dissolve the phage pellet by vortexing, and incubate for 10 min at 68 °C.

Protocol 6. *Continued*

11. Extract the solution with 100 μl of phenol/chloroform, and recover the phage DNA by ethanol precipitation.

12. Dissolve the DNA in 100 μl of TE buffer pH 8.0.

---

a Agarose, not agar, should be used. Phage DNA prepared from agar plates is often resistant to restriction enzyme digestion.

Protocol 7. Excision rescue of cDNA-containing plasmid from λGEX5 clones

Equipment and reagents

- *Not*I reaction mixture [for 10 reactions, 45 μl of water, 10 μl of 10 × *Not*I buffer (NEB or equivalent) and 5 μl (5–25 units) of *Not*I; prepare the mixture on ice just before use]
- 2X YT medium (1.6% bacto-tryptone, 1% bacto-yeast extract, 0.5% NaCl)

- ligase reaction mixture [for 10 reactions, 40 μl of water, 7 μl of 10 × ligase buffer (*Protocol 2*), and 3 μl (300–1200 units) of T4 DNA ligase; prepare the mixture on ice just before use]
- XL1-Blue competent cells (Stratagene)

Method

1. Place 4 μl of each phage DNA solution from *Protocol 6* in a microfuge tube. Add 6 μl of a *Not*I reaction mixture, incubate at 37°C for 1 h, and then at 70°C for 20 min.

2. Centrifuge the tubes briefly, and then transfer 2 μl of the *Not*I-digested DNA into a well of a U-bottom, 96-well microtitre plate.

3. Add 5 μl of ligase reaction mixture. Mix by vortexing the plate, and incubate for 30 min at 16°C.

4. Add 20 μl of XL1-Blue competent cells directly to the ligated DNA, mix, and incubate on ice for 5 min. Heat for 90 sec at 42°C, add 200 μl of 2X YT medium, and incubate for 30 min at 37°C with gentle shaking.

5. Transfer an appropriate volume (200 μl per 90 mm plate or 50–100 μl per 35 mm plate in a 6-well plate) of the cell suspension on to an agar plate containing 100 μg/ml ampicillin.

6. Incubate overnight (>10 h) at 37°C. Most of the clones thus obtained usually contain plasmids of the expected structure (i.e. pGEX-PUC-3T containing a cDNA between the *Sse*8387I sites), but it may be a good idea to pick up two colonies from each plate for back-up.

6. Initial characterization of candidate clones

First of all, cDNA clones from positive plaques need to be checked with respect to the following criteria:

- Can the cDNA-encoded polypeptide really act as a substrate for the protein kinase?
- Does the polypeptide sequence represent at least a part of any naturally existing protein?

The former point can be examined by *in vitro* phosphorylation assay of each of the purified GST fusion proteins. The latter question arises since a polypeptide translated from a wrong reading frame or non-coding region of cDNA could be a good substrate by chance. Fortunately, clones that encode such artefactual products can be effectively excluded from the list of candidates by checking their product sizes on SDS–PAGE (see Section 6.1).

6.1 Purification and analysis of GST fusion proteins encoded by candidate clones

GST-fused recombinant proteins encoded by the rescued plasmids can be expressed in small bacterial cultures, purified by GSH–agarose affinity beads, and then tested for phosphorylation by the protein kinase. An SDS–PAGE analysis of the phosphorylated GST fusion proteins offers information about their sizes and extent of phosphorylation.

Protocol 8. Analysis of GST fusion proteins produced by candidate clones

Equipment and reagents

- sonicator with a microtip
- TLB (50 mM Hepes–NaOH, pH 7.4, 150 mM NaCl, 10% glycerol, 1% Triton-X100, 1 mM EGTA, 1 mM DTT, 1 mM PMSF)
- GSH–agarose beads (Sigma, or equivalent)
- lysozyme solution (5 mg/ml in 0.1 M Tris–HCl, pH 8.0)
- 2 × SB (0.1 M Tris–HCl, pH 6.8, 4% SDS, 20% glycerol, 40% 2-mercaptoethanol, 0.04% bromophenol blue)

Method

1. Pick a single colony and inoculate 3 ml of 2 × YT medium containing 0.1% glucose and 100 μg/ml ampicillin. Incubate overnight (12–18 h) at 37°C with vigorous shaking.

2. Place 0.15 ml of the overnight culture into a microfuge tube containing 50 μl of 60% sterile glycerol. Mix well and store at −70°C as a glycerol stock.

3. Place 1.35 ml of the overnight culture into a microfuge tube and prepare plasmid DNA by a miniprep method suitable for DNA sequencing.

4. Add 7.5 μl of 100 mM IPTG to the remaining culture (1.5 ml), and incubate for 3–4 h at 37°C.

5. Transfer the culture into a microfuge tube, centrifuge at 10000 *g* for

---

Protocol 8. *Continued*

30 sec at 4°C. Remove the supernatant and resuspended the bacterial pellet in 100 μl of ice-cold TE (pH 7.5).

6. Add 10 μl of freshly prepared lysozyme solution, mix, and leave on ice for 5 min.

7. Add 0.5 ml of TLB and lyse the cells by sonication. Centrifuge the tube at 12 000 *g* for 15 min at 4°C and transfer the supernatant to a fresh tube.

8. Add 100 μl of 50% (v/v) GSH–agarose beads and incubate for 1 h at 4°C with rocking.

9. Wash the beads twice with 0.5 ml of ice-cold TLB and once with 1 ml of MAPK reaction buffer (see *Protocol 4*). Add to the washed beads 50 μl of the same buffer and mix by vortexing.

10. Transfer 20 μl of the suspension containing about 10 μl of the beads into a fresh tube and add 5 μl of MAPK reaction buffer containing 250 μM unlabelled ATP, 0.5 μCi of [γ-^{32}P]ATP, and 10–50 ng of ERK1 MAP kinase. Incubate the reaction mixture (25 μl) for 30–60 min at 30°C with occasional agitation.

11. Add 25 μl of 2 × SB, boil for 3–5 min, and then analyse 10–30 μl of the samples by SDS–PAGE on a 12.5% polyacrylamide gel.

12. Stain the gel with Coomassie blue, dry and expose to an X-ray film for autoradiography.

In our screen for ERK1 substrates, we screened about 3×10^5 independent clones of a cDNA library prepared from HeLa cells, and obtained 120 positive clones from the secondary screening (5). *Figure 5* shows a typical example of *in vitro* phosphorylation analysed by SDS–PAGE. Although almost all of the recombinant proteins were phosphorylated by ERK1 MAP kinase *in vitro*, more than half of them were GST proteins with a very short fusion partner (S5, S6, S8, S17, and S19). Sequencing analysis revealed that most of the extremely small fusion partners were artefactual products derived from out-of-frame ligation of cDNAs. Therefore, out of 120 clones, we selected 32 clones that expressed GST fusion proteins larger than 32 kDa in total size (i.e. with a fusion partner of >5 kDa) for further characterization (5). Every recombinant product of these 32 clones was a good *in vitro* substrate for ERK1, and is likely to represent some part of a naturally existing protein (see Section 6.2). Generalizing from this case, it is advisable to select only clones that produce a GST fusion protein whose size is larger than, for example, 32 kDa, which means that clones encoding cDNA-derived polypeptides of <5 kDa should be excluded from the substrate candidates. Although the frequency of appearance of such artefactual recombinants may depend on

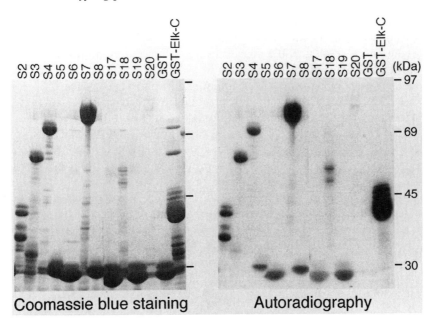

Figure 5. *In vitro* phosphorylation analysis of positive clones in a screen for ERK1 substrates. Positive phage clones (S2–S8 and S17–S20) obtained in the screening were converted into plasmid clones, and then their GST fusion proteins were expressed and purified using GSH–agarose. The resulting GST fusion proteins together with GST and GST–Elk-C were phosphorylated by ERK1 MAP kinase in the presence of [γ-^{32}P]ATP and analysed by SDS–PAGE. Left: Coomassie blue staining of the gel. Right: autoradiogram of the same gel. Note that GST itself was not phosphorylated at all, whereas the artefactual recombinants (S5, S6, S8, S17, and S19) with similar small sizes and quantities were relatively weakly phosphorylated.

the protein kinase, it is probably desirable to exclude artefacts by checking the size of each GST fusion protein prior to sequencing.

6.2 Sequencing analysis

The 5′- and 3′-nucleotide sequences of the candidate clones can be determined using oligonucleotide primers designed to anneal just outside the cloning sites. Amino acid sequences of the fusion proteins can be easily deduced from the 5′-nucleotide sequence. The nucleotide and amino acid sequences should be subjected to identity/homology searches using appropriate programs such as BLAST (19). If a particular cDNA encodes a polypeptide sequence that is identical or similar to that of any known protein, this information may offer clues with regard to further analysis. Some of them may be very plausible candidates for physiological targets, whereas some others may be most unlikely (e.g. secreted proteins). The nucleotide sequences of some clones may be found in an EST (expressed sequence tag) database, which may offer additional nucleotide sequence missing in the partial cDNA. 'Walking' in EST databases via overlapping sequences often results in the elucidation of a long contiguous sequence. If a consensus phosphorylation motif is known for the protein kinase in question, then it is worth scanning the predicted sequence of the cDNA product for potential phosphorylation sites.

In our screen for ERK1 substrates, sequencing analysis revealed that 14 clones out of the 32 candidates mentioned in Section 6.1 corresponded to fragments of known proteins, including two known physiological ERK substrates, p90^{RSK2} and c-Myc (5). We also showed that one of the novel clones encoded a new MAP kinase-activated protein kinase and that this protein kinase, named MNK1, was an *in vivo* target for ERK1 MAP kinase (5). Although we have not characterized other clones in detail, most of them contained potential MAP kinase recognition sites, Ser–Pro or Thr–Pro, in their amino acid sequences. In another screen for cyclin E/Cdk2 substrates, 54 positive clones were identified from 1.0×10^6 independent clones of the HeLa cDNA library. At least two of them are likely to be *in vivo* Cdk substrates, one is caldesmon and the other is a novel protein, named PRC1, that plays a role in cytokinesis (6).

7. Analysis of physiological function of the identified substrates

If a large number of clones are isolated in a screen, it may be necessary to decide which clone(s) are to be further characterized. In the case of clones showing identity or homology to some known sequence, this information can help to inform this decision. In the case of novel proteins, however, it is not easy to choose one out of the candidates. Therefore, some additional experiments need to be done to guide selection. A simple experiment would be to test whether each GST–fused substrate protein can bind the protein kinase by an *in vitro* binding assay. If a substrate physically associates with the kinase to form a complex, it may be worth analysing further. Alternatively, analysis of expression patterns of the candidates by Northern or *in situ* hybridization may be helpful. If a clone shows a characteristic expression pattern such as cell type specificity, cell cycle dependence, developmental regulation, or inducible expression, this may suggest some physiological relationship between the kinase and substrate.

Finally, it is important to remember that a protein identified by phosphorylation screening remains simply an '*in vitro*' substrate until proven to be a physiological target of the protein kinase used, and this issue can be clarified only by *in vivo* analysis. It is important to determine whether the protein is phosphorylated under conditions where the protein kinase in question is known to be activated. The final goal is, of course, to elucidate the function of the newly identified substrates and the physiological significance of their phosphorylation. Although there is no general strategy for this purpose, isolation and sequence determination of a full-length cDNA clone, as well as production of antibodies, are essential steps. Various experiments such as overexpression or ectopic expression, decreasing expression by antisense methods, or inhibition of protein function by antibody microinjection, may suggest possible roles for the protein *in vivo*. Also, it may be informative to examine the subcellular localization and kinetics of synthesis and degradation of the protein. Furthermore, determination of the sites phosphorylated *in vitro* and *in vivo*, and mutation of these sites, would help to elucidate the biological significance of phosphorylation in regulating the function of the protein kinase target.

8. Application to other protein kinases

The phosphorylation screening approach is, in principle, generally applicable to all protein kinases. A major problem that may arise in some cases, as discussed in Section 2, is a high background caused by strong phosphorylation of an endogenous protein(s) derived from *E. coli* or λ phage. The severity of this background noise may depend on the substrate specificity of the protein kinase used. Although we did not encounter this problem in the screens for ERK1 MAP kinase and cyclin E/Cdk2 substrates, which are both proline-directed protein kinases, a preliminary experiment with the catalytic subunit of cAMP-dependent protein kinase showed indistinguishable signals between positive and negative plaques. The high background problem may be overcome if it is possible to develop a GSH-derivatized filter on to which the GST–fused recombinants can selectively be immobilized. The use of this affinity filter system should, in principle, reduce the background problem caused by bacterial proteins. No such GSH filter is yet commercially available, but in preliminary experiments we have coupled GSH to cellulose filters and have found that the λGEX5/GSH–cellulose filter system seemed to work well for ERK1. However, further development is required for practical, large-scale usage.

Another modification for reducing the problem of a high background may be possible. Some protein kinases prove to form a relatively stable complex with their substrates through a domain distinct from the phosphorylation site, as is the case for JNK and c-Jun (20–22), and for certain MAP kinases and MAP kinase kinases (23, 24). As discussed in Section 7, this phenomenon might be of use to select some clones out of a great number of substrate candidates in the phosphorylation screening. Carrying this idea one step further, a modified screening method may be applicable to some protein kinases. In this alternative protocol, plaque filters are first pre-incubated with the protein kinase in the absence of ATP to allow the kinase to bind recombinant substrates. Then the filters are briefly washed to remove the excess, unbound protein kinase (different stringencies of washing could usefully be tried), and subsequently incubated in the presence of $[\gamma\text{-}^{32}\text{P}]$ATP to allow phospho-transfer to occur within the kinase–substrate complex. We have not tested this alternative method, but it may reduce background signals derived from non-specific phosphorylation, and would be a way of identifying substrates that have a high affinity binding site. In fact, phosphorylation screening may work well because, with the relatively low protein kinase concentration used in the screen, only high affinity substrates are phosphorylated efficiently.

Application of phosphorylation screening to protein–tyrosine kinases may be bedeviled by the high background problem, because in general protein–tyrosine kinases are rather non-specific *in vitro*. The use of anti-phosphotyrosine antibodies to screen the phosphorylated λgt11 cDNA expression library, selecting only the strongest signals, has proved to be a successful way of identifying Src protein–tyrosine kinase substrates (7). An alternative solution would be to use an additional far-Western selection step utilizing a phosphotyrosine-binding domain. For example, plaque filters are first incubated with a protein–tyrosine kinase in the presence of unlabelled ATP. After washing, the filters are next incubated with an epitope-tagged or radiolabelled protein containing a particular SH2 or PTB (phosphotyrosine-binding) domain to allow them to bind, and then positive plaques are visualized by immunodetection or autoradiography. This protocol can be a substrate screen based not only on the phosphorylating specificity of the protein kinase, but also on the binding specificity of the phosphotyrosine-binding modules used. In fact, Kavanaugh and co-workers showed that a screen of a λgt11 cDNA library with solid-phase phosphorylation by platelet-derived growth factor receptor tyrosine kinase, followed by binding of ^{32}P-labelled Shc PTB domain, resulted in cloning of a c-ErbB2 (Neu) cDNA, which contained tyrosine phosphorylation-dependent PTB-binding sites (25). Similar 'double selection strategies' utilizing phosphorylation-dependent protein–protein interactions might be applicable to some protein–serine/threonine kinases, in combination with a protein module probe such as 14-3-3 proteins (26), Pin1 (27, 28), and the MPM-2 monoclonal antibody (29), each of which is shown to specifically interact with a phosphoamino acid-containing peptide motif.

We expect that the phosphorylation screening technique will be applicable, with further refinement, to other protein kinases including protein–tyrosine kinases and receptor-type protein kinases.

Acknowledgements

We are deeply appreciative of the support and advice of our colleagues, especially Wei Jiang for helpful discussions during the development of this method.

References

1. Songyang, Z., Blechner, S., Hoagland, N., Hoekstra, M. F., Piwnica-Worms, H., and Cantley, L. C. (1994). *Curr. Biol.*, **4**, 973.
2. Zhao, J., Dynlacht, B., Imai, T., Hori, T., and Harlow, E. (1998). *Genes Dev.*, **12**, 456.
3. Yang, X., Hubbard, E. J., and Carlson, M. (1992). *Science*, **257**, 680.
4. Waskiewicz, A., Flynn, A., Proud, C. G., and Cooper, J. A. (1997). *EMBO J.*, **16**, 1909.
5. Fukunaga, R. and Hunter, T. (1997). *EMBO J.*, **16**, 1921.
6. Jiang, W., Jimenez, G., Wells, N.J., Hope, T.J., Wahl, G.M., Hunter, T., and Fukunaga, R. (1998). *Mol. Cell*, **2**, 877.
7. Lock, P., Abram, C. L., Gibson, T., and Courtneidge, S. A. (1998). *EMBO J.*, **17**, 4346.
8. Smith, D. B. (1993). *Meth. Mol. Cell. Biol.*, **4**, 220.
9. Valtorta, F., Schiebler, W., Jahn, R., Ceccarelli, B., and Greengard, P. (1986). *Anal. Biochem.*, **158**, 130.
10. Glover, C. V. C. and Allis, C. D. (1991). In *Methods in enzymology* (ed. T. Hunter and B. M. Sefton), Vol. 200, p. 85. Academic Press, San Diego.
11. Young, R. A. and Davis, R. W. (1983). *Proc. Natl. Acad. Sci. USA*, **80**, 1194.
12. Huynh, T. V., Young, R. A., and Davis, R. W. (1985). In *DNA cloning, volume I: a practical approach* (ed. D. M. Glover), p. 49. IRL Press, Oxford.
13. Sambrook, I., Fritsch, E. F., and Maniatis, T. (ed.) (1989). *Molecular cloning: a laboratory manual*, 2nd edn. Cold Spring Harbor Laboratory Press, NY.
14. Ausubel, F. M., Brent, R., Kingston, R. E., Moore, D. D., Seidman, J. G., Smith, J. A., and Struhl, K. (ed.) (1997). *Current protocols in molecular biology* (Supplement 37). John Wiley & Sons, Inc., NY.
15. Bullock, W. O., Fernandez, J. M., and Short, J. M. (1987). *BioTechniques*, **5**, 376.
16. Marais, R., Wynne, J., and Treisman, R. (1993). *Cell*, **73**, 381.
17. Williams, N. G., Paradis, H., Agarwal, S., Charest, D. L., Pelech, S. L., and Roberts, T. M. (1993). *Proc. Natl. Acad. Sci. USA*, **90**, 5772.
18. Short, J. M., Fernandez, J. M., Sorge, J. A., and Huse, W. D. (1988). *Nucl. Acids Res.*, **16**, 7583.
19. Altschul, S. F., Gish, W., Miller, W., Myers, E. W., and Lipman, D. J. (1990). *J. Mol. Biol.*, **215**, 403.
20. Hibi, M., Lin, A., Smeal, T., Minden, A., and Karin, M. (1993). *Genes Dev.*, **7**, 2135.
21. Kallunki, T., Su, B., Tsigelny, I., Sluss, H. K., Dérijard, B., Moore, G., Davis, R., and Karin, M. (1994). *Genes Dev.*, **8**, 2996.
22. Gupta, S., Barrett, T., Whitmarsh, A. J., Cavanaugh, J., Sluss, H. K., Dérijard, B., and Davis, R. J. (1996). *EMBO J.*, **15**, 2760.
23. Bardwell, L., Cook, J. G., Chang, E. C., Cairns, B. R., and Thorner, J. (1996). *Mol. Cell. Biol.*, **16**, 3637.
24. Fukuda, M., Gotoh, Y., and Nishida, E. (1997). *EMBO J.*, **16**, 1901.
25. Kavanaugh, W. M., Turck, C. W., and Williams, L. T. (1995). *Science*, **268**, 1177.
26. Muslin, A. J., Tanner, J. W., Allen, P. M., and Shaw, A. S. (1996). *Cell*, **84**, 889.
27. Ranganathan, R., Lu, K. P., Hunter, T., and Noel, J. P. (1997). *Cell*, **89**, 875.
28. Lu, P.J., Zhou, X.Z., Shen, M., and Lu, K.P. (1999). *Science*, **283**, 1325.
29. Davis, F. M., Tsao, T. Y., Fowler, S. K., and Rao, P. N. (1983). *Proc. Natl. Acad. Sci. USA*, **80**, 2926.

14

Analysis of protein kinase interactions using the two-hybrid method

RONG JIANG and MARIAN CARLSON

1. Introduction

The yeast two-hybrid system is a genetic approach that detects protein–protein interactions (1, 2). This method relies on the functional reconstitution of transcriptional activators, such as Gal4, in the nucleus of yeast cells. The interaction of two-hybrid proteins, one bearing a DNA-binding domain and the other bearing a transactivation domain, results in transcriptional activation of a reporter gene. Certain features of the two-hybrid system make it particularly suitable to analyse protein kinase interactions. First, it is highly sensitive and can detect transient, as well as stable, interactions. This probably results from the fact that the two-hybrid system uses gene expression as a readout to amplify the signal generated by protein–protein interactions. The two-hybrid method has uncovered interactions not easily revealed with other methods. For example, the interaction between the Raf kinase and Ras is readily detectable by the two-hybrid method, but not by co-immunoprecipitation (3, 4). Secondly, as the two-hybrid system analyses interactions within the environment of a eukaryotic cell, many types of post-translational modifications critical for interaction are preserved. While yeast is the popular host for this method, a mammalian version of the two-hybrid system has also been developed (5, 6). Thirdly, the two-hybrid method detects bridged as well as direct interactions, which facilitates the dissection of protein interactions within kinase complexes. Such analyses have revealed important roles for scaffold proteins, bridging proteins, and adaptor proteins in kinase signalling. Finally, many kinases contain separable catalytic and regulatory domains, making protein kinases themselves amenable to the structure/function analysis offered by the two-hybrid system.

The interaction of a protein–serine/threonine kinase, Snf1, with its activator Snf4, served as the prototype for the two-hybrid system (1, 7, 8). The identification of a Snf1-interacting protein by a two-hybrid interaction screen provided the first evidence that the two-hybrid method can be used to identify kinase-interacting proteins (9). Subsequently, the two-hybrid method has been applied to the study of a large number of diverse protein kinases, including both serine/threonine- and tyrosine-specific classes. Major applications can be summarized as follows:

(a) Screening for kinase-interacting proteins such as kinase subunits, substrates, regulators, and scaffold proteins.

(b) Structural dissection of protein kinases, such as mapping of interacting sequences and identification of residues critical for interaction. In some cases, information garnered from structural analysis directly points to regulatory mechanisms. For example, an interaction between the regulatory domain and the catalytic domain within a protein kinase itself may indicate an autoinhibitory regulation.

(c) Functional analysis of kinase interactions. This includes isolation of mutants with altered interaction profiles, such as interaction-defective or interaction-enhanced mutants, and compensatory mutants that restore a disrupted interaction. These analyses have been very useful in addressing the functional relevance of an observed interaction.

2. Identification of protein kinase-interacting proteins by the two-hybrid system

2.1 Getting started with the yeast two-hybrid system

2.1.1 Materials and reagents

The yeast two-hybrid system has three components:

- a 'bait' hybrid, which is a fusion between a test protein, for example, a kinase of interest, and a DNA-binding domain (BD) commonly derived from yeast Gal4 (GBD), or from *E. coli* LexA

- a 'prey' hybrid, which is a fusion between an interacting protein and a transactivation domain (AD), commonly that of Gal4 (GAD), or VP16 from the type II herpes simplex virus

- a *Saccharomyces cerevisiae* (budding yeast) reporter strain

A reporter strain normally has a reporter gene construct integrated in the chromosome, which contains, in its promoter region, binding sites matching the BD moiety, i.e. UAS_{GAL} (upstream activating sequence) for GBD hybrids, or $lexA_{op}$ (operators) for LexA hybrids. Two types of reporter genes are generally used, a *lacZ*-based reporter for blue/white screens, and the yeast amino acid markers (*URA3*, *HIS3*, or *LEU2*) for selection. For identifying new interacting proteins, the latter has the advantage of cutting the workload. However, the existence of simple assays for *lacZ* expression makes it the reporter of choice for quantitative analysis of defined protein–protein interactions.

Three improved versions of the original two-hybrid method (2) have been designed, and all appear to work comparably (4, 10, 11). A detailed description of these variants, including vectors and reporter strains, has been published (12). Standard protocols for growth and transformation of yeast cells can be found in other sources (13, 14).

2.1.2 Assaying two-hybrid interactions

In the few cases tested, there appears to be a general correlation between the strength of interactions assayed in the two-hybrid system, and that of the same interactions measured by *in vitro* binding experiments (15, 16). However, examples also exist in which *in vitro* studies yield different results from the two-hybrid system (17). Caution should be exerted when comparing different two-hybrid interactions, as the interaction between a pair of hybrid proteins is likely to be affected by multiple parameters. These include the expression level, the correctness of protein folding, the efficiency of nuclear entry, the ability to bind DNA, and the effect on general growth and transcription of yeast hosts. Comparison of the two-hybrid interaction of a wild-type protein with that of a point mutant is most likely to provide meaningful results. Two protocols are listed here for assaying two-hybrid interactions using *lacZ* reporters.

Protocol 1. β-Galactosidase filter assay (modified from ref. 18)[a]

Reagents

- Z buffer [for 1 litre, mix 16.1 g Na2HPO4.7H2O, 5.5 g NaH2PO4.H2O, 0.75 g KCl, 0.25 g MgSO4.7H2O. pH should be 7.0; adjust if necessary. Do not autoclave; filter sterilize instead. Add β-mercaptoethanol (β-Me, 13.5 ml per 5 ml Z buffer) fresh each time]
- X-gal (5-bromo-4-chloro-3-indolyl β-D-galactopyranoside; 20 mg/ml in DMF (N, N'-dimethylformamide); store in dark at –20°C
- nitrocellulose filter (Millipore cat. no. HATF 08225)
- Whatman 3MM paper

Method

1. Using flat-ended toothpicks, pick yeast colonies from transformation plates and make patches (0.5 cm in diameter) or streaks, on selective plates to maintain the plasmids. Grow cells overnight at 30°C.[b]

2. Wear gloves when handling nitrocellulose filters.[c] To transfer cells to filters, lay a filter on to the plate and lightly press down on the filter with finger tips to ensure good contact by removing air pockets. Lift the filter with forceps, starting from the edge. Put the filter in a clean Petri dish with the cells facing up.

3. Put the dish with lid on into a –80°C freezer for 30 min.[d] The filter should freeze inside the dish. Filters can stay frozen up to a few days before the next step.

Protocol 1. *Continued*

4. Let filters thaw at room temperature for 2 min. Take the filter out of the dish. Add the reaction cocktail to the dish (5 ml Z buffer, 13.5 ml β-Me, and 75 ml X-gal).

5. Soak two pieces of Whatman 3MM paper, previously cut to the size of the dish, in the buffer. Pour away any extra buffer. Lay the thawed filter on top of the soaked papers.

6. Wrap the plate with parafilm and leave at 30°C to develop colour. Check plates as the reaction proceeds; colour may develop after a few minutes or after overnight incubation. Stop the reaction by taking out the filters and drying them in a fume hood.

[a] The filter assay, which is performed on broken cells, is more sensitive than growing cells on X-gal plates (19).
[b] Yeast cells can be grown directly on nitrocellulose filters laid on plates. This eliminates the need to filter-lift cells from plates. Cells can be suspended in sterile H_2O (final OD_{600} 0.1), and cell suspensions (15 μl) can be spotted on filters by using multichannel pipettes. Incubate overnight at 30°C and proceed as in the protocol.
[c] There is no need to sterilize filters.
[d] An alternative way to freeze cells is to dip the filter in liquid nitrogen for a few seconds; however, extra care must be taken to avoid breaking the filter when taking it out of the liquid nitrogen, as the filter becomes extremely brittle.

Protocol 2. β-Galactosidase liquid assay (20)[a]

Reagents

- Z buffer with β-Me (27 μl β-ME per 10 ml of Z buffer; add β-Me fresh to Z buffer before use)
- ONPG (o-nitrophenol-β-D-galactoside; 4 mg/ml in H_2O; made fresh)
- chloroform
- 0.1% SDS
- 1 M Na_2CO_3

Method

1. Grow yeast at 30°C in selective liquid medium to mid-log phase (A_{600} in the range of 0.2–0.8). For 3 ml cultures, the 15 ml culture tubes are convenient.

2. Spin down 1 ml of A_{600} = 0.5, or equivalent amount of cells, in small glass tubes in a clinical table-top centrifuge at 1000 g for 5 min at room temperature. Aspirate off medium. Be careful not to dislodge the pellets.

3. Add 1 ml cold Z buffer with β-Me. In a fume hood, add 3 drops of chloroform and 2 drops of 0.1% SDS, with a glass pipette.

4. Vortex at room temperature for 10 sec at maximum speed to break open the cells. Equilibrate the tubes at 28°C in a water bath for 5 min.

5. Start the reaction by adding 0.2 ml ONPG. Mix by gentle vortexing. Start timing the reaction. To process multiple samples, add ONPG to successive tubes at 10 sec intervals.

6. When the reaction mixture turns yellow, stop the reaction by adding 0.5 ml Na_2CO_3. Gently mix. Record the reaction time.

7. Spin down cell debris (1000 g, 5 min). Check A_{420} of the supernatant. Calculate the β-galactosidase activity[b] in Miller units (21) using the formula: $(A_{420} \times 2000)/\text{time(min)}$.

[a] Instead of using permeabilized cells, β-galactosidase activity can be assayed using yeast protein extracts. This is advantageous when analysing weak two-hybrid interactions. See *Protocol 3* on how to make yeast protein extracts. The final β-galactosidase activity should be normalized to the total protein levels of extracts.
[b] For comparison, a few units of β-galactosidase activity are sufficient to produce a blue colour in filter lift assays after overnight incubation at 30 °C (9).

2.1.3 Analysing two-hybrid interactions in any yeast genetic background by using plasmid-based reporters

In yeast, classic genetic analysis can be combined with the two-hybrid system to study the effect of chromosomal gene mutations on the interaction of a defined interacting pair. However, for this purpose, the GBD-based two-hybrid system with an integrated reporter has many drawbacks. Either the mutation of interest must be introduced into the reporter strain, or the reporter construct must be integrated into a strain carrying the mutation. Moreover, the GBD-based system requires the reporter strain to have *gal4* and *gal80* mutations (in order to remove the endogenous GBD as well as the galactose induction mechanism). An alternative approach is to carry out two-hybrid assays using plasmid-borne LexA reporters, such as those developed in ref. 16, for which the status of the *GAL4* and *GAL80* genes is irrelevant. This strategy allows application of the two-hybrid system in any yeast strain or genetic background (see ref. 22). Recent development of two-hybrid vectors containing antibiotic resistance markers for selection in *S. cerevisiae*, such as the *Streptoalloteichus hindustanus* bleomycin gene, encoding resistance to zeocin, and the *E.coli* kanamycin-resistance gene, encoding resistance to geneticin (23), should aid this application.

2.2 Comments on two-hybrid screens

The success of a two-hybrid screen for kinase-interacting proteins depends on the nature of the bait and the quality of the library. The following points need to be considered when starting a two-hybrid screen or a two-hybrid analysis of known interactions:

(a) Most BD and AD fusions are N-terminal. It is possible that the BD or AD moiety could perturb the structure of the adjacent N-terminal region of the protein kinase, or an interacting protein. If this is suspected, C-terminal BD or AD fusions should be made instead (see ref. 24 for C-terminal AD fusion vectors).

(b) Both hybrid proteins need to be inside the nucleus for their interaction to activate reporter gene expression. Most fusion moieties also contain a nuclear localization signal (NLS), although many hybrid proteins appear to diffuse into the nucleus even without an NLS (25).

(c) The DNA-binding ability of a LexA hybrid can be tested using a transcriptional repression assay (26). A positive result also indicates a nuclear entry of the bait hybrid.

(d) The DNA-binding domain hybrid should not activate transcription by itself, or at least not strongly. For two-hybrid screens, this problem of weak self-activators can be overcome by using the *HIS3* reporter, and including a minimal amount of 3-amino-1,2,4-triazole (3AT), a drug that acts to inhibit histidine biosynthesis, in the medium (10). Strong self-activators can be used as BD hybrid bait only if transactivation regions are removed. Alternatively, a strategy has been designed in which the activator protein is fused to AD and the library is constructed in the BD fusion hybrid. Using this method, the *Drosophila* RBF was isolated as an interacting protein of E2F1, a transcription activator (27). When examining two-hybrid interactions of known proteins, interaction can sometimes be distinguished from weak self-activation by quantitative assays of the β-galactosidase activity of a *lacZ* reporter (28). However, it is always possible that the presence of the AD hybrid indirectly enhances the otherwise weak activation by such a BD hybrid.

(e) Make sure hybrid proteins are expressed in yeast. Western blot analysis can be carried out by preparing yeast protein extracts or by boiling cells directly before loading a gel (*Protocols 3* and *4*, respectively). Antibodies to GBD, LexA, GAD, and VP16 are commercially available (Clontech; Santa Cruz Biotech). It is also possible to perform kinase assays on immunoprecipitates to ensure that a kinase hybrid maintains its catalytic function in yeast. Mammalian tyrosine kinases may well preserve their catalytic function when expressed in yeast [see point (i) below for a discussion on tyrosine kinases]. For yeast kinases, a test for functional complementation of the cognate mutation usually suffices. If degradation of hybrid proteins is a persistent problem, a protease mutation (e.g. *pep4*) can be introduced into the reporter strain.

(f) If the expression of hybrid proteins is toxic to yeast, try using vectors with low copy number (for centromeric GBD and GAD fusion vectors see ref. 29; for a centromeric LexA fusion vector, see ref. 30), with attenuated promoter strength (e.g. the 'mild' *ADH1* promoters in pGAD10, pGADGH, pGADGL vectors, Clontech), or with inducible promoters which allow conditional expression of fusion proteins [galactose-inducible

(11, 23); Cu^{2+}-inducible (31); methionine-repressible (32)]. The use of centromeric vectors, which are maintained at one or two copies per cell, has the added advantage of minimizing variations in hybrid protein levels caused by the fluctuation of copy numbers, which is common for the multicopy, 2μ vectors. For protein kinases, toxicity may arise due to promiscuous phosphorylation of yeast proteins, and therefore a kinase 'dead' mutant may reduce the toxicity. Occasionally, yeast strains that are tolerant to the toxic expression of hybrid proteins can be selected and used in two-hybrid screens (33).

(g) For protein kinases, certain mutant alleles may exhibit improved inter-actions (34). Using such mutants as bait for two-hybrid screens could allow recovery of interactors that are otherwise hard to detect. For instance, a catalytically deficient kinase may bind substrates normally but fail to release them from the active site, thus trapping substrates. In addition, kinase mutants that cause a dominant interfering phenotype due to sequestration of other proteins *in vivo* (35), are also good baits for two-hybrid screens.

(h) For membrane-bound receptor kinases, use intracellular domains, which usually contain the catalytic function, as bait. Examples include the TGF-β type I receptor serine/threonine kinase (36) and the insulin receptor tyrosine kinase (37). In addition, a two-hybrid analysis has recently been carried out using the extracellular domain of a receptor (31).

(i) Except for the dual-specificity protein kinases, budding yeast has no tyrosine-specific protein kinases, which poses a problem for interactions that depend on tyrosine phosphorylation. However, introducing the necessary tyrosine kinase, together with the two-hybrid proteins, has been shown to circumvent this potential problem (30). In addition, the demonstration of a two-hybrid interaction between the cytoplasmic domain of the insulin receptor tyrosine kinase and a substrate in yeast (37), which is strictly dependent on the receptor tyrosine kinase activity, suggests that budding yeast permits tyrosine phosphorylation and allows characterization of interactions involving protein tyrosine kinases.

(j) After potentially interacting clones are identified, it is crucial to show the specificity of the interaction by testing interactions with several irrelevant baits and vector controls (38). In addition, with exceptions (see ref. 39), coding sequences of potential interactors should be in-frame with the AD fusion moiety (as most cDNA libraries are made as AD fusions).

(k) To test possible interactions between known proteins, it is advisable to make both BD and AD-hybrid constructs for a given kinase and its interactor, and test which combination results in the best interaction. For some proteins, a two-hybrid interaction depends on which protein is in the BD or AD hybrid.

Protocol 3. Preparation of protein extracts from yeast cells

Reagents

- extraction buffer (50 mM Tris–HCl, pH 7.5, 1 mM EDTA, 150 mM NaCl, 1% Triton X-100); cocktail of protease inhibitors may be added; store at 4°C
- octanol (Sigma)
- selective synthetic complete (SC) liquid medium
- glass beads, ~0.45 mm in size (Sigma cat. no. G-8772)

Method

1. Grow yeast cells to mid-log phase (A_{600} ~0.5) in 50 ml SC medium to maintain the selection for plasmids. The doubling time for wild-type yeast is about 2 h in SC medium.
2. Pellet the cells in the cold, decant the medium, and resuspend cells in 10 ml cold extraction buffer. Transfer the cell suspension to a 15 ml plastic culture tube and repellet the cells. Cell pellets can be stored at −80°C for up to a few months.
3. Resuspend cells in 250 μl cold extraction buffer. Add 2 μl of octanol to prevent foaming during vortexing. Add glass beads just up to the liquid surface.
4. Vortex at maximum speed 10 times for 30 sec each at 4°C. Chill tubes on ice for 10 sec in between.
5. Centrifuge at 4°C to settle beads (2000 rpm, 5 min in a Sorvall SS-34 rotor) and transfer the upper milky layer to a cold microcentrifuge tube. Spin in the microcentrifuge at 15 000 g for 15 min to remove the cell debris; the resulting supernatant is the protein extract, with a concentration around 20 μg/μl.
6. For Western blotting analysis, 50–150 μg of protein should suffice.

Protocol 4. Preparation of yeast cells for Western blotting analysis

Reagents

- 2 × SDS–PAGE sample buffer

Method

1. Collect 10 × A_{600} units of yeast cells by spinning at room temperature, and wash cells once with 1.5 ml sterile H_2O. Resuspend cells in 100 μl of 2 × sample buffer, and boil for 10 min.
2. Spin down (15 000 g, 15 min) at room temperature in a micro-centrifuge.

> 3. Load one-third of the supernatant on an SDS–polyacrylamide gel. Avoid touching the viscous lower layer, which contains DNA and other cell debris. The result is a little messier than that obtained with *Protocol 3*.

3. Analysis of protein kinase interactions

3.1 Mapping interaction regions

To characterize an identified interaction, it is often informative to know where the interaction regions reside in each partner. In some instances, such interaction regions can become apparent if several overlapping clones of an interacting protein are isolated from a two-hybrid screen (40). Generally, the two-hybrid system offers a convenient way to delineate interacting regions. A panel of constructs containing deletions in the coding sequence fused to BD or AD can be made, introduced into a yeast reporter, and tested for interaction against a full-length partner hybrid. Defined, in-frame amino-terminal deletions can also be created by PCR. For analysis of the carboxyl-terminus, double-stranded, nested deletions can be derived directly from hybrid constructs using exonuclease III. With careful planning, a library of progressive C-terminal deletions can be obtained for virtually any target gene. The key is to include two unique restriction enzyme sites at the 3'-end of the target sequence: a *proximal* site that leaves a blunt or 5'-overhanging end after digestion (e.g. *Sma*I, *Not*I), followed by a *distal* site that creates 3'-overhangs (longer than 3 bases, e.g. *Kpn*I, *Pst*I). The presence of both sites enables the access of the Exo III enzyme (for detailed procedures, see Pharmacia nested deletion kit).

Once specific interaction domains are mapped, the two-hybrid method can be used to identify residues critical for interaction by coupling mutagenesis to the two-hybrid method (see Section 4.1). The identified interacting sequences are good agents for dominant interference studies: an isolated interaction domain, when overexpressed, may interfere with the function of the full-length protein. In addition, the interacting sequences provide a tool for probing conformational changes of kinase complexes (41, 42). Finally, knowing the interacting sequences may help design inhibitory peptides to disrupt pharmacologically important kinase interactions.

3.2 Bridged interactions

In theory, interactions detected in the two-hybrid system reflect either direct protein–protein interactions, or interactions that are mediated by other proteins. For example, proteins X and Y could interact via a bridging protein B. In this regard, X and Y do not interact without B, while B interacts with both X and Y directly. To bridge the interaction, B must contain distinct binding sites for X and Y, such that it binds X and Y simultaneously. Therefore, the two-hybrid bridging experiment can be used to demonstrate trimeric complex formation *in vivo*. Two-hybrid analysis has revealed such bridging proteins in diverse signalling systems involving protein kinases. In the sexual response pathway of *Schizosaccharomyces pombe*, the Byr2 protein kinase is capable of bridging a two-hybrid interaction between Ste4 and Ras1, which otherwise do not interact (43). Mammalian Raf kinase can mediate inter-actions between Ras and MEK (a MAPK kinase), as shown by two-hybrid bridging experiments (3).

It should be noted, however, that the finding that two proteins only interact in the presence of a third protein is not sufficient to conclude that the third protein is physically bridging the interaction. A third protein could modify one of the interacting partners, therefore allowing it to interact with the other. For example, phosphorylation of a protein by tyrosine kinases creates binding sites for SH2 domain-containing proteins. Therefore, the interaction requires the catalytic activity, rather than the physical presence, of the kinases, and a catalytically 'dead' kinase will not support the interaction. In contrast, a bridging kinase might still support an interaction even it is catalytically deficient.

When examining two-hybrid interactions of yeast proteins in yeast, one point to bear in mind is that endogenous bridging proteins could be taking part in the interaction. It is also possible that yeast could encode bridging proteins that function with mammalian proteins. Therefore, it is important to confirm that a two-hybrid interaction is direct, by *in vitro* binding analyses.

3.3 Scaffold proteins

Analyses of two-hybrid interactions within kinase complexes have uncovered a unique class of proteins called scaffold proteins (44). Scaffold proteins are molecules that bind to and facilitate interactions of other proteins in the context of protein complexes. Scaffold proteins may bridge an interaction but, unlike bridging proteins, they may not be absolutely required for other proteins to interact with one another. A good example is Ste5, a yeast protein in the pheromone response pathway. By using the two-hybrid system, several groups have shown that Ste5 interacts, through distinct binding sites, with all three sequential members of a MAP kinase module. Interactions between members of the MAP kinase cascade can occur in the absence of Ste5, indicating that the Ste5 scaffold facilitates but is not absolutely required for interactions (24, 45, 46). In the Snf1 pathway, the conserved Sip1/Sip2/Gal83 family of scaffold proteins each bind to Snf1 and Snf4 independently via dis-tinct domains, thus facilitating the interaction between Snf1 and Snf4 in the complex. The triple knock-out of *SIP1/SIP2/GAL83* impairs formation of a complex containing Snf1 and Snf4 (22). However, Snf1 and Snf4 also interact directly with one another (41).

3.4 Self-association of protein kinases

Some protein kinases exist or function in the form of dimers or oligomers, for example, ERK2, Raf (47, 48). The two-hybrid system offers a convenient method to assess kinase self-association by testing the interaction between BD and AD hybrids of the same kinase.

3.5 Interactions within a protein kinase

Many protein kinases, both serine/threonine and tyrosine specific, are regulated by autoinhibitory mechanisms (49, 50). One such mechanism involves a separate, regulatory domain of a protein kinase that binds to and inhibits the catalytic domain of the same molecule. Often regulatory domains are modular, i.e. they behave as independently foldable units. Thus, regulatory domains and catalytic domains can be isolated and separately fused to BD or AD, and two-hybrid interaction assays can be performed. Such analyses have revealed that the Snf1 and Byr2 kinases are regulated by autoinhibition (41, 42).

4. Mutational analysis of two-hybrid interactions

4.1 Mutagenesis of interaction domains

To confirm the authenticity of an interaction domain identified through deletion analysis, it is often necessary to obtain point mutations within the domain that disrupt the interaction. Such mutations also identify the amino acid residues critical for interaction. Compared with deletion mutations, loss-of-interaction point mutants are invaluable in functional studies in that they are less likely to affect the global folding or the stability of a protein. If such mutations are associated with a phenotype, it could well mean that the interaction is functionally important. In addition, point mutations allow a finer dissection of regulatory mechanisms. For example, if two proteins interact with the same minimal domain of a third protein, point mutation analysis could determine whether the two proteins recognize the same or different residues.

To obtain point mutations that disrupt an interaction, the interacting domain sequence is first randomly mutagenized, and the resulting mutant sequence pool is ligated into a BD or AD vector to make a library. The two-hybrid screen is then carried out to look for mutants that have lost interactions with a target protein. PCR-based mutagenesis of a defined region is commonly used, taking advantage of the inherent error-prone nature of the *Taq* DNA polymerase. The *Taq* enzyme lacks a 3' to 5' exonuclease activity, which serves as a proof-reading function in other enzymes. As a result, it has an error rate of 10^{-4} per nucleotide incorporated (51). The fidelity of *Taq* polymerase can be further compromised by the addition of a low amount of Mn^{2+} in the reaction, and by lowering the ratio of dATP to the remaining nucleotides (52). This is particularly useful to increase the mutation frequency for small regions. A protocol is given below (*Protocol 5*).

Protocol 5. PCR-mediated mutagenesis of defined regions (modified from refs 52 and 53)

Reagents

• PCR reagents

• $MnCl_2$, 10 mM

Method

1. Mix PCR reagents in 100 μl volume (final concentration, 6.0 mM $MgCl_2$, 1 mM each of dGTP, dCTP, dTTP, 0.2 mM dATP, 1 μM each primer. Use 100 ng DNA template and 2.5 units of the *Taq* polymerase). Set up several parallel tubes to ensure independent mutational hits.

2. Amplify templates for 10 initial cycles with a profile that fits the length of the DNA template and the annealing temperature of primers.

3. Add Mn^{2+} to a final concentration of 0.6 mM. Continue PCR for an additional 20 cycles.

4.2 Screen for loss-of-interaction mutants

Two methods are commonly used to introduce the mutant sequences into a reporter strain. The first one involves constructing a mutant library by digesting and ligating mutagenized sequences into a suitable AD or BD vector, and by transforming *E. coli* with the ligation mixture. Transformants are scraped off plates and plasmid DNA isolated and subsequently used to transform bait-bearing reporter yeast. A second method (54) uses a gap repair procedure. In this method, PCR generates mutagenized DNA fragments with ends homologous to those of a gapped vector. The PCR fragments are co-transformed with the gapped vector DNA into yeast directly. Homologous recombination in yeast repairs the gap of the vector with the mutagenized DNA sequence, forming intact plasmids that can now be selected on plates. This procedure obviates the need to clone the mutant sequences.

To identify mutant partners that fail to interact with a bait, two-hybrid screens are carried out. For *lacZ*-based reporters, white colonies are to be picked and analysed further for potential loss-of-interaction mutants. The use of a *URA3* reporter, which allows growth of potential interaction-defective mutants on plates containing the drug 5-fluoro-orotic acid, greatly facilitates the isolation of potential mutants (53). To eliminate truncation mutants, potential interaction-defective mutants are checked for the expression of full-length protein by Western blot analysis. As an alternative method, the truncation mutants may be selected against by making a C-terminal fusion with a selectable marker, such as a drug resistance gene or a yeast amino acid marker, to the coding sequence of interest.

To recover the library plasmid containing the mutant prey, the bait plasmid is first segregated from yeast, and total DNA prepared from the prey-bearing yeast is used to transform *E. coli*. Protocols are listed here for segregating plasmids and for isolating yeast DNA (*Protocols* 6 and 7, respectively). *Protocol 8* describes a method to selectively rescue the library plasmid in a single step. This procedure is based on the fact that the sequences of the bait and prey plasmids are known; therefore, the bait plasmid can be selectively destroyed by cleavage at a unique site that is not present in the library plasmid. Such a unique restriction site (e.g. the 8 base-pair cutter *Not*I) can be conveniently introduced during subcloning. This method takes 2 days to complete, compared with at least a week for standard procedures. For two-hybrid screens of new interacting proteins whose sequences are unknown and potentially contain *Not*I sites, other methods such as complementation of an *E. coli leuB* strain by yeast *LEU2* plasmids, or an *E. coli trpC* strain by yeast *TRP1* plasmids, can be used (12).

Protocol 6. Segregation of plasmids from yeast cells by growth on non-selective medium[a]

Equipment and reagents
- non-selective synthetic complete (SC) plates
- replica block and sterile velvets

Method

1. As an example, suppose that the yeast has two plasmids: a *TRP1*-containing plasmid to be segregated, and a *LEU2* plasmid to be kept in the yeast. The key is to grow the yeast in medium supplemented with tryptophan but lacking leucine.[b]

2. Make three yeast patches (each 0.5 cm in diameter) on a synthetic complete plate lacking leucine (SC–Leu) but containing all the other necessary auxotrophic requirements.[c] Grow overnight at 30 °C.

3. Repeat patching on the SC–Leu plate, using the cells from the previous overnight plate. Grow overnight at 30 °C. This process can be repeated a few times more for 'die-hard' plasmids.

4. From the second set of overnight patches, streak out for single colonies on SC–Leu plates. Incubate at 30 °C until colonies are formed (~2 days).

Rong Jiang and Marian Carlson

Protocol 6. *Continued*

5. Replica plate colonies to SC–Leu and SC–Leu–Trp plates. Grow overnight at 30°C. Those colonies that are present on the SC–Leu plates but fail to grow on the SC–Leu–Trp plates have lost the *TRP1* plasmid.

[a] The frequency of mitotic loss of yeast centromeric (CEN) plasmids with autonomous DNA replication sequence (ARS) is ~1% when grown non-selectively (55). The loss frequency for 2 μ plasmids, from which most two-hybrid vectors are derived, is even higher.
[b] The loss of a plasmid can be selected if the parental vector contains a drug-sensitive marker, for example, $CYH2^S$ or $CAN1^S$, and the host bears a corresponding drug-resistance gene. Those cells that have lost the bait plasmid will grow on plates containing cycloheximide or canavanine, respectively.
[c] Instead of patching cells on plates, liquid cultures in non-selective synthetic medium can also be used for segregating plasmids.

Protocol 7. Preparation of yeast DNA for plasmid recovery (56)

Reagents

- selective synthetic complete (SC) liquid medium
- cracking buffer (10 mM Tris–HCl, pH 8.0, 1 mM EDTA, 2% Triton X-100, 1% SDS; store at room temperature)
- TE buffer (10 mM Tris–HCl, pH 8.0, 1 mM EDTA)
- glass beads, ~0.45 mm (Sigma cat. no. G-8772)

Method

1. Grow 3 ml overnight culture at 30°C in selective SC medium to maintain selection for the plasmid.

2. Pellet cells in screw-capped microcentrifuge tubes. Wash once with 1 ml H_2O.

3. Add 0.1 ml cracking buffer, 0.1 ml phenol (saturated with TE buffer pH 8.0)/chloroform/isoamyl alcohol (25:24:1), and 300 mg glass beads. It is convenient to add glass beads using a scoop made from a 0.5 ml microcentrifuge tube, trimmed such that it holds 300 mg of beads. Be sure to clear away any beads stuck in the screw threads before closing the tube; otherwise, leaks may arise and cause a phenol spill.

4. Place tubes in a vortex mixer microcentrifuge tube adapter, and vortex at full speed at room temperature for 10 min to disrupt the yeast cells.

5. Add 0.1 ml TE buffer to each tube. Mix by gentle vortexing. Spin for 5 min at 15 000 g.

6. Transfer the aqueous layer, which contains yeast genomic DNA and the plasmid DNA, into a new tube. Precipitate the DNA by mixing with 0.5 ml of 95% ethanol at room temperature. Spin down the pellet (10 000 g, 5 min). Rinse with cold 75% ethanol and dry the pellet.

328

7. Dissolve the pellet in 0.1 ml of sterile H_2O. Take 2 μl to transform *E. coli* by electroporation.

Protocol 8. Single-step selective plasmid rescue

Method

1. The plasmid that is not wanted must contain a unique restriction enzyme site (i.e. *Not*I), which is not present in the plasmid to be rescued from the yeast.

2. Isolate total DNA from plasmid-bearing yeast (*Protocol 7*).

3. Digest half of the DNA with the unique restriction enzyme for 2 h to ablate the unwanted plasmid. Heat inactivate the enzyme at 65°C for 20 min.

4. Transform *E. coli*.

4.3 Isolation of compensatory mutants

A loss-of-interaction mutation alters the nature of an interacting interface to such an extent that the resulting mutant protein loses affinity for its partner. However, it is sometimes possible to restore the disrupted interaction by isolating compensatory mutations in the partner. To achieve this, the interacting sequence of the partner is mutagenized as before, and a two-hybrid screen is then performed to seek mutant partners that now interact with the mutated bait. Some compensatory mutants may interact only with the mutant but not with the wild-type bait, which suggests that compensation takes place at the original contact point. A different class of compensatory mutants may interact with both the mutant and the wild-type bait, suggesting that compensation occurs by creating a new contact point between the two mutant proteins; in this case, the interaction with the wild-type bait should be stronger than with the mutant bait.

Isolating compensatory mutations serves two purposes. First, it provides *in vivo* evidence that the two proteins are in direct contact (41, 57). Secondly, a concurrent restoration of a phenotype strongly suggests a causal relationship between the disrupted interaction and the phenotype (57).

4.4 Role of phosphorylation in kinase interactions

The catalytic function of a protein kinase may be required for its own interaction with other proteins in two different ways. First, many kinases are autophosphorylated *in vivo*. For some, such a phosphorylation has been shown to be important for interaction with regulatory proteins. Examples include cAMP-dependent protein kinase (58), and the insulin receptor and Flt-1 receptor tyrosine kinases (37, 59). Secondly, a protein kinase may phosphorylate a regulator such as a phosphatase, which then binds to the kinase and down-regulates its activity (60). To address the role of catalytic activity in the requirement for kinase interaction, a catalytically 'dead' mutant can be used in two-hybrid assays. A commonly used mutation is the substitution of the invariant lysine residue in the ATP-binding pocket of the catalytic domain.

In some cases mutation of the catalytic activity enhances interaction of the kinase with regulators. For example, the K84R substitution in Snf1 elevates interaction with the Reg1 regulatory subunit of protein phosphatase 1 (34). Such catalytically deficient mutants may prove useful as baits to trap kinase-interacting proteins whose interactions are otherwise too weak to detect.

Many kinases are regulated via transphosphorylation by other kinases. It has been shown that several kinase interactions are dependent on such modifications. A good example is the human cyclin-dependent kinase Cdc2: phosphorylation of the conserved T loop threonine-161 is required for the binding of cyclins (61, 62). For a given kinase, if the sites of phosphorylation have been identified, their effect on kinase interactions can be assessed in the two-hybrid system using phosphorylation site mutants. In the case of the yeast Snf1 kinase, mutation of the conserved T loop threonine (T210) abolished the interaction with its activating subunit, Snf4, suggesting that phosphorylation of T210 is required for conformational changes in the kinase complex that allow the binding of Snf4 to Snf1 (34).

5. Analysis of simultaneous multiple interactions using a double two-hybrid method

Many proteins function *in vivo* in the context of large complexes in which one protein simultaneously interacts with multiple factors (63). Such a multiplicity of protein–protein interactions provides a challenge to the conventional two-hybrid methodology, which analyses interactions of two proteins at a time. A strategy, called the double two-hybrid method, has been developed that allows multiple interactions to be analysed simultaneously (41, see *Figure 1*). In this method, two different baits, LexA-X (protein X) and GBD-Y (protein Y), are introduced together with a GAD-M (protein M) into a strain bearing two integrated reporters, $lexA_{op}$-lacZ and *GAL1-HIS3*. Transformants are tested for blue colour and His phenotype, thereby allowing both X–M and Y–M interactions to be assessed in the same cell. In addition, the double two-hybrid method can be used to isolate alleles in M which differentiate X–M and Y–M interactions. For example, a library of mutant M sequence fused to

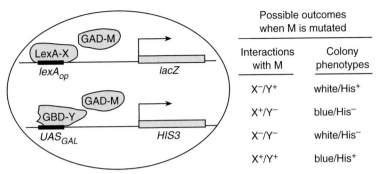

	Possible outcomes when M is mutated	
	Interactions with M	Colony phenotypes
	X^-/Y^+	white/His$^+$
	X^+/Y^-	blue/His$^-$
	X^-/Y^-	white/His$^-$
	X^+/Y^+	blue/His$^+$

Figure 1. The double two-hybrid method (40). In the yeast nucleus, interactions between LexA-X and GAD-M, and between GBD-Y and GAD-M can be simultaneously analysed. Mutant alleles of M may disrupt one or both interactions. Possible outcomes are listed.

GAD can be transformed into the reporter containing LexA-X and GBD-Y baits. All four possible combinations of interactions, X^-/Y^+, X^+/Y^-, X^-/Y^-, and X^+/Y^+ (where '–' indicates no interaction and '+' an interaction), can be resolved in a single screen by the resulting white/His$^+$, blue/His$^-$, white/His$^-$, and blue/His$^+$ phenotypes, respectively. Similar systems using *URA3* and *lacZ* reporters have now been described (64, 65).

6. Concluding remarks

The two-hybrid system is a powerful genetic method both for the identification of proteins that interact with protein kinases and for the analysis of inter- and intra-molecular interactions exhibited by protein kinases. However, findings must be interpreted with caution, as 'false-positive' interactions are not uncommon. Artefactual interactions can result from the overexpression of proteins, the expression of an incomplete coding sequence, the presence of the fused BD or AD moiety, or the expression of a protein out of its normal cellular context (as is always the case for mammalian proteins expressed in yeast). These potential problems necessitate the use of complementary genetic or biochemical experiments to confirm the authenticity of an observed two-hybrid interaction, and to address its functional relevance. None the less, the two-hybrid method offers a unique approach to the analysis of protein interactions, and the rapidly expanding use of this method attests to its great utility in the analysis of protein kinases.

References

1. Fields, S. and Song, O. (1989). *Nature*, **340**, 245.
2. Chien, C., Bartel, P. L., Sternglanz, R., and Fields, S. (1991). *Proc. Natl. Acad. Sci. USA*, **88**, 9578.
3. Van Aelst, L., Barr, M., Marcus, S., Poverino, A., and Wigler, M. (1993). *Proc. Natl. Acad. Sci. USA*, **90**, 6213.

4. Vojtek, A. B., Hollenberg, S. M., and Cooper, J. A. (1993). *Cell*, **74**, 205.
5. Dang, C. V., Barrett, J., Villa-Garcia, M., Resar, L. M. S., Kato, G. J., and Fearon, E. R. (1991). *Mol. Cell Biol.*, **11**, 954.
6. Luo, Y., Batalao, A., Zhou, H., and Zhu, L. (1997). *BioTechniques*, **22**, 350.
7. Celenza, J. L. and Carlson, M. (1986). *Science*, **233**, 1175.
8. Celenza, J. L. and Carlson, M. (1989). *Mol. Cell Biol.*, **9**, 5034.
9. Yang, X., Hubbard, E. J. A., and Carlson, M. (1992). *Science*, **257**, 680.
10. Durfee, T., Becherer, K., Chen, P.-L., Yeh, S.-H., Yang, Y., Kilburn, A., Lee, W.-H., and Elledge, S. (1993). *Genes Dev.*, **7**, 555.
11. Gyuris, J., Golemis, E., Chertkov, H., and Brent, R. (1993). *Cell*, **75**, 791.
12. Bartel, P. L. and Fields, S. (ed.) (1997). *The yeast two-hybrid system*. Oxford University Press, Oxford.
13. Rose, M. D., Winston, F., and Hieter, P. (ed.) (1990). *Methods in yeast genetics: a laboratory course manual*. Cold Spring Harbor Laboratory Press, Cold Spring Harbor, New York.
14. Guthrie, C. and Fink, G. R. (ed.) (1991). *Guide to yeast genetics and molecular biology. Methods in enzymology*, Vol. 194. Academic Press, London.
15. Li, B. and Fields, S. (1993). *FASEB J.*, **7**, 957.
16. Estojak, J., Brent, R., and Golemis, E. A. (1995). *Mol. Cell Biol.*, **15**, 5820.
17. Fagan, R., Flint, K. J., and Jones, N. (1994). *Cell*, **78**, 799.
18. Breeden, J. and Nasmyth, K. (1985). *Cold Spring Harbor Symposia on Quantitative Biology*, **L**, 643.
19. Dagher, M. C. and Filhol-Cochet, O. (1997). *BioTechniques*, **22**, 916.
20. Guarente, L. (1983). In *Methods in Enzymology* (ed. R. Wu, L. Grossman, and K. Moldave), Vol. 101, p. 181. Adademic Press, London.
21. Miller, J. H. (ed.) (1972). *Experiments in molecular genetics*. Cold Spring Harbor Laboratory, Cold Spring Harbor, NY.
22. Jiang, R. and Carlson, M. (1997). *Mol. Cell Biol.*, **17**, 2099.
23. Huang, J. and Schreiber, S. L. (1997). *Proc. Natl. Acad. Sci. USA*, **94**, 13396.
24. Printen, J. A. and Spague, G. F. (1994). *Genetics*, **138**, 609.
25. Golemis, E. A. and Brent, R. (1992). *Mol. Cell Biol.*, **12**, 3006.
26. Golemis, E. A. and Brent, R. (1997). In *The yeast two-hybrid system* (ed. P.L. Bartel and S. Fields), p. 43. Oxford University Press, Oxford.
27. Du, W., Vidal, M., Xie, J.-E., and Dyson, N. (1996). *Genes Dev.*, **10**, 1206.
28. Kuchin, S., Yeghiayan, P., and Carlson, M. (1995). *Proc. Natl. Acad. Sci. USA*, **92**, 4006.
29. Chevray, P. and Nathans, D. (1992). *Proc. Natl. Acad. Sci. USA*, **89**, 5789.
30. Osborne, M. A., Dalton, S., and Kochan, J. P. (1995). *BioTechnoloby*, **13**, 1474.
31. Ozenberger, B. A. and Young, K. H. (1995). *Mol. Endocrinol.*, **9**, 1321.
32. Tirode, F., Malaguti, C., Romero, F., Attar, R., Camonis, J., and Egly, J. M. (1997). *J. Biol. Chem.*, **272**, 22995.
33. Bai, C., Sen, P., Hoffmann, K., Ma, L., Goeble, M., Harper, J. W., and Elledge, S. J. (1996). *Cell*, **86**, 263.
34. Ludin, K., Jiang, R., and Carlson, M. (1998). *Proc. Natl. Acad. Sci. USA*, **95**, 6245.
35. Van den Heuvel, S. and Harlow, E. (1993). *Science*, **262**, 2050.
36. Wang, T., Danielson, P. D., Li, B.-Y., Shah, P. C., Kim, S. D., and Donahoe, P. K. (1996). *Science*, **271**, 1120.
37. O'Neill, T. J., Craparo, A., and Gustafson, T. A. (1994). *Mol. Cell Biol.*, **14**, 6433.

38. Bartel, P., Chien, C.-T., Sternglanz, R., and Fields, S. (1993). *BioTechniques*, **14**, 920.
39. Lesage, P., Yang, X., and Carlson, M. (1996). *Mol. Cell. Biol.*, **16**, 1921.
40. Yang, X., Jiang, R., and Carlson, M. (1994). *EMBO J.*, **13**, 5878.
41. Jiang, R. and Carlson, M. (1996). *Genes Dev.*, **10**, 3105.
42. Tu, H., Barr, M., Dong, D. L., and Wigler, M. (1997). *Mol. Cell Biol.*, **17**, 5876.
43. Barr, M., Tu, H., Van Aelst, L., and Wigler, M. (1996). *Mol. Cell Biol.*, **16**, 5597.
44. Pawson, T. and Scott, J. (1997). *Science*, **278**, 2075.
45. Marcus, S., Polverino, A., Barr, M., and Wigler, M. (1994). *Proc. Natl. Acad. Sci. USA*, **91**, 7762.
46. Choi, K.-Y., Satterberg, B., Lyons, D. M., and Elion, E. A. (1994). *Cell*, **78**, 499.
47. Khokhlatchev, A. V., Canagarajah, B., Wilsbacher, J., Robinson, M., Atkinson, M., Goldsmith, E., and Cobb, M. H. (1998). *Cell*, **93**, 605.
48. Farrar, M. A., Alberol-Ila, J., and Perlmutter, R. M. (1996). *Nature*, **383**, 178.
49. Soderling, T. R. (1990). *J. Biochem. Biol.*, **265**, 1823.
50. Hubbard, S. R., Wei, L., Ellis, L., and Hendrickson, W. A. (1994). *Nature*, **372**, 746.
51. Zhou, Y., Zhang, X., and Ebright, R. H. (1991). *Nucl. Acids Res.*, **19**, 6052.
52. Leung, D., Chen, E., and Goeddal, D. (1989). *Technique*, **1**, 11.
53. Vidal, M., Braun, P., Chen, E., Boeke, J. D., and Harlow, E. (1996). *Proc. Natl. Acad. Sci. USA*, **93**, 10321.
54. Lehming, N., McGuire, S., Brickman, J. M., and Ptashne, M. (1995). *Proc. Natl. Acad. Sci. USA*, **92**, 10242.
55. Stinchcomb, D. T., Mann, C., and Davis, R. W. (1982). *J. Mol. Biol.*, **158**, 157.
56. Hoffman, C.S. and Winston, F. (1987). *Gene*, **57**, 267.
57. White, M., Nicolette, C., Minden, A., Polverino, A., Van Aelst, L., Karin, M., and Wigler, M. (1995). *Cell*, **80**, 533.
58. Levin, L. R., Kuret, J., Johnson, K. E., Powers, S., Cameron, S., Michaeli, T., Wigler, M., and Zoller, M. J. (1988). *Science*, **240**, 68.
59. Cunningham, S. A., Waxham, M. N., Arrate, P. M., and Brock, T. A. (1995). *J. Biol. Chem.*, **270**, 20254.
60. Camps, M., Nichols, A., Gillieron, C., Antonsson, B., Muda, M., Chabert, C., Boschert, U., and Arkinstall, S. (1998). *Science*, **280**, 1262.
61. Ducommun, B., Brambilla, P., Felix, M.-A., Franza Jr B. R., Karsenti, E., and Draetta, G. (1991). *EMBO J.*, **10**, 3311.
62. Desai, D., Wessling, H. C., Fisher, R. P., and Morgan, D. O. (1995). *Mol. Cell. Biol.*, **15**, 345.
63. Alberts, B. (1998). *Cell*, **92**, 291.
64. Xu, C. W., Mendelsohn, A. R., and Brent, R. (1997). *Proc. Natl. Acad. Sci. USA*, **94**, 12473.
65. Inouye, C., Dhillon, N., Durfee, T., Zambryski, P. C., and Thorner, J. (1997). *Genetics*, **147**, 479.

15

Analysis of protein kinase interactions using biomolecular interaction analysis

FRIEDRICH W. HERBERG and BASTIAN ZIMMERMANN

1. Introduction

Novel technologies employing surface plasmon resonance (Biacore AB) or resonance mirrors (IAsys, Lab Systems) can be used for the characterization of protein-protein, protein-DNA and ligand-receptor interactions (for review see refs 1–3). These techniques directly measure the binding of a molecule in the soluble phase (the 'analyte', also called the 'ligate') to a 'ligand' molecule immobilized on a sensor surface. Thereby the association (k_a) and dissociation rate constants (k_d) are determined separately. With a known concentration of the analyte, apparent equilibrium binding constants (K_D or K_A) can be calculated. The chemistries for coupling molecules to the surface via amine, sulfhydryl, carboxyl, and other groups are well defined and reproducible. Non-covalent, site-directed immobilization can be achieved by avidin–biotin interaction, by the use of antibodies, or by binding fusion proteins to specific surfaces. Surfaces containing lipid monolayers are also available. This technology can be used for monitoring interactions, in real time without the use of labels, between two or more molecules, e.g. between proteins, peptides, nucleic acids, or pharmaceuticals. Interactions involving lipid vesicles, bacteria, viruses, and mammalian cells can also be measured. Questions that can be addressed include the following:

- which components interact, and under what conditions?
- how fast do they bind and dissociate, and how strongly do they interact?
- how much *active* interactant do I have in my sample?
- how is the interaction influenced by molecular modifications (e.g. phosphorylation and other post-translational modifications)?
- do different components influence each other's binding to a common interactant, for example by steric, allosteric, or co-operative effects?

- what effect do additional effectors, co-factors (e.g. metal ions, nucleotides, pH), have on the interaction of the interactants?

Molecules do not necessarily need to be purified, although for the accurate detection of apparent association and dissociation rate constants in a defined system, well characterized fractions are preferable. The functional characteristics of biological interactions, such as kinetics and affinity, are analysed by monitoring the association and dissociation rate constants. The method can also be used qualitatively, e.g. (i) to determine relative binding affinities; (ii) for ligand fishing; (iii) to detect expression of recombinant proteins; and (iv) to characterize binding topology (epitopes, antibodies). Determination of concentrations in a highly specific manner is another application.

1.1 Basics of optical biosensors, or how to utilize an evanescent field

Both systems available, Biacore or IAsys, are evanescent field detectors, but the basis of the physics and the detection principles are quite different. In the following a brief introduction is given for sensors based on surface plasmon resonance (SPR, Biacore) and waveguiding technology (resonant mirror biosensor, IAsys) for the reader interested in technical details.

1.1.1. Biacore

The Biacore system consists of a light source emitting near-infrared light, a sensor microchip, an automated liquid-handling system (constant flow), and a diode array position-sensitive detector. Biomolecular binding events cause changes in the refractive index at the surface layer, which are detected as changes in the surface plasmon resonance signal. In general, the refractive index change for a given change of mass concentration at the surface layer, is practically the same for all proteins and peptides (4), and is similar for glycoproteins, lipids, and nucleic acids. This value is plotted as response units (RU), where 1000 RU corresponds to 1 ng protein/mm^2 sensor surface (4). 10 000 RU corresponds to 0.01 refractive index units (RIU) which is 0.1° arc in the SPR angle.

Surface plasmon resonance
The detection principle of the Biacore relies on surface plasmon resonance (SPR), an optical phenomenon that arises when light illuminates thin conducting films under specific conditions. The resonance is a result of interaction between electromagnetic vectors in the incident light and free electron clouds called surface plasmons in the conductor. If the interface is coated with a thin layer of a conducting material (gold is used in the case of the Biacore), the phenomenon of SPR can arise. A resonant coupling between the incident light energy and surface plasmons in the conducting film occurs at a specific angle of incident light, absorbing the light energy and causing a characteristic drop in the reflected light intensity at that angle (*Figure 1*).

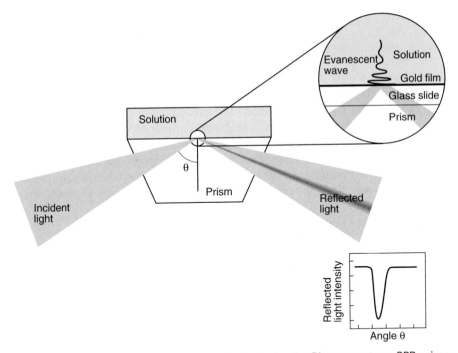

Figure 1. Surface plasmon resonance as the basis for the Biacore system. SPR arises when light is totally internally reflected from a metal-coated interface between two media of different refractive index (a glass prism and solution). If the incident light is focused on the surface in a wedge, the drop in intensity at the resonance angle appears as a 'shadow' in the reflected light wedge, which is detected by a diode array detector. (Courtesy of F. Schindler, BiacoreAB.)

The resonance angle is sensitive to a number of factors, including the wavelength of the incident light, and the nature and thickness of the conducting film. Most importantly for this technology, the angle depends on the refractive index of the medium into which the evanescent wave propagates. When other factors are kept constant, the resonance angle is a direct measure of the refractive index of the medium. Only the angle on which SPR occurs is altered and detected with the diode array detector; the intensity of the 'shadow' in the reflected light is unchanged.

The evanescent wave decays exponentially with distance from the interface, and effectively penetrates the lower refractive index medium to a depth of approximately one wavelength (5). Under the conditions used in this technology, this distance is about 300 nm. In consequence, SPR only detects changes in refractive index very close to the surface.

1.1.2. IAsys

The IAsys is an evanescent field detector based on resonance mirror technology (3). The system consists of a laser light source (670 nm), scanning about a $10°$ arc across the device, a cuvette system, and a charged coupled device detector. The signal derived can be plotted in arc seconds. Thereby, 1 ng protein/mm^2 corresponds to 163 arc seconds (6).

The resonance mirror

In a construction similar to the SPR device, the resonance mirror is a structure of two dielectric layers of glass. The resonance mirror consists of a high index waveguide separated from a high index prism block by an intervening, low index coupling layer (*Figure 2*). Light is totally internally reflected from the sensor surface using a prism. Below the sensor layer, where the interaction takes place, a dielectric resonant layer of high refractive index (about 100 nm thick) is placed. This is separated from the prism by a low index layer (coupling layer) around 1 mm thick, this being sufficiently thin that light can couple into the resonant layer via the evanescent field. Efficient coupling occurs only for certain incident angles, where phase matching between the incident beam and the resonant modes of the high index layer is achieved. At the resonance point, light couples into the high index layer and propagates along within this layer before coupling back into the prism, emerging to strike the detector. The angle of excitation of resonance is very sensitive to changes at the sensing interface, so that small changes are monitored by measuring shifts in the excitation angle.

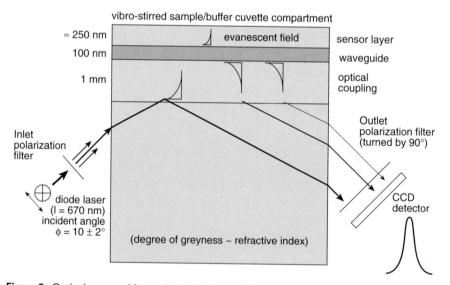

Figure 2. Optical waveguide as the basis for the IAsys system. At a certain angle during the movement of the laser, mode coupling of laser light into the waveguide occurs, followed by polarization shift and reflection of the laser light. The time of the light flash occurrence (maximum of resonance scan) at the detector is correlated with the corresponding laser position. This resonance laser position changes proportionally with the refractive index of the sensor layer. (Courtesy of H. Trutnau, Affinity Sensors.)

With SPR, the resonance position is seen as a dip in the reflected light intensity, as a component of the light is absorbed in the metal layer. With the resonance mirror being a low loss system, no dip is seen, but there is a shift in the phase of the reflected light on resonance.

2. Immobilization strategies

2.1 Principle strategies

Immobilization of the ligand to the surface of the sensor chip is the crucial step in interaction analysis. The ligand molecule has to be immobilized in such a way that it is in an active conformation and is not hindered sterically in its interaction with the analyte. The ligand can either be immobilized directly to the surface, or it can be captured by a previously immobilized capturing molecule (e.g. an antibody or a receptor ligand). After preparation, the surface can often be used many times, although the degree of re-use depends upon the stability of the molecule immobilized, and the regeneration method used to remove the bound analyte.

The immobilized ligand is the basis for the specificity in the interaction analysis. Several factors must be considered in deciding the best immobilization conditions, and these are discussed in detail in Sections 2.1.1–2.1.7 below.

2.1.1 The immobilization method

The immobilization of ligands can be performed by one of two different strategies:

- covalent, random coupling of ligands using well-defined chemical coupling methods
- site-directed, specific, and high affinity immobilization via capturing molecules

Carboxymethylated (CM) surfaces enable chemical coupling of ligands or capturing molecules via native amine ($-NH_2$), sulfhydryl ($-SH$), aldehyde ($-CHO$), or carboxyl ($-COOH$) groups. Covalent immobilization techniques require specific buffer conditions (pH, ionic strength) to electrostatically attract the ligand to the matrix and to favour the chemical reaction. Other sensor surfaces are supplied pre-immobilized with capturing molecules used for rapid binding of tagged ligands, e.g. streptavidin (SA) surfaces enabling immobilization of biotinylated ligands with high affinity. The use of capturing molecules usually does not require specific buffer conditions, because electrostatic pre-concentration is generally not relevant for immobilization that relies on high-affinity binding of ligand to capturing molecules. The choice of the immobilization method depends strongly on the biochemical properties of the ligand (e.g. its pI, its stability at low pH or low ionic strength) and the kind of analysis to be performed.

2.1.2 The choice of ligand versus analyte

The choice of which interactant should be the ligand, i.e. the immobilized component, and which should be the analyte, i.e. the component in the soluble phase, is a decision of key importance. If the investigation is to compare the interaction of several molecules with a common component, it is usually most convenient to use the common component as the ligand. The ideal ligand has the following properties:

- *specificity:* a high specificity exhibited by the ligand ensures a selective interaction analysis even when the bulk solution is a complex mixture
- *purity:* the ligand should be as pure as possible, both for reproducible immobilization and to avoid non-specific binding
- *size:* if the components are markedly different in size, it is often better to immobilize the smaller component, because the response is proportional to the mass of the binding analyte
- *regeneration stability:* the ligand should be able to withstand the conditions used for the regeneration, i.e. the removal of the bound analyte from the sensor surface, so that the surface with the immobilized ligand can be re-used in a series of analyses

2.1.3 Retention of biological activity

The primary requirement for ligands immobilized on sensor surfaces is that functional interaction with the analyte is retained. The biological activity may be adversely affected by the coupling reaction, and steric restrictions may also arise. Amine coupling is widely applicable, but most macromolecular ligands have a diversity of potential attachment points using this method. It may be more advantageous to use a coupling method where the attachment site is more rigorously defined, in order to obtain a homogeneous ligand population on the surface. Other coupling methods may also be necessary if the ligand contains reactive amines, or other nucleophilic groups, at the analyte-binding site. The cAMP-dependent protein kinase, for example, contains a lysine (K72) in the active site that is essential for the catalytic activity of the enzyme. This site is very reactive, and is involved in ATP binding and phosphotransfer. However, it can be protected by including 1 mM ATP and 2 mM $MgCl_2$ in the immobilization buffer. By protecting K72, the surface activity (stoichiometry of binding) of the catalytic subunit to the regulatory subunit (R-subunit as ligand) could be increased from 33 to 88% (7). The theoretical surface activity can be calculated according to $S = MW_L R_A / MW_A R_L$, where S is the stoichiometry, subscript L defines ligand (immobilized protein), subscript A defines analyte (injected protein), R represents the response in RU, and MW is the molecular weight of ligand or analyte.

2.1.4 Effect of buffer conditions

The carboxymethylated dextran matrix is negatively charged at pH values above about 3, and positively charged ligands are electrostatically attracted to the surface, enabling efficient immobilization from relatively dilute ligand solutions. Electrostatic pre-concentration is critically affected by the ionic strength and the pH of the immobilization buffer. The ligand must have a net positive charge to utilize this effect, and the pH of the buffer must therefore be below the isoelectric point (pI) of the ligand. A low ionic strength of the ligand solution also favours electrostatic interactions between the surface matrix and the ligand. The ionic strength of the ligand solution should therefore be kept as low as possible, with the buffer concentration around 10 mM, provided that the ligand is stable under these conditions. A trial immobilization without activating the surface ('pre-concentration run') can be performed to check the efficiency of electrostatic pre-concentration. The buffer and other additives for the ligand solution should be chosen with respect to the chemistry of the immobilization reaction. Buffers containing groups that can react with the activated surface should be avoided (e.g. primary amine buffers like Tris for amine coupling). The use of capturing molecules enhances the specificity of the immobilization and the buffer conditions are less critical in these cases.

2.1.5 Effect of ligand concentration

Pre-concentration increases with increasing ligand concentration, as long as the ligand carries a significant net positive charge (pH<pI), but this effect reaches a maximum at quite low concentrations. At pH values closer to the pI of the ligand there is no clear maximum level, and higher ligand concentrations are required to attain maximum pre-concentration as the protein becomes less positively charged. For a given pH, the ideal ligand concentration is the lowest value that gives maximum pre-concentration. A suitable concentration is usually in the range 10–200 μg/ml, so that less than one microgram of pure ligand might be sufficient for an immobilization. Ligands which are not electrostatically attracted to the matrix may require very high (mM) concentrations during immobilization via primary amines. Here, high pH values (pH 8–9) favour the coupling efficiency.

2.1.6 Level of immobilization

Optimization of the amount of immobilized ligand is an important factor in quantitative measurements. The sensitivity and range of an analytical interaction analysis is determined by the immobilization level in relation to the binding constant of the interaction. For purely qualitative applications it is useful to immobilize a high amount of ligand, to get a clear answer as to whether any interaction takes place between ligand and analyte. However, for kinetic measurements the level of immobilized ligand should be kept low (typically 50–500 RU or even lower). The amount of ligand to be immobilized also depends on the relative size of the ligand and the analyte, since the response signal correlates with the molecular mass. For a larger ligand a given level in RU represents a smaller number of analyte-binding sites, giving a correspondingly smaller maximum binding capacity in RU. On the other hand, a larger analyte will give a larger response than a small analyte for the same number of binding sites.

2.1.7 Regeneration conditions

The removal of the bound analyte from the immobilized ligand, referred to as the regeneration, is an important step, since dissociation of the analyte might take a long time. It should be considered before deciding which immobilization method will be applied and which component should be immobilized. The regeneration conditions (see *Table 1*) depend on the nature of the ligand, and must be checked for each experiment. The ligand should be able to withstand the regeneration conditions and retain its full biological activity. On the other hand, incomplete removal of the analyte from the surface, detected by an upward drift of the baseline between different measurements, may affect the binding kinetics. This may lead to an accumulation of material on the surface and reduces the response for a given concentration of analyte in a subsequent analysis.

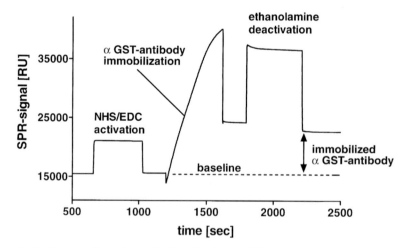

Figure 3. Amine coupling of an αGST-antibody to a CM-dextran surface. The surface is activated by NHS/EDC, the αGST-antibody is immobilized to the activated surface, and the surface is deactivated by ethanolamine further removing all non-specifically bound antibody. The amount of immobilized antibody can be calculated from the difference between the baseline before the immobilization and after deactivation of the surface, in response units (RU).

Table 1. Maximum concentrations of agents which may be useful for removal of the bound analyte or or for the regeneration of the sensor surface using different immobilization techniques (use only short pulses of injection for 1 min). Keep in mind that most proteins will not withstand any of those agents given in the table at maximum concentrations

	Amine coupling	Thiol coupling	Antibody capture	Streptavidin/avidin	NTA–surface
Acetonitrile*	(20%)	(20%)		(20%)	(20%)
DMSO*	(8%)	(8%)		(8%)	(8%)
DTE in HBS-buffer	10 mM	n.r.	n.r.	10 mM	10 mM
EDTA	0.35 M	0.35 M	0.35 M	0.35 M	0.35 M
Ethanol	(70%)	(70%)		(70%)	(70%)
Ethanolamine	(100 mM)	(100 mM)	(100 mM)	(100 mM)	(100 mM)
Ethylene glycol in HBS-buffer	50%	50%	50%	50%	50%
Formamide	(40%)	(40%)		(40%)	(40%)
Formic acid	(20%)			(20%)	(20%)
Glycine pH 2.2	(100 mM)		100 mM	(10 mM)	(100 mM)
HCl	(10 0mM)		(100 mM)	(10 mM)	(100 mM)
Imidazole pH8	300 mM	300 mM	300 mM	300 mM	300 mM
MgCl$_2$	4 M	4 M	4 M	4 M	4 M
NaOH	(100 mM)	n.r.	n.r.	(100 mM)	(100 mM)
NaCl	2 M	2 M	2 M	2 M	1 M
Octyl glycosid	40 mM	40 mM	40 mM	40 mM	40 mM
SDS	(0.5%)		0.05%	(0.5%)	(0.5%)
Surfactant P20	0.5%	0.5%	0.5%	0.5%	0.5%
Urea	(8 M)	(8 M)	n.r.	(3 M)	(8 M)?

n.r. = not recommended, () = limited by stability of ligand, *try to avoid.

2.2 Covalent coupling methods
2.2.1 Amine coupling
Amine coupling is recommended as the first choice for most applications. This immobilization method is usually easy to perform, and only requires uncharged primary amines (e.g. lysine residues or free N-termini) on the ligand. The reaction is favoured by high pH, but the optimal pH depends on the nature of the ligand (see Section 2.1.4). The covalent link introduced by amine coupling is stable, and the conditions for surface regeneration after binding of the analyte are only limited by the stability of the immobilized ligand. Amine coupling should be avoided if:

• the pI of the ligand is below 3.5, since the low pH required for electrostatic pre-concentration causes protonation of the primary amino groups and reduces the coupling efficiency (see Section 2.1.4);

• the active site of the ligand includes reactive amino or other nucleophilic groups, and the immobilization therefore causes loss of biological activity;

• the coupling sites of the ligand interfere with the binding of the analyte; in this case a site-directed immobilization should be preferred.

Protocol 1. Amine coupling (*Figure 3*)

Reagents
• running buffer, e.g. HBS (10mM Hepes, pH 7.4, 150mM NaCl, 3.4mM EDTA, 0.005% surfactant P20, filtered and degassed)
• immobilization buffer (see Section 2.1.4 concerning choice of immobilization buffer)
• NHS (100 mM *N*-hydroxysuccinimide)
• ligand solution (1–100 μg/ml ligand in an appropriate immobilization buffer)
• EDC [400 mM *N*-ethyl-*N'*-(dimethylaminopropyl)-carbodiimide]
• ethanolamine hydrochloride (1M, pH 8.5)

Method[a]
1. Set flow rate to 5 μl/min.
2. Mix equal volumes of NHS and EDC (final volume 100 μl).
3. Inject the EDC/NHS mixture for 7 min to activate the CM surface.
4. Inject the ligand solution using the manual inject command; observe the increase in the response signal to yield the desired immobilization level.
5. Inject ethanolamine for 7 min to deactivate excess reactive groups and to remove non-covalently bound material from the surface.

[a]All coupling protocols are optimized for immobilization using Biacore but may be applicable to other biosensors.

2.2.2 Thiol coupling

Thiol coupling provides an alternative to amine coupling, and is recommended for ligands where amine coupling cannot be used or is unsatisfactory, e.g. for acidic proteins or peptides and other small ligands. The efficiency of thiol coupling is usually very high, so that the conditions for immobilization are often less critical than with amine coupling. Thiol-coupled ligands cannot be used together with strong reducing agents, since the disulfide bridge linking the ligands to the surface is unstable under such conditions. If reducing conditions are needed for the interaction, the stability of the immobilized ligand must be tested first. Basic solutions (pH>9) should be avoided for the regeneration of the surface, to prevent a loss of immobilized ligand. There are two ways by which thiol coupling can be performed, the *ligand thiol* and the *surface thiol* procedure. The ligand thiol procedure uses intrinsic thiol groups in the ligand, i.e. cysteines, which are exchanged with a reactive disulfide group on the sensor surface. In contrast, the surface thiol procedure uses thiol groups on the sensor surface exchanging with a reactive disulfide group introduced into carboxyl or amino groups of the ligand. This method can be used for ligands that do not contain intrinsic thiol groups (i.e. no cysteines) and allows coupling of different functional groups of the ligand. Amino groups can be modified with NHS-activated heterobifunctional reagents such as SPDP or SMPT, carboxyl groups with PDEA. Modification with PDEA is especially useful for immobilization of acidic proteins, since the reaction raises the pI of the protein and thus favours electrostatic pre-concentration on the sensor surface matrix.

Protocol 2.   Ligand thiol coupling

Reagents

- running buffer, e.g. HBS (see *Protocol 1*)
- PDEA pH 8.5 [80 mM 2-(2-pyridinyldithio)-ethaneamine hydrochloride (PDEA) in 0.1M borate buffer pH 8.5; freshly prepared]
- immobilization buffer (see Section 2.1.4 concerning choice of immobilization buffer)
- NHS (see *Protocol 1*)
- ligand solution (10–200 μg/ml ligand in an appropriate immobilization buffer
- EDC (see *Protocol 1*)
- cysteine/NaCl (50 mM L-cysteine, 1M NaCl in 0.1M formate buffer pH 4.3; freshly prepared)

Method

1. Set the flow rate to 5 μl/min.

2. Mix equal volumes of NHS and EDC.

3. Inject the EDC/NHS mixture for 2 min to activate the CM surface.

4. Inject the freshly prepared PDEA solution for 4 min to introduce the reactive disulfide group onto the surface.

Protocol 2. *Continued*

5. Inject the ligand solution for 7 min.

6. Inject the cysteine/NaCl solution for 4 min to deactivate the surface and to remove non-covalently bound material.

Protocol 3. Surface thiol coupling

Reagents

- ligand solution (1 mg/ml in 0.1M MES buffer, pH 5.0)
- fast desalting column (NAP-10 column or equivalent)
- running buffer, e.g. HBS (see *Protocol 1*)
- NHS (see *Protocol 1*)
- EDC (see *Protocol 1*)
- cystamine dihydrochloride (40mM in 0.1M borate buffer pH 8.5)

- DTT or DTE (dithioerythritol) (0.1M in 0.1M borate buffer pH 8.5)
- immobilization buffer (see Section 2.1.4 concerning choice of immobilization buffer)
- ligand solution (10–200 µg/ml in an appropriate immobilization buffer
- PDEA/NaCl pH 4.3 (20 mM PDEA, 1M NaCl in 0.1M sodium formate buffer pH 4.3; freshly prepared)

Method A. Ligand modification

1. Add PDEA to a final concentration of 25 mM to 1 ml of ligand solution (1 mg/ml in 0.1M MES buffer, pH 5.0) and cool on ice.

2. Add 50 µl EDC.

3. Mix and allow to react for 1 h on ice.

4. Remove the reagents by gel filtration on a desalting column equilibrated with an appropriate immobilization buffer.

Method B. Surface thiol coupling

1. Set flow rate to 5 µl/min.

2. Dilute the modified ligand (10–200 µg/ml) in immobilization buffer.

3. Mix equal volumes of NHS and EDC.

4. Inject the EDC/NHS mixture for 2 min to activate the CM surface.

5. Inject the cystamine solution for 3 min.

6. Inject the DTE or DTT solution for 3 min to reduce cystamine disulfides to thiols.

7. Inject the modified ligand solution for 7 min.

8. Inject 20mM PDEA/NaCl pH 4.3 for 4 min to deactivate excess thiol groups, and to remove non-covalently bound material from the surface.

2.2.3 Aldehyde coupling

Aldehyde coupling of ligands may be useful in specific cases, and provides an alternative method for covalent linking of polysaccharides, glycoconjugates, oxidized glycoproteins, and other biomolecules containing aldehyde groups (8).

2.3 Site-directed immobilization strategies

The site-directed immobilization of ligands has the advantage of orienting the ligand molecules in a homogenous manner by binding to only one specific interaction site. Biotinylated ligands and recombinant proteins containing specific fusion tags like glutathione-*S*-transferase (GST) or poly(His) tags can be immobilized via capturing molecules, e.g. streptavidin/avidin, specific antibodies, or nickel-saturated NTA, respectively. In these cases only the tag will be linked to the surface, and the ligand will be accessible for binding to the analyte.

2.3.1 Antibody surfaces

The immobilization of ligands on specific antibody surfaces provides an excellent tool for site-directed immobilization. Using antibodies against a fusion portion of a recombinant protein, e.g. the widely used fusion proteins with GST, ensures an immobilization via epitopes that should not be involved in binding to the analyte. The influence of the fusion partner itself on the kinetics should be examined in control experiments using different fusion tags, by cleaving the fusion partner from the protein, or by using different immobilization techniques and comparing the resulting rate constants. Anti-Fc antibodies can be used as capturing molecules for specific antibodies, used for example in epitope mapping experiments, or as an immobilization matrix for another ligand antibody in 'sandwich assays'. Because of the specificity of the immobilization, the captured molecules do not have to be as pure as for other immobilization procedures. The regeneration is also facilitated because the whole complex of captured ligand and analyte can be removed from the antibody surface using low pH (10 mM glycine pH 1.8–4) or by low concentrations of SDS. However, not every antibody can be used for this application. The antibody must have a sufficiently high affinity for the ligand, at least two orders of magnitude larger than the affinity of the interaction to be determined. More precisely, the off-rate of the ligand from the antibody surface has to be much slower than that of the interaction itself.

Protocol 4.   Interaction analysis using antibody surfaces

Reagents

- running buffer, e.g. HBS (see *Protocol 1*)
- NHS (see *Protocol 1*)
- EDC (see *Protocol 1*)
- immobilization buffer (10 mM Na acetate pH5.0)
- ethanolamine hydrochloride (1 M, pH 8.5)

- antibody solution (30 µg/ml purified anti-ligand antibody in immobilization buffer)
- regeneration solution (10mM glycine, pH 2.2, or 0.05% SDS)
- ligand solution (typically 10 µg/ml ligand in running buffer)

Method A. Immobilization of the capturing antibody (see Figure 3)

1. Set the flow rate to 5 µl/min.

Protocol 4. *Continued*

2. Mix equal volumes of NHS and EDC (final volume 100 μl).

3. Inject the EDC/NHS mixture for 7 min to activate the CM surface.

4. Inject the antibody solution for 7 min.

5. Inject ethanolamine for 7 min to deactivate excess reactive groups and to remove non-covalently bound material from the surface (the immobilization level for the antibody is typically about 8000 RU using these conditions).

6. Inject regeneration solution for 2 min to remove non-covalently bound material.

7. Immobilize the same amount of antibody on a second (control) surface.

Method B. Capturing the ligand and regeneration of the antibody surface

1. Set the flow rate to 5 μl/min.

2. Equilibrate the surface in running buffer.

3. Inject the ligand solution for, typically, 5 min.

4. Check the immobilization level according to the recommendations (see Section 2.1.6).

5. Adjust the immobilization level by changing the ligand concentration or the contact time of the ligand to the surface.

6. Change the flow rate to 20 μl/min (use the flow rate that is used in the subsequent interaction analysis).

7. Wait until the response is stable; observe the baseline for at least 5 min to have a measure for the loss of ligand from the surface.

8. Inject regeneration solution for 2 min.

9. Repeat the optimization of the immobilization and the regeneration conditions on a control surface using a suitable ligand as a negative control (e.g. recombinant GST for the anti-GST antibody surface).

Method C. Interaction analysis

1. Immobilize the ligand on the specific surface using the optimized immobilization conditions.

2. Immobilize the control protein on the control surface using the optimized immobilization conditions.

3. Set the flow rate as desired for the interaction analysis.

4. Perform the interaction analysis using several concentrations of analyte on the specific and the control surface.

5. Regenerate both surfaces by injecting regeneration solution for 2 min (if the regeneration is not satisfactory, the addition of 30% ethylene glycol to the regeneration solution, or an injection of 0.05% SDS for 1 min, are suggested).

6. Repeat the immobilization for another cycle of interaction analysis.

See *Figure 5* for an example of this approach involving the interaction of a GST-R-subunit with the C-subunit of cAMP-dependent protein kinase.

2.3.2 Immobilization of biotinylated ligands

Sensor surfaces with pre-immobilized streptavidin or avidin are multi-functional immobilization matrices for biotinylated ligands. They are ideal for the capture of biotinylated DNA fragments, as well as for biotinylated peptides and proteins. Even vesicles containing biotinylated lipids can be immobilized, so that interactions of membrane-bound proteins can be examined (9). The coupling procedure can be performed with high efficiency without relying on electrostatic pre-concentration on the sensor surface (see Section 2.1.4). However, the ligand always has to be modified by biotinylation. Avidin and streptavidin are tetrameric proteins, containing four identical subunits of about 15 kDa. Each subunit contains one high-affinity binding site for biotin ($K_D \approx 10^{-15}$ M^{-1}), but this extremely high affinity may change when the biotin is conjugated to another molecule. The binding characteristics may also be influenced depending on the biotinylation chemistry, or the site of biotinylation. An extensive discussion of the chemistry and practice of biotinylation can be found in *Methods in enzymology*, Vol. 184 (1990). Avidin is a basic glycoprotein (pI ≈ 10) and the pI, in combination with the presence of carbohydrate moieties, may result in a higher non-specific binding which may be useful in specific cases. Streptavidin has a lower pI (pI 5–6) than avidin, and lacks the glycoprotein constituents, and should therefore be preferred for most applications. The high affinity of this interaction withstands a wide range of regeneration conditions, only limited by the stability of the captured ligand. For short oligonucleotides, 1 mM HCl is recommended, whereas for longer oligonucleotides, or for the removal of bound protein from nucleic acids, 1 M NaCl in 50 mM NaOH is often preferable (for other conditions see Section 2.1.7). Urea (3 M) has also proved to be a useful regeneration reagent.

2.3.3 Poly(His) fusion proteins

Recombinant poly(His)-tagged proteins expressed in heterologous expression systems are widely used, and easily purified via Ni^{2+}–NTA metal-affinity chromatography. The fusion tag, usually consisting of six consecutive histidines, is very small and usually does not interfere with the structure and function of the recombinant protein. Ni^{2+}–NTA surfaces can be used as an immobilization matrix for poly(His)-tagged proteins, but the affinity and selectivity of the fusion tag to the surface must be checked in control experiments. Even if the immobilization is very easy and can be carried out without changing the pH or the ionic strength during the coupling procedure, there are some limitations. The affinity of the poly(His) tags to the Ni^{2+}–NTA matrix can be affected by moieties adjacent to the tag in the protein, or by the buffer conditions, e.g. the pH, the ionic strength, or additives like EDTA or bivalent metal ions (Mg^{2+}, Ca^{2+}). The affinity of the poly(His)-tagged ligand for the surface should be significantly higher than the affinity of the analyte to the ligand, therefore the stability of the immobilization should be checked. Fusion proteins modified with more than one poly(His) tag display higher affinity to the Ni^{2+}–NTA surface (10). Side chains of cysteine, tyrosine, tryptophan, and lysine residues on the surface of a protein may also participate in binding to chelated metal, although the affinity of these interactions is typically lower than that obtained with poly(His) tags. The regeneration of the Ni^{2+}–NTA surface may either be performed by stripping the nickel from the surface with high concentrations of EDTA (350 mM), or by using conditions that selectively dissociate the analyte, leaving the ligand on the surface for the next cycle of analyte binding (see Section 2.1.7).

Protocol 5. Immobilization of poly(His)-tagged proteins

Reagents

- running buffer (10mM Hepes, 0.15M NaCl, 50 μM EDTA, 0.005% surfactant P20, pH 7.4, filtered and degassed; the low concentration of EDTA neutralizes contaminating metal ions, but is dilute enough not to strip the nickel from the surface)
- nickel solution (500 μM $NiCl_2$ in running buffer)

- ligand solution [avoid EDTA and bivalent metal ions in the buffer; the buffer conditions can be varied in ionic strength and pH to prevent non-specific binding; low concentrations of imidazole in the buffer (10–20 mM) may help]
- regeneration solution (10 mM Hepes, 0.15 M NaCl, 0.35 M EDTA, 0.005% surfactant P20, pH 8.3)

Method[a]

1. Set flow rate to 20 μl/min.

2. Inject regeneration solution for 1 minute; repeat this step if necessary.

3. Inject nickel solution for 1 minute to saturate the NTA[b] surface with nickel (the baseline increases by about 40 RU).

4. Change the flow rate to 10 μl/min or less; the contact time in the immobilization step can be varied to control the amount of bound ligand.

5. Inject the purified poly(His)-tagged ligand.

6. Repeat the injection of the nickel solution (20 μl/min, 1 min) to remove ligands bound with low affinity; check the stability of the bound ligand by monitoring the baseline.

[a] A protocol describing the production of NTA surfaces based on carboxymethylated dextran matrices is given by Hochuli *et al.* (1987) *J. Chromatogr.* **411**, 177, the linkage is described in ref. 11.
[b] NTA is manufactured by QIAGEN GmbH and is under license from Hoffmann–LaRoche Ltd and QIAGEN GmbH.

2.4 Immobilization of low molecular weight ligands

The immobilization of low molecular weight ligands (<1000 Da) enables the design of specific surfaces that can be used for interaction studies, or as immobilization matrices for other binding partners. Since these ligands often have to be chemically modified to allow immobilization, the effect of modification has to be determined in complementary assays where the biological functionality or binding characteristics can be tested. High density surfaces of small ligands which bind with high affinity to the analyte tend to exhibit mass transfer-limited binding and rebinding behaviour, but are well suited for concentration determinations of the analyte, or as immobilization matrices. Kinetic analysis, however, should be performed using low density surfaces. The immobilization process cannot be followed directly, since the response obtained from small molecules is low. The level of immobilization can sometimes be assessed indirectly by binding a high molecular weight analyte. A surface with a lower degree of carboxylation, might be valuable to obtain lower immobilization levels (see other sensor surfaces described in Section 2.5.2).

2.5 Other sensor surfaces

2.5.1 Hydrophobic surfaces

Membrane-associated interactions can be examined using hydrophobic surfaces, e.g. surfaces composed of long-chain alkanethiol molecules that form hydrophobic layers. The tendency to bind proteins non-specifically is very high, but can be reduced when the surface is coated with a lipid monolayer directed with its hydrophilic side towards the solution phase. Liposomes adsorb spontaneously to the hydrophobic surface to form a supported lipid monolayer in which membrane-associated ligands can be included. The process of coating typically takes 0.5–3 hours depending on liposome composition, the diameter of the liposomes, the eluent buffer, and the temperature (12). The adsorbed monolayer is normally stable and withstands conditions typically used for regeneration of the embedded or associated ligands. Even 100 mM NaOH or 100 mM HCl can be used, given a stable ligand. However, detergents and organic solvents will alter or destabilize the lipid monolayer and should only be used to strip the monolayer from the surface. Another possibility to investigate membrane-associated processes is the immobilization of vesicles containing biotinylated lipids on streptavidin/avidin, or on anti-biotin antibody surfaces, measuring the interaction of ligands that are included in the vesicle preparation (13, 14). Overall, the homogeneity of the vesicles, achieved for example using an extrusion apparatus (15), is the critical point for reliable measurements with lipid vesicles.

2.5.2 Sensor surfaces with altered density or different matrices

Applications for altered sensor surfaces could be:

- binding of bulky ligands like viruses or whole cells
- basic proteins or crude samples where the risk of non-specific binding to the dextran matrix should be kept to a minimum
- kinetic studies where questions relating the effects of the dextran matrix are examined

A flat, carboxymethylated surface without a dextran matrix, where the carboxyl groups are directly attached to the surface layer, provides a better functionality in binding viruses or whole cells, and allows investigation of the influence of the dextran matrix on kinetic determinations. Since the evanescent field decays exponentially from the surface of the chip, the signal-to-noise ratio is better for large particles. The lack of the dextran matrix leads to a lower hydrophilicity that may increase non-specific binding for some samples, but this may be counteracted by coating with BSA (e.g. 0.2 mg/ml in running buffer). The immobilization yield is only 10% of that obtained on a carboxymethylated dextran matrix of 100 nm, under comparable conditions. A short carboxymethylated dextran may give reduced non-specific binding in work with serum samples, and may be valuable in testing the influence of the dextran matrix on kinetic determinations. A dextran matrix with a low degree of carboxylation is useful for analytes with a high positive charge, or for culture supernatants and cell homogenates, because non-specific binding is lowered in these cases due to the reduced charge of the surface. This surface might also be a valuable tool for the immobilization of low molecular weight ligands because the lower surface density of the ligand reduces the risk of mass transport and rebinding effects.

3. Interaction analysis

Biomolecular interaction analysis allows determination of association and dissociation rate constants separately. Apparent K_Ds can then be calculated from known concentrations of analyte. The interaction analysis in the following example is described for a Biacore 2000 instrument using components of the cAMP-dependent protein kinase system as an example. The general considerations, however, can be applied to any other system. A large number of potentially interesting interaction patterns within the protein kinase family are beyond the scope of this chapter. However, some interaction analyses are presented here as examples, and potential pitfalls and misinterpretations are discussed. Each kinetic analysis should consist of a sequence of subsequent injections with different analyte concentrations on the same sensor surface. These data sets are then used to generate data, e.g. calculating apparent rate constants and equilibrium binding constants.

3.1 Interaction analysis using the cAMP-dependent protein kinase as a model

3.1.1 Amine coupled C-subunit

In *Figure 4* the interaction of 250 nM cAMP-free R^I_α-subunit, prepared according to ref. 16, with C-subunit, immobilized on a CM 5 chip, as described in *Protocol 1* in 10 mM sodium phosphate, pH 6.2, containing 1 mM ATP and 2 mM $MgCl_2$, is shown. The R-subunit was exchanged into running buffer (20 mM MOPS, 150 mM KCl, 1 mM DTT, pH 7.0, 0.005% P20) also containing 1 mM ATP and 5 mM $MgCl_2$, using a PD10 column (Pharmacia). To minimize mass transfer effects (see Section 3.3), a lower level of C-subunit was coupled for kinetic analysis (data not shown). Non-specific binding and bulk changes were subtracted on the basis of blank runs using a non-activated sensor chip surface. To obtain reliable association and dissociation rate constants, the R-subunit at several concentrations was injected over surfaces with immobilized C-subunit. After injections of the R-subunit, the C-subunit surface was regenerated physiologically (up to 100 times) by injection of 10 µl of 100 µM cAMP, 2.5 mM EDTA in running buffer (*Figure 4*; 7, 17, 18). The binding stoichiometry (for calculation see Section 2.1.3) with saturating concentrations of R-subunit was calculated to 55% assuming 1:1 binding.

Site-directed immobilization via ligand thiol with a mutant protein of the C-subunit (C199A) containing only one cysteine in position 343 yielded a surface stoichiometry of about 5%, indicating steric problems, although this mutant protein had wild-type-like properties.

3.1.2 R-subunit coupled as a GST fusion protein

Immobilization of the αGST antibody was performed as described in the legend to *Figure 3*. A total of 6000 RU of the antibody was immobilized. Another mutant form of the R-subunit (GSTΔ1-45/R209K) was bound by injecting 20 µl at 5 µg/ml. 900 RU were bound to the antibody surface. With a baseline drift of 1 RU/min, this surface was relatively stable, allowing investigation of the binding of several concentrations of C-subunit and regeneration with cAMP, as described above, without re-addition of the R-subunit. However, the stoichiometry of binding was only 20%. *Figure 5* shows the interaction of the C-subunit at concentrations between 12 and 250 nM with the immobilized

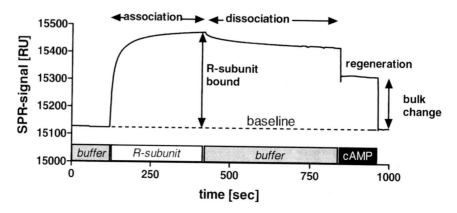

Figure 4. Biomolecular interaction analysis. Interaction of cAMP-free R-subunit with C-subunit, immobilized via primary amines on a CM surface. 250 nM cAMP-free R-subunit was injected in running buffer containing 1 mM ATP, 5 mM $MgCl_2$. Dissociation was initiated by switching to the same buffer without R-subunit. Finally, the surface was regenerated with running buffer containing 100 μM cAMP and 2.5 mM EDTA. A bulk change was observed due to the higher ionic strength of the regeneration buffer.

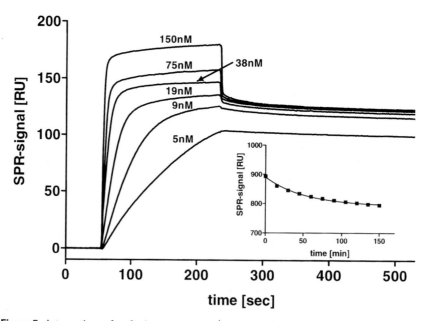

Figure 5. Interaction of a fusion-captured R^I-subunit (GSTΔ1–45/R209K) with a recombinant C-subunit. The baselines are normalized; the inset shows the amount of R^I-subunit still captured by the αGST antibody after each run.

354

GSTΔ1-45/R209K, after normalization of the baseline. The inset in *Figure 5* shows the drift after each cycle of about 15 minutes. Apparent rate and binding constants are listed in *Table 2*.

3.1.3 R-subunit coupled as a poly(His) fusion protein

A NTA sensor chip was used to couple poly(His)-tagged R-subunit (R^{II}-R213K, an analogue mutant to R^{I}-R209K, see *Protocol 5*) and the sensor surface was prepared according to Section 2.3.3. The C-subunit was injected on this surface in concentrations varying from 18 nM to 300 nM. Although the on-rates were comparable with those of the R^{I}-R209K mutant, the off-rates were one order of magnitude faster. Whether this faster off-rate is due to the accelerated kinetics of the R^{II}-R213K mutant protein, or to a faster dissociation of the poly(His) fusion complex, remains to be clarified. Apparent rate and binding constants are given in *Table 2*.

Table 2. Apparent rate and binding constants for the interaction of type I and type II R-subunits with the C-subunit

Immobilized ($_{im}$)	Apparent K_D (nM)	Apparent k_{ass} (M^{-1}s^{-1})±SE	Apparent k_{diss}(s^{-1})±SE
R^{I}R209K$_{im}$/C-SU	0.073	$2.6 \times 10^6 \pm 3.4 \times 10^5$	$1.9 \times 10^{-4} \pm 5.9 \times 10^{-6}$
His$_6$-R^{II}R213K$_{im}$/C-SU	0.75	$5.7 \times 10^5 \pm 8.1 \times 10^4$	$4.3 \times 10^{-4} \pm 9.2 \times 10^{-5}$
GST-R^{I}R209K$_{im}$/C-SU	0.055	$2.2 \times 10^6 \pm 3.8 \times 10^5$	$1.2 \times 10^{-4} \pm 4.2 \times 10^{-5}$
C-SU$_{im}$/R^{I}R209K	0.036	1.4×10^6	5.1×10^{-5}

3.1.4 Comparison of different immobilization strategies

Although the reproducibility of each single interaction study is extremely good, the absolute numbers derived from the association and dissociation phases have to be evaluated carefully. This becomes apparent when data of 'vice versa' experiments are compared. Here, first one of the interacting partners is immobilized (R-subunit) and the binding of the other partner (C-subunit) is monitored, and then the other way around. By comparing the numbers it becomes clear that, although the association and dissociation rate constants differ significantly, the apparent K_D is very similar (*Table 2*). The differences in the rate constants might be due to different degrees of freedom for the immobilized partner, or for several other reasons, as discussed below. SPR should always be accompanied by other methods, preferably in free solution, to check if the apparent numbers reflect a realistic interaction *in vivo* (see Section 6). This has been done for the R–C interaction using analytical gel filtration (19), activity assays (20), calorimetric studies (M. Doyle, personal communication), or fluorescence studies (S.S. Taylor, personal communication).

3.2 Non-specific binding

Non-specific binding can be due, for example, to the interaction of analyte molecules with the dextran matrix or other components of a sensor chip, or with immobilized non-specific interaction partners. It can also be due to non-specific protein–protein interactions. Non-specific binding will be high especially when using crude lysates. The following are some procedures that can be adopted to tackle this problem:

(a) Substract values obtained using a control surface. Bear in mind that the *same amount* of protein or DNA, or whatever is immobilized, should be on the reference cell, since when using a 'refractive index change detector' the same sample volume should be occupied in the control as in the detection cell.

(b) When making measurements with unpurified cell lysates, use a control surface with an unrelated ligand; this is even better than using a control lysate without the analyte molecule of choice, because of the tremendous heterogeneity of proteins at high concentration in lysates.

(c) If non-specific binding to antibodies occurs, run a control surface with an unrelated antibody.

(d) When working with immobilized DNA or peptides, use random sequence DNA or an unrelated peptide at the same concentration on a different surface.

(e) Use cleaner ligand and analyte preparations, exchange buffer with fast buffer exchange columns (e.g. PD10, NAP5, Pharmacia), or spin columns for DNA purification (BioRad) to exchange analyte against running buffer, including P20.

(f) Use high salt (500 mM KCl or even higher concentrations), or detergent.

(g) Perform the analysis in the presence of soluble CM–dextran (Fluka) at 1 mg/ml in the running buffer.

(h) Exchange the ligand and analyte if possible ('vice-versa' experiment).

(i) Use a different type of surface or immobilization chemistry.

Never forget to check for non-specific binding, because in most cases you will obtain a binding curve whether it is the interaction you are looking at or not.

3.3 Mass transfer-limited reactions

3.3.1 What are mass transfer limitations?

Mass transfer limitations occur when the observed binding rate is limited by the rate of transfer of analyte to and from the matrix. This happens if the actual interaction-controlled rate constants are much faster than the diffusion

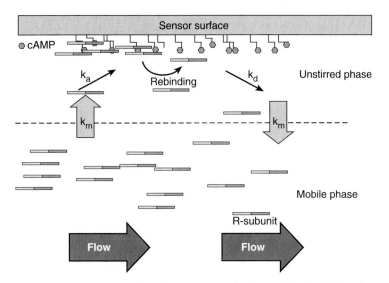

Figure 6. Mass transfer: interaction of R-subunit with immobilized cAMP. The analyte (R-subunit) has to jump from a zone of laminar flow to the unstirred phase, limited by the mass transfer coefficient k_m. Association takes place with the rate constant k_a, dissociation with k_d. Rebinding will be more likely to occur, since analyte molecules are not efficiently removed from the unstirred phase.

limitations for the interaction. The effect of mass transfer will slow down both the association phase and the dissociation phase. This is because the analyte molecules have to 'jump' from a zone of laminar flow to the unstirred layer, where the interaction takes place in the association phase (*Figure 6*). In turn, analyte molecules are not transported fast enough away from the surface in the dissociation phase, which also makes them additionally available for rebinding (see Section 3.4). *Figure 6* shows these effects schematically for the interaction of the cAMP-free R-subunit with a high density cAMP surface (see also Section 4.3).

3.3.2 How to recognize mass transfer effects

Figure 10 shows the concentration determination of R-subunit on a cAMP surface, and is an example for pure mass transfer-limited interaction. No curvature of the plot in the association phase is detectable (see inset in *Figure 10*) and the ln dR/dt versus time plots show a straight line (data not shown). In *Figure 7* data are simulated under mass transfer limitation, and purely interaction-controlled conditions. Mass transfer can affect data severely (21, 22), and it is most important to recognize these effects. This can be done by performing the following controls:

(a) Test your interaction at three different flow rates (i.e. 5, 15, and 50 µl/min, *Figure 8*) or at different stirrer speeds (IAsys).

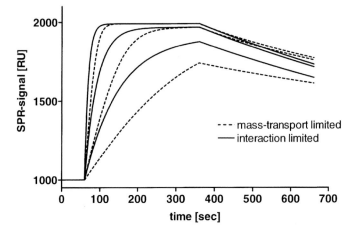

Figure 7. Mass transfer-limited reactions: comparison of mass transport- (dashed line) and interaction-limited interactions (solid line) using simulated data. Three different concentrations of an analyte are shown, binding either under mass transport-limited conditions that are characterized by lowered rate constants for both association and dissociation phases, or under interaction-limited conditions.

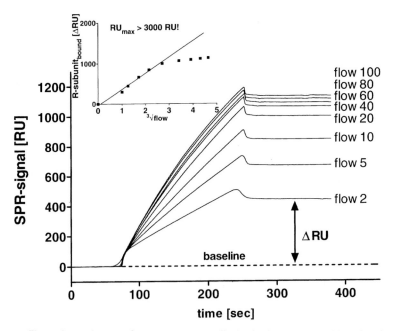

Figure 8. Flow dependence of mass transport-limited binding. 17 nM cAMP-free R^I-subunit were injected for 3 min on a cAMP surface using flow rates as indicated. The figure shows a plot of SPR response versus time. A replot of response after injection versus the cube root of the flow is given in the inset. Clearly, a linear dependence of the cube root of the flow is detected using low flow rates.

(b) Use different surface densities. In all cases the kinetics should look similar; if not, there may be a mass transfer limitation.

(c) Perform data analysis with linearized plots of ln $[abs(dR/dt)]$ versus time; mass transfer-limited reaction will not show any change over time.

(d) If you use global fit analysis, repeat fitting using local parameters for k_a, k_d, and R_{max}. These data will scatter significantly if mass transfer limitations occur. However, there might be additional reasons for data scatter.

3.3.3 How to minimize mass transfer effects

Several strategies can be used to minimize mass transfer limitations:

(a) Reduce the surface density, because mass transfer is directly dependent on the interaction-controlled association rates, and the concentration of the available ligand sites on the surface, in relation to the mass transfer-limited association rates, according to the equation $k_a[B]/k_m$; [B] being the ligand concentration. For that reason, mass transfer limitations are most effectively reduced by working at low ligand concentrations. Since the diffusion coefficient decreases with increasing molecular weight, large analytes (or more exactly analytes with higher frictional coefficients) are more likely to be subject to mass transfer limitations. The diffusion coefficient is also dependent on the solution viscosity, which is in turn strongly dependent on the temperature and solution composition. The temperature and the solution composition (e.g. the presence of any viscogens in the buffer) can also be changed, within limits.

(b) Mass transfer decreases with the cube root of the flow, meaning an increase in the flow rate increases the rate of transfer of the analyte to the surface. This is shown in an experiment where a fixed concentration of cAMP-free R-subunit was injected on a high density surface of 8-AHA-cAMP, using flow rates between 1 and 100 μl/min. *Figure 8* shows the plots of the response versus time for the indicated flow rates. A replot of the response versus the cube root of the flow clearly demonstrates that, up to a flow rate of about 20 μl/min, a purely mass transfer-limited reaction is detected. At higher flow rates the binding becomes more and more interaction-controlled, detected as the deviation from the straight line (inset).

(c) High efficiency stirring systems (vibro stirrer, IAsys), may be useful against mass transfer-limited reactions, although this has not yet been demonstrated conclusively. In principle, a stirred system is also subject to mass transfer limitations, since analyte molecules have to move from the stirred phase to the unstirred layer in the matrix of the sensor.

More recently, software packages have become available with sophisticated models implementing mass transfer limitations (23).

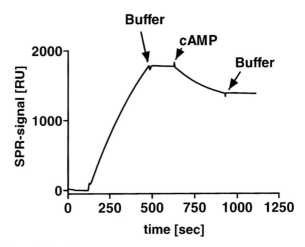

Figure 9. Rebinding. Effect of a soluble ligand (cAMP) on the rebinding behaviour of the analyte. cAMP-free R^{II} subunit was injected onto a cAMP surface at a flow rate of 10 μl/min. The initial dissociation phase was performed in running buffer, followed by running buffer + 5 mM cAMP, as indicated on the plot. A 30-fold faster dissociation rate constant is observed in the presence of ligand.

3.4 Rebinding

Rebinding occurs when an analyte molecule, dissociating from the immobilized ligand, rebinds to another ligand molecule (*Figure 6*). This can strongly affect the dissociation rate constants, as seen in *Figure 9*. There are several ways to minimize rebinding:

(a) Use a high concentration of analyte, and an injection time long enough to saturate the surface. At least in the initial dissociation phase, most of the ligand molecules are occupied with analyte, making it less likely that rebinding can occur.

(b) Use a low density surface.

(c) Use a high flow rate, or stirrer speeds, in the dissociation phase.

(d) Add soluble ligand at the same concentration as the immobilized ligand to the dissociation phase. This might be difficult if fusion proteins are used, since these will interact with their specific immobilization matrix.

(e) Use a different immobilization matrix.

3.5 Further problems

In this section some further problems in biomolecular interaction analysis are listed, without detailed discussion. Heterogeneity of the ligand and analyte, either derived from the preparation procedure, or from the biological patterns of the analyte and ligand, is a frequent problem and is difficult to overcome. Heterogeneity is seen in protein interactions when multiple binding sites are

present. Control experiments using different methods, and a more sophisticated data analysis are necessary to address these cases (see Section 5).

Steric problems can arise when, after initial binding of analyte molecules to
the surface of a sensor chip, the entry of analyte molecules into the dextran
matrix is inhibited. Matrices with low densities of dextran or no dextran have
been tested (24). Steric hindrance which abolishes the degree of freedom
necessary to allow the interaction to take place may be reduced by engineering a linker. This could be the aminohexyl chain for cAMP, as described
above, the biotinylation of peptides, which has been demonstrated to be very
useful to maximize biological activity, or the modification of a CM surface
with ε-aminocaproic acid as a linker using standard amine coupling chemistry
at pH 8.5.

4. Applications of biomolecular interaction analysis

Since there is a very large number of applications using biosensors for the
characterization of protein kinase interactions, only a few can be mentioned
here. Biomolecular interaction analysis has been used for unravelling signal
transduction pathways and analysing macromolecular complexes (2, 25), and
to scrutinize the importance of dimerization of the human growth hormone,
using numerous mutants of the extracellular binding domain (26). A large
number of protein kinase interactions has been reported in order to verify the
results of yeast two-hybrid screens (27, 28). The method has also been combined with high throughput screening technologies such as phage display (see
ref. 29 for review).

4.1 Signalling domain interactions

SPR analysis has been applied to study the interaction of signalling domains
like that of phosphotyrosine-containing motifs with Src homology 2 (SH2)
domains [e.g. in the adapter protein Grb2 (30)] or the interaction of specific c-
Src phosphorylation sites in the epidermal growth factor receptor (EGFR)
with the SH2 domain of c-Src (31). Binding constants for SH3 domains and
PDZ domains with their respective binding partner (32–34) or zinc finger/
DNA interactions (35) have also been investigated using SPR.

4.2 Ca^{2+}/calmodulin-dependent interactions

Ca^{2+}/calmodulin-dependent binding processes of protein kinases have been
examined by SPR. The investigation of the Ca^{2+}-induced selective binding of
a Ca^{2+}/calmodulin complex to a covalently immobilized, 26-amino acid
peptide from myosin light chain kinase revealed an ion selectivity. Binding
was observed for Ca^{2+}, whereas Mg^{2+}, K^+, and Li^+ could not induce any
binding (36). A Ca^{2+}-dependent myristyl switch was described for the

interaction of recoverin with phospholipid vesicles (9). Regulatory domains of Ca^{2+}/calmodulin-dependent protein kinase were also identified by SPR (37).

4.3 Ligand surfaces

Another application of the cAMP system is the immobilization of ligands, e.g. the use of cAMP to capture the R-subunit. cAMP surfaces are very robust, allowing the use of SDS for regeneration. Therefore, even crude lysates can be tested. This immobilization technique is especially useful when interaction with other binding partners of the R-subunit is studied. For example, the binding of specific PKA anchoring proteins (AKAPs) to the R-subunit has been investigated (Herberg and Tasken, in preparation). AKAP binding has been previously studied with immobilized peptides using SPR or resonance mirrors (38), but not with intact proteins because of the lack of a regeneration procedure. By removing the entire R-subunit/AKAP complex with 0.05% SDS, the cAMP surface could be used again. Resonance mirrors have been used to characterize the interaction of peptide motifs of the C-subunit in interaction with the R-subunit (39).

4.3.1 Concentration determination

Concentration determinations can be performed either directly or as a competition experiment. *Figure 10* shows the mass transfer-limited interaction (see also Section 3.3) of cAMP-free R-subunit with a high density 8-AHA-cAMP surface on a CM5 chip. Five different concentrations of R-subunit were subsequently injected over the surface using identical flow rates and injection times. The response, taken at a fixed point, or the slope of each association, plotted against the concentration of the R-subunit was used as a regression curve to determine an unknown concentration of R-subunit. Surprisingly, the R-subunit could not be competed away readily with cAMP. Even an injection for 4 minutes using 10 mM cAMP would displace only 10 % of the R-subunit from the cAMP surface. This indicates extremely slow off-rates. However, the R-subunit could be displaced using the C-subunit. Binding of the C-subunit did not increase the response due to the additional mass added, but instantaneously took off the R-subunit from the surface (data not shown). This effect was amplified by the addition of MgATP and cAMP, suggesting a physiological competition of the C-subunit with the bound R-subunit on the cAMP matrix (Herberg *et al.*, in preparation).

4.3.2 Solution competition assays

Solution competition is a useful method for the concentration determination of a ligand in the soluble phase, after pre-incubation with the analyte before injection to the surface with the same ligand. This approach is also very useful for the determination of equilibrium binding constants in solution (see Section 5; ref. 40). In the case of PKA a constant amount of R-subunit, pre-incubated with different concentrations of cAMP or analogues of cAMP, was

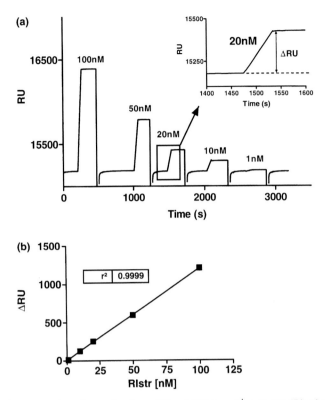

Figure 10. Concentration determination. (A): cAMP-free R^I subunit (RIstr) was injected over an 8-AHA-cAMP surface with concentrations as indicated. Each injection was performed for 2 min at a flow rate of 5 μl/min. The inset (middle) demonstrates that the interaction is purely mass transport controlled. (B) Shows the absolute response taken after each injection plotted against the concentration of the R^I-subunit.

injected onto a cAMP surface using constant injection times and volumes. Only the residual binding was determined, and a regression curve could be generated. This kind of approach was used to measure concentrations of cAMP in unpurified lysates (C. Schultz and F.W. Herberg, unpublished results).

4.4 Characterization and verification of interacting partners using antibodies

If the interaction of a multiple complex as an analyte with a ligand surface is performed, it is often difficult to determine which of the components in the complex is binding to the ligand. Antibodies offer an excellent tool to analyse protein complexes pre-formed on a sensor surface. We have examined the binding of components of the cardiac troponin complex with different phosphoforms of the cardiac troponin I (cTnI), phosphorylated by PKA (41). In

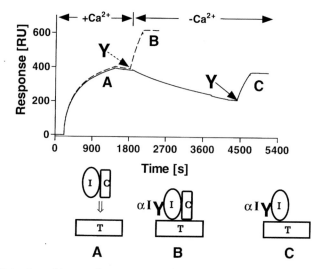

Figure 11. Detection of interacting partners using antibodies. 1010 RU cardiac troponin T (cTnT) labelled with 3 mol of biotin/mol of protein were immobilized on a streptavidin surface (30). 9 µM cTnC/cTnI complex pre-formed in the presence of Ca^{2+} were injected for 1000 sec in a first experiment (dotted line, A). After a dissociation phase of 400 sec in the presence of 0.1 mM Ca^{2+} (indicated as $+Ca^{2+}$), 5 µg cTnI-specific antibody (Y) were injected on the ternary holocomplex (B). In a second experiment the same antibody was injected after a dissociation phase of 45 min in the presence of 2 mM EGTA (indicated as $-Ca^{2+}$). Once again, the cTnI-specific antibody (Y) was injected (C) and only a slightly reduced response of antibody binding was observed, indicating that in the absence of Ca^{2+} the TnC is released much faster from the complex than TnI.

the experiment shown here, a complex of cTnI was formed with cardiac troponin C (cTnC) in the presence of Ca^{2+} and then the complex was injected over cardiac troponin T (cTnT), coupled via biotin/streptavidin to a sensor chip. During the dissociation phase Ca^{2+} was omitted from the buffer, and a faster dissociation rate constant was observed. Antibodies against cTnI were then used to clarify if cTnI or cTnC remained bound to cTnT in the absence of Ca^{2+}. *Figure 11* describes how this analysis was performed, with the result that cTnC dissociates first from cTnT, and is followed by cTnI with a much slower rate (S. Reiffert and F.W. Herberg, unpublished results).

5. Data evaluation

5.1 Calculations of rate and equilibrium binding constants

The calculation of apparent association and dissociation rate constants is probably the most difficult part of the whole operation. Before using the data it must be ensured that proper controls have been performed, especially when subtracting non-specific binding. In most cases it is appropriate to subtract results of blank runs using surfaces containing the same ligand in a biologically inactive form. It is not recommended to use simply non-specific protein surfaces (e.g. BSA) as a control, since the non-specific interactions might be different (see also Section 3.2).

Bulk shifts due to buffer variations should be removed, although software packages are capable of subtracting these aberrations. Nevertheless, in our experience the best results were obtained with very pure analytes, exchanged into running buffer using gel filtration or fast buffer exchange columns.

Once the raw data have been prepared, appropriate software may be applied. The user can choose between different generations of software supplied by the manufacturers (Biacore or IAsys), software packages not especially written for biosensors (for example, Graphpad Prism or Sigmaplot), or his/her own algorithms.

It is left open which is the 'best' evaluation procedure for interaction patterns, which are sometimes rather complex. Initial rate analysis might be appropriate for some cases (42), while non-linear analysis usually is capable of fitting more complex models to the data obtained (43). More complex global fit analysis gives a better overall control for the calculated rate constants, since a single set of rate constants is fitted to a complete set of binding curves performed at different concentrations. Grouping data sets that share common parameter values provides a basis for the reaction models tested, and improves the statistical behaviour of parameter estimates. There are two software packages using global fit analysis currently available, i.e. *Clamp* by David Myszka and Thomas Morton (44) and *Biaevaluation 3.0* (Biacore). Both are based on the non-linear Levenburg–Marquardt algorithm (45), which optimizes parameters by minimizing the sum of squared residuals. The residuals are the difference between the calculated curve and the experimental data. The algorithm is an iterative process aiming at a local minimum, which may or may not be the true minimum. Control runs should be performed by starting the analysis with different parameters.

In many cases a simple 1:1 Langmuir binding model might not fit the data, and different models assuming more complex interaction patterns may have to be applied. However, as the complexity of the model increases, the ability to fit the equations to an experiment will improve automatically! This is simply because there is more scope for varying parameters to generate a close fit. Therefore, assumptions about the mechanism of interaction should be decided on before applying a more complex model. Complex systems are extremely difficult to interpret, and even sophisticated evaluation software cannot substitute for careful experimental design. However, in some cases the biological interactions involving protein kinases are not a simple 1:1 interaction, and a more complicated model has to be applied, e.g. when studying receptor kinase dimerization. *Clamp* and *Biaevaluation 3.0* give a choice of models, with more complex situations on the ligand and the analyte side of the analysis (24).

The binding of an analyte to a ligand under constant flow can be regarded as a pseudo-first-order reaction, since the concentration of the analyte is constant in the flow cell. This is not absolutely true, especially when using a cuvette system (46, 47); the depletion of analyte can have a significant effect on the analyte concentration. The same might also be true for flow systems; owing to mass transfer limitations, the concentration of analyte might be reduced close to the dextran matrix, where interaction with the immobilized ligand takes place (48). This inherent problem can produce the same kind of deviations from pseudo-first-order binding processes. Therefore, global fitting may potentially result in conclusions as doubtful as those derived from conventional linear analysis of data (49). Using a series of different concentrations of analyte, the association rate constants can be calculated. In the footnotes below, the terms for calculation of association (*Equation 11*) and dissociation rate constants (*Equation 14*) are derived which can be analysed using linear or non-linear data processing.

Equations for the calculation of association and dissociation rate constants are given below:

The association and dissociation rate for binary complex formation is described by:

$$A + B \underset{k_d}{\overset{k_a}{\longleftrightarrow}} AB \tag{1}$$

where k_a and k_d are the rate constants. The association rate is given by:

$$d[AB]/dt = k_a[A][B] \tag{2}$$

and the dissociation rate by:

$$-d[AB]/dt = k_d[AB] \tag{3}$$

At equilibrium, association and dissociation are equal, so:

$$K_D = [A][B]/[AB] = k_d/k_a \tag{4}$$

$$K_A = [AB]/[A][B] = k_a/k_d \tag{5}$$

where K_D and K_A are the equilibrium dissociation and association binding constants, respectively.

If mass transfer is much faster than interaction-controlled association, the analyte at the surface is maintained at the same concentration as in the bulk phase, and the rate equation can be written as

$$d[AB]/dt = k_a[A][B] - k_d[AB] \tag{6}$$

The concentration of unoccupied ligand [B] is the difference between the total amount of ligand on the surface, $[B]_0$, and the amount of complex, [AB]:

$$[B] = [B]_0 - [AB] \tag{7}$$

Substitution of this expression for [B] in *Equation 6* gives:

$$d[AB]/dt = k_a[A]([B]_0-[AB])-k_d[AB] \qquad (8)$$

If the total amount of ligand $[B]_0$ is expressed in terms of the maximum analyte binding capacity of the surface, all concentration terms can then be expressed as SPR response (RU), eliminating the need to convert from mass to molar concentration:

$$dR/dt = k_a C(R_{max}-R)-k_d R \qquad (9)$$

where dR/dt is the rate of change of the SPR signal, C is the concentration of the injected analyte, R_{max} is the maximum analyte binding capacity of the sensor surface in RU, and R is the SPR signal at time t.

Equation 9 can be rearranged to give:

$$dR/dt = k_a CR_{max}-(k_a C + k_d)R \qquad (10)$$

Kinetic constants can be calculated by linear regression, using the pseudo-first-order rate equation (*Equation 10*). Plots of dR/dt versus R have a slope of k_s. When k_s is plotted against C, the resulting slope is equal to the k_a.

Equation 10 can be solved analytically if C is assumed to be constant (which is a valid approximation using dispersion-free injection of the sample). Separating variables and integrating yields:

$$R_t = k_a CR_{max}/(k_a C + k_d)\times (1-e^{-(k_a C+k_d)t}) \qquad (11)$$

This equation describes the response at any time during association, and can be used for non-linear regression analysis of single curves.

Assuming a 1:1 complex between analyte and immobilized ligand, the corrected association phase may be described as:

$$R_t = k_a CR_{max}/(k_a C + k_d)\times(1-e^{-(k_a C+k_d)t}) + R_{bulk} + R_{drift}t \qquad (12)$$

The terms R_{bulk} and R_{drift} consider the bulk concentration of the sample refractive index and the drift of the baseline, respectively.

Following *Equation 3* and the considerations leading to *Equation 9*, the dissociation phase can be described as:

$$dR/dt = -k_d R \qquad (13)$$

assuming that the reassociation of released analyte (rebinding) is negligible. Dissociation kinetics were checked for rebinding effects by co-injecting non-biotinylated ligand in the same concentrations as the immobilized ligand during the dissociation phase.

Separating variables and integrating yields:

$$R_t = R_0 e^{-kd\,(t-t0)} \qquad (14)$$

where R_t is the response at time t and R_0 the time at an arbitrary starting point t_0. Introducing correction variables yields:

$$R_t = R_0 e^{-kd\,(t-t0)} + R_{drift}(t - t_0) + R_{residue} \qquad (15)$$

where $R_{residue}$ is a term for the non-dissociating analyte which remains bound to the surface during buffer flow.

6. Complementary techniques

The strategy of using multiple methods is fundamental to the characterization of biomolecular interactions, because systematic errors and inaccuracies can be ruled out. Independent methods, preferably in free solution, such as those discussed in Sections 6.1–6.4, should be applied. The different values yielded for a specific parameter should be compared, rather than trusting one exclusive measurement from only one method.

6.1 Analytical gel chromatography

Analytical gel chromatography, based on the molecular size-dependent partitioning of a protein between the stationary and mobile phase of a sizing column, can be used to check the purity of the interactants used in biomolecular interaction analysis, potentially revealing heterogeneity or aggregation in the preparation. Equilibrium binding constants and stoichiometries of complexes can be determined, as well as the dependence of the interaction on the buffer composition. Hydrodynamic properties such as Stoke's radii, giving information about the shape of proteins or protein complexes, can also be obtained (50).

6.2 Analytical ultracentrifugation

The characterization of protein–protein interactions in solution can also be performed using analytical ultracentrifugation. There are two general approaches: equilibrium sedimentation is thermodynamic, and yields information about molecular mass and assembly state, whereas velocity sedimentation is hydrodynamic and yields information about molecular mass and shape. With equilibrium sedimentation, equilibrium binding constants in the micro- to nanomolar range can be determined.

6.3 Calorimetric methods

Two techniques based on calorimetry are used for the determination of affinities, i.e. differential scanning calorimetry (DSC) and isothermal titration calorimetry (ITC). DSC has been shown to be a useful method to determine relatively weak binding interactions, such as cytidine-2′-monophospate binding to ribonuclease A (51). ITC is regarded as a thermodynamically robust method for investigating protein–protein interactions. Reactants can be studied in their native form without the use of labels or attachment to a matrix. ITC measures equilibrium constants directly, and does not use kinetic rate constants as the basis for binding constants. However, it yields a large amount of thermodynamic information. The intrinsic binding enthalpy for a given protein–protein interaction can be calculated from the primary heat signal (52), and, when combined with affinity, yields the binding entropy change. ITC is also a measure of the amount of conformational change which is coupled to the binding investigated (53). The main drawback of ITC is that high concentrations of reactants are needed.

6.4 Mass spectrometry

The accurate determination of the molecular mass of the interactants is an important step in interaction analysis. This is now routinely done using mass spectrometry, which allows the identification of the components and its post-translational modifications such as phosphorylation, myristylation and glycosylation (see Chapter 6). Electrospray mass spectrometry (ES–MS) is the method of choice for high resolution mass determination but the choice of buffer is critical. The buffer is less critical for matrix-assisted laser desorption/ionization mass spectrometry (MALDI–MS). The precision and accuracy of both methods are good, even for large proteins. The approach of using MS to identify molecules eluted from the sensor chip surface opens new potential for on-line integration of the two techniques. MALDI–MS can also be integrated with biomolecular interaction analysis using SPR to identify molecules bound to the sensor chip surface, as demonstrated by Krone *et al.* (54). The sensor chip itself can be prepared directly in the solid phase sample matrix for MALDI–MS, making the interface with SPR particularly easy.

Acknowledgements

We thank Bernd Haase, Uwe Roder, and Franz Schindler from Biacore AB, Hans Trutnau from Affinity Sensors, Ariane Maleszka, Silke Reiffert, Mark Kibschull, Claudia Hahnefeld and Ludwig Heilmeyer (Ruhr-Universität Bochum), and Attila Toth (University of Debrecen, Hungary) for the supply of data and for valuable discussions.

References

1. Szabo, A., Stoltz, L., and Granzow, R. (1995) *Curr. Opin. Struct. Biol.* **5**, 699.
2. Raghavan, M and Bjorkman, P.J. (1995) *Structure* **3**, 331.
3. Cush, R., Cronin, J.M., Steward, J., Maule, C.H., Molloy, J., and Goddard, N.J. (1993) *Biosens. Bioelectron.* **8**, 347.
4. Stenberg, E., Persson, B., Roos, H., and Urbaiczky, C. (1991) *J. Coll. Interface Sci.* **143**, 513.
5. Kovacs, G.D. (1982) In *Electromagnetic surface modes* (ed. A.D. Boardman), 143pp., Wiley, New York.
6. Edwards, P.R., Gill, A., Pollard-Knight, D.V., Hoare, M., Buckle, P.E., Lowe, P.A., and Leatherbarrow, R.J., (1995) *Anal. Biochem.* **231**, 210.

7. Herberg, F.W., Dostmann, W.R., Zorn, M., Davis, S.J., and Taylor, S.S. (1994) *Biochemistry* **33**, 7485.

8. O'Shannessy, D. and Wilchek, M. (1990) *Anal. Biochem.* **191**, 1.

9. Lange, C. and Koch, K.W. (1997) *Biochemistry*, **36**, 12019.

10. Nieba, L., Nieba-Axmann, S.E., Persson, A., Hamalainen, M., Edebratt, F., Hansson, A., Lidholm, J., Magnusson, K., Karlsson, A.F., and Pluckthun, A. (1997) *Anal. Biochem.* **252**, 217.

11. Gershon, P.D. and Khilko, S. (1995) *J. Immunol. Meth.* **183**, 65.

12. Kalb, E., Frey, S., and Tamm, L.K. (1992) *Biochim. Biophys. Acta* **1103**, 307.

13. Masson, L., Mazza, A., and Brousseau, R. (1994) *Anal. Biochem.* **218**, 405.

14. Soulages, J.L., Salamon, Z., Wells, M.A., and Tollin, G. (1995) *Proc. Natl. Acad. Sci. USA* **92**, 5650.

15. MacDonald, R.C., MacDonald, R.I., Menco, B.P., Takeshita, K., Subbarao, N.K., and Hu, L.R. (1991) *Biochim. Biophys. Acta* **1061**, 297.

16. Buechler, Y.J., Herberg, F.W., and Taylor, S.S. (1993) *J. Biol. Chem.* **268**, 16495.

17. Herberg, F.W., Taylor, S.S., and Dostmann, W.R. (1996) *Biochemistry* **35**, 2934.

18. Herberg, F.W., Zimmermann, B., McGlone, M., and Taylor, S.S. (1997) *Protein Sci.* **6**, 569.

19. Herberg, F.W. and Taylor, S.S. (1993) *Biochemistry* **32**, 14015.

20. Hofmann, F. (1980) *J. Biol. Chem.* **255**, 1559.

21. Glaser, R.W. (1993) *Anal. Biochem.* **213**, 152.

22. Kortt, A.A., Gruen, L.C., and Oddie, G.W. (1997) *J. Mol. Recognit.* **10**, 148.

23. Myszka, D.G., He, X., Dembo, M., Morton, T.A., and Goldstein, B. (1998) *Biophys. J.* **75**, 583.

24. Karlsson, R. and Falt, A. (1997) *J. Immunol. Meth.* **200**, 121.

25. Schuster, S.C., Swanson, R.V., Alex, L.A., Bourret, R.B., and Simon, M.I. (1993) *Nature* **365**, 343.

26. Cunningham, B.C. and Wells, J.A. (1993) *J. Mol. Biol.* **234**, 554.

27. Knibiehler, M., Goubin, F., Escalas, N., Jonsson, Z.O., Mazarguil, H., Hubscher, U., and Ducommun, B. (1996) *FEBS Lett.* **391**, 66.

28. Chiorini, J., Zimmermann, B., Yang, L., Smith, R., Ahearn, A., Herberg, F., and Kotin, R. (1998) *Mol. Cell. Biol.* **18**, 5921.

29. Malmborg, A.C. and Borrebaeck, C.A. (1995) *J. Immunol. Meth.* **183**, 7.

30. Muller, K., Gombert, F.O., Manning, U., Grossmuller, F., Graff, P., Zaegel, H., Zuber, J.F., Freuler, F., Tschopp, C., and Baumann, G. (1996) *J. Biol. Chem.* **271**, 16500.

31. Lombardo, C.R., Consler, T.G., and Kassel, D.B. (1995) *Biochemistry* **34**, 16456.

32. Felder, S., Zhou, M., Hu, P., Urena, J., Ullrich, A., Chaudhuri, M., White, M., Shoelson, S.E., and Schlessinger, J. (1993) *Mol. Cell. Biol.* **13**, 1449.

33. Ladbury, J.E., Lemmon, M.A., Zhou, M., Green, J., Botfield, M.C., and Schlessinger, J. (1995) *Proc. Natl. Acad. Sci. USA* **92**, 3199.

34. Kim, E., DeMarco, S.J., Marfatia, S.M., Chisti, A.H., Sheng, M., and Streler, E.E. (1998) *J. Biol. Chem.* **273**, 1591.

35. Yang, W.P., Wu, H., and Barbas, C.F., III (1995) *J. Immunol. Meth.* **183**, 175.

36. Ozawa, T., Kakuta, M., Sugawara, M., Umezawa, Y., and Ikura, M. (1997) *Anal. Chem.* **69**, 3081.

37. Tokumitsu, H., Wayman, G.A., Murumatsu, M., and Soderling, T.R. (1997) *Biochemistry* **36**, 12823.

15: Biomolecular interaction analysis

38. Burton, K.A., Johnson, B.D., Hausken, Z.E., Westenbroek, R.E., Idzerda, R.L., Scheuer, T., Scott, J.D., Catterall, W.A., and McKnight, G.S. (1997) *Proc. Natl. Acad. Sci. USA* **94**, 11067.

39. Sahara, S., Sato, K., Kaise, H., Mori, K., Sato, A., Aoto, M., Tokmakov, A.A., and Fukami, Y. (1996) *FEBS Lett.* **384**, 138.

40. Nieba, L., Krebber, A., and Pluckthun, A. (1996) *Anal. Biochem.* **234**, 155.

41. Reiffert, S.U., Jaquet, K., Heilmeyer, L.M.G., Jr, and Herberg, F.W. (1998) *Biochemistry* **37**, 13516.

42. Edwards, P.R. and Leatherbarrow, R.J. (1997) *Anal. Biochem.* **246**, 1.

43. Morton, T.A., Myszka, D.G., and Chaiken, I.M. (1995) *Anal. Biochem.* **227**, 176.

44. Myszka, D.G. and Morton, T.A. (1998) *Trends Biochem. Sci.* **23**, 149.

45. Press, W.H., Teukolsky, S.A., Vetterling, W.T., and Flannery, B.P. (1992) *Numerical recipes in C.* Cambridge University Press, Cambridge.

46. Hall, D.R., Gorgani, N.N., Altin, J.G., and Winzor, D.J. (1997) *Anal. Biochem.* **253**, 145.

47. O'Shannessy, D. and Winzor, D.J. (1996) *Anal. Biochem.* **236**, 275.

48. Hall, D.R., Cann, J.R., and Winzor, D.J. (1996) *Anal. Biochem.* **235**, 175.

49. Schuck, P. and Minton, A.P. (1996) *Anal. Biochem.* **240**, 262.

50. Ackers, G.K. (1975) In *The proteins* (ed. H. Neurath and R. Hill), Vol. 1. Academic Press, Inc., New York.

51. Brandts, J.F. and Lin, L.N. (1990) *Biochemistry* **29**, 6927.

52. Doyle, M.L., Louie, G., Dal, M.P., and Sokoloski, T.D. (1995) In *Methods in enzymology*, Vol. 259, p. 183. Academic Press, New York.

53. Spolar, R.S. and Record, M.T., Jr (1994) *Science* **263**, 777.

54. Krone, J.R., Nelson, R.W., Dogruel, D., Williams, P., and Granzow, R. (1997) *Anal. Biochem.* **244**, 124.

371

16

Analysis of protein kinase specificity using oriented peptide libraries

ZHOU SONGYANG and LEWIS C. CANTLEY

1. Introduction

This chapter focuses on the use of oriented peptide libraries to study the specificities of protein kinases. This approach not only predicts an optimal sequence from a single experiment, with no prior knowledge of *in vivo* phosphorylation sites required, but it also provides information about the relative importance of each position for selectivity, and about which amino acids are tolerated. In the following sections we will explain peptide library design strategies, detail the experimental techniques for peptide synthesis and selection, and discuss how the data obtained are interpreted.

2. Application of the peptide library approach

2.1 Protein kinases

Protein kinases involved in cell signalling are generally classified into three categories depending on their abilities to phosphorylate serine or threonine residues, tyrosine residues, or both. Protein kinases share regions of conserved sequence in the catalytic domain (the kinase domain), suggesting a common ancestry (1). However, different protein kinases have evolved to function distinctly in response to diverse cellular stimuli. Because protein phosphorylation is reversible, it allows living cells to rapidly reset to the basal state after stimulation, and the process is crucial in the regulation of a plethora of intracellular biological activities. An understanding of the specificities of protein kinases for downstream targets is critical to our understanding of signal transduction.

The specificity of a protein kinase in the cell is governed both by its intracellular localization and by its intrinsic specificity for protein substrates. The intrinsic specificity of the kinase domain is perhaps the most important factor. Studies of several protein–serine/threonine kinases, including cAMP-dependent protein kinase (PKA), indicated that the kinase domain recognizes a short primary sequence around the phosphorylation site (2, 3).

2.2 Determining protein kinase specificity: shortcomings of traditional approaches

On the basis of the sequences of known *in vivo* substrates, and of studies of amino acid substitutions made either via chemical synthesis of peptide substrates, or via site-directed mutagenesis of protein substrates, the specificities of a few protein–serine/threonine kinases have been determined (see Chapter 10). However, this conventional approach has several drawbacks. First, it is expensive and time consuming to synthesize and assay all of the possible substitutions. Since around 10 amino acids of the peptide/protein substrate are likely to contact the active-site cleft of a protein kinase (3), there are up to 20^{10} (10^{13}) different peptides to consider. In addition, this method cannot be used in cases where the *in vivo* targets of protein kinases have not yet been identified. Taking advantage of a technique that simultaneously synthesizes large numbers of degenerate peptides on a solid matrix (beads or pins) (4, 5), alternative strategies based on combinatorial chemistry have been used to address these problems. The specificity of the protein kinase can be deduced by sequencing the peptides on the beads that give a phosphorylation signal. However, these approaches have the drawback that phosphorylation of a peptide on a solid matrix may be different from that in solution. The peptide substrates identified by these methods may therefore not be physiologically relevant.

Based on a strategy developed to examine the binding specificity of SH2 domains, we have developed an oriented peptide library technique to rapidly determine the optimal sequences for protein kinase targets (6, 7). The kinase of interest is used to phosphorylate a soluble mixture of billions of distinct peptides, which have identical length and orientation and have a single phosphorylatable amino acid in the centre. The small fraction of phosphorylated peptides is quantitatively separated from the bulk of the non-phosphorylated peptides, and the mixture is then sequenced. The abundance of amino acids at each degenerate position surrounding the phosphorylation site is compared with that at the same position in the starting mixture. The result should indicate the preference for a particular amino acid at each position.

3. Oriented peptide libraries for protein kinases

The method discussed below has been used successfully to determining the substrate specificities of several protein kinases (7, 8). Similar to the strategy adopted for our earlier technique for determining binding motifs for SH2 domains (6), the procedure quantitatively separates the degenerate phospho-peptide products from the bulk of the non-phosphorylated peptides.

3.1 Design of degenerate peptide libraries for protein kinases

Peptide libraries for protein kinases can be classified depending on the number of fixed recognition residues. If little is known about the specificity of a given kinase, completely degenerate libraries (primary libraries) can be used to identify key residues for substrate recognition. In primary libraries, the phosphorylatable amino acids (Tyr, Ser, and Thr) are fixed to orient the library, but other positions contain a random mixture of amino acids. Because the proportion of peptides that can be efficiently phosphorylated by the kinase of interest is often low in primary libraries, the sequencing of the phosphorylated peptide mixture may be carried out on rather small amounts of material, and will be subject to a high background. However, secondary libraries can then be constructed based on the results of the primary library screens. In these secondary libraries, key residues identified as being required for phosphorylation in the primary screen are fixed, as well as the phosphorylatable amino acids themselves. Screening is then repeated using the secondary library to define more fully the specificity of the protein kinase.

3.1.1 Proper orientation of the peptide library

Orientation is easily achieved for protein kinase libraries, because the phosphorylatable amino acids (Tyr, Ser, and Thr) can be used to orient the library. The following primary library (Ser degenerate library) constructed for analysis of protein–serine/threonine kinases can be used as an example. The sequence of the library was: Met–Ala–Xxx–Xxx–Xxx–Xxx–Ser–Xxx–Xxx–Xxx–Xxx–Ala–Lys–Lys–Lys, where Xxx represents all amino acids except Trp, Cys, Tyr, Ser, and Thr (7). To ensure that the only potential site of phosphorylation was the serine at residue 7, the three phosphorylatable amino acids were omitted at the degenerate positions. Having the degenerate amino acids at certain positions only, and fixing the phosphorylatable amino acid at position 7, means that the phosphorylated peptides are sequenced in phase. To design a secondary library, a second residue in addition to the phosphorylatable amino acid is fixed. For example, the following secondary library (Ser–Pro library) was constructed for the analysis of cyclin-dependent kinases: Met–Ala–Xaa–Xaa–Xaa–Xaa–Ser–Pro–Xaa–Xaa–Xaa–Ala–Lys–Lys–Lys, where Xaa indicates all amino acids except Trp and Cys (7).

3.1.2 Selection of amino acids at the degenerate positions

For primary libraries, phosphorylatable amino acids (Tyr, Ser, and Thr) are usually avoided in the degenerate positions. For secondary libraries, Tyr, Ser, and Thr can be included at degenerate positions because phosphorylation of these residues at the degenerate position can usually be neglected. In the case of the Ser–Pro library, the chance of serine and proline being adjacent at the degenerate positions is quite small (~1.5%). Thus, peptides phosphorylated at positions other than the fixed one do not interfere with the sequencing of oriented peptides. Cysteine and tryptophan are omitted from the degenerate positions in all peptide libraries to avoid problems with oxidation and problems during sequencing, respectively. Cysteine residues can be added and studied after the optimal peptides have been obtained. If 15 different amino acids are present in any one of the eight degenerate positions (for primary libraries), the total theoretical degeneracy of this library is $15^8 = 2\,562\,890\,625$. A primary peptide library for tyrosine kinases (Tyr–kinase substrate library) was constructed in a similar fashion except that residue 7 was tyrosine (8).

3.1.3 Other considerations

(a) The length of peptide libraries can vary, but the number of degenerate positions should not exceed 15. A library of 15 degenerate positions already has 20^{15} different molecules. For a peptide with a molecular weight of 2 kDa, this translates into roughly 100 milligrams. We started with eight degenerate positions, because we argued that four residues N-terminal and C-terminal to the phosphorylation site would be likely to include the region involved in substrate recognition, based on the recognition motifs that had already been determined for protein kinases by other approaches.

(b) Placing a short leading sequence before the degenerate positions is generally beneficial. In the case of the Ser degenerate library (Met–Ala–Xaa–Xaa–Xaa–Xaa–Ser–Xaa–Xaa–Xaa–Xaa–Ala–Lys–Lys–Lys), the Met–Ala sequence at the N-terminus of the peptide libraries not only provides two amino acids to verify that peptides from this mixture are being sequenced, but also allows quantification. Similarly, the Ala at residue 12 makes it possible to quantify and estimate the amount of peptide lost during sequencing. The poly(lysine) tail not only prevents wash-out during sequencing, but also improves the solubility of the mixture (no solubility problems occurred at neutral pH and 5 mg/ml concentration). Importantly, the poly(lysine) sequence also allows the peptides to stick to phosphocellulose paper (P81 paper) facilitating the separation of [γ-^{32}P]ATP (see *Protocol 3*).

4. Use of oriented peptide libraries

4.1 Peptide library synthesis

In our laboratory the synthesis of degenerate peptide libraries is accomplished according to standard BOP/HOBt coupling protocols using a Peptide BioSynthesizer (ABI 431A, Perkin–Elmer). Only general guidelines, rather than a detailed method, are given in *Protocol 1*, because the user will wish to adapt the manufacturer's protocols that come with their own peptide synthesizer.

Protocol 1. Synthesis of degenerate peptide libraries

Equipment and reagents
- FMOC-blocked amino acids and other reagents for synthesis
- peptide synthesizer

Method
1. At the degenerate positions, add simultaneously equal moles of different FMOC-blocked amino acids (except for cysteine) at a fivefold excess over the coupling resin. The ratio of input FMOC-blocked amino acids may need to be adjusted on different synthesizers to achieve even distribution of degenerate amino acids.
2. Deprotect the peptides and cleave them off the resin using trifluoroacetic acid.
3. Lyophilize the supernatants from step 2 to obtain crude peptide library powder mixtures.
4. Sequence 1–2 μg of the peptide libraries, e.g. using a peptide sequencer, to confirm that all amino acids are present in similar amounts (within a factor of 3) at the degenerate positions.

4.2 Protein kinases

4.2.1 Preparation of protein kinases

Protein kinases can be obtained from many different sources. Most commonly, they are overexpressed as recombinant proteins in either bacteria or eukaryotic cells (see Chapter 10). Expression in insect cells (Sf9 cells) using baculovirus vectors is a good system for most protein kinases, and it often yields active enzymes. Expressed protein kinases can be purified using conventional chromatography, affinity chromatography, or immunoprecipitation (see Chapter 9). Purification of recombinant protein kinases has become routine and protocols can be found in ref. 9. The amount of enzymes required to phosphorylate peptide libraries for sequencing depends on their specific activities. In general, we use microgram quantity of protein kinases in most experiments.

Protocol 2. Phosphorylation of peptide library

Reagents
- Ser/Thr kinase buffer (50 mM Tris, pH 7.4, 10 mM $MgCl_2$, 1 mM DTT)
- $[\gamma\text{-}^{32}P]ATP$
- Tyr kinase buffer (50 mM Tris, pH 7.4, 10 mM $MnCl_2$, 1 mM DTT)

Protocol 2. *Continued*

Method

1. Perform the kinase reactions with either soluble or immobilized kinases.
2. Add the protein kinase to 300 μl of solution containing 1 mg of degenerate peptide mixture, 100 μM ATP with a trace of [γ-^{32}P]ATP (roughly 6×10^5 cpm) in appropriate kinase buffer. Also perform, as a control, a mock reaction in which no kinase is added.
3. Incubate the mixture at 25–30°C for 3 h. The aim is to phosphorylate around 1% of the peptide mixture.
4. Terminate the reaction by adding acetic acid to a final concentration of 15%.

4.3 Phosphopeptide separation

To study the substrate specificity of a protein kinase using peptide libraries, it is critical to separate efficiently those peptides that have been phosphorylated from the large amount of unphosphorylated peptides. In general, phosphorylated peptides represent less than 1% of the total peptide mixture. To achieve efficient separation, a ferric ion–iminodiacetic acid (Fe^{3+}–IDA) column is used; the ferric ion binds the phosphate moiety relatively specifically (10, 11). However, ATP should be removed first since it may block binding of phosphopeptides to the Fe^{3+}–IDA column.

4.3.1 Removal of ATP using an anion-exchange column

Since ATP is negatively charged at pH values less than 5, it binds to a diethylaminoethyl (DEAE) anion-exchange column. Because all of the peptides in the library contain a poly(lysine) tail, they have a net positive charge in 30% acetic acid and do not bind to the column (12). Therefore, a DEAE column can be used to remove ATP quickly.

Protocol 3. Separation of peptides from ATP using an anion-exchange column

Equipment and reagents

- DEAE–Sephacel (Sigma)
- acetic acid (30% v/v)

- centrifugal vacuum concentrator (e.g. Savant Speedvac)

Method

1. Wash 1–1.5 ml of packed DEAE–Sephacel beads with 10 ml of acetic acid. Centrifuge at 1000 *g* for 2 min and remove the supernatant. Repeat the washing once more.
2. Suspend the beads in 10 ml of acetic acid and pour the slurry into a 15 ml disposable column. Allow the acetic acid to drip through by gravity until the beads run dry.

3. Layer the peptide supernatants (from *Protocol 2*, step 3) carefully onto the beads and allow all the liquid to run into the column.

4. Elute the column with 30% acetic acid. Discard the first 600 μl of flow through and collect the next 1 ml.

5. Lyophilize the collected fraction on a centrifugal vacuum concentrator.

Using this protocol, the peptide mixtures are in the void volume due to their poly(lysine) tail, while the ATP and the denatured protein kinases are retained on the column. In pilot experiments, the eluates from the column were analysed for peptide, phosphopeptide, [γ-^{32}P]ATP, and [^{32}P]phosphate by adsorption on to phosphocellulose (P81) paper, thin-layer chromatography, or SDS–PAGE. The first revealed that the first 600 μl represented the void volume of the column and the next 1 ml contained both phosphorylated and non-phosphorylated peptides that were free of [γ-^{32}P]ATP. The radioactivity of this fraction provides an initial estimate of the fraction of the total peptide mixture that has been phosphorylated.

4.3.2 Purification of phosphopeptides on a ferric ion–iminodiacetic acid column

An Fe^{3+}–IDA column is used for separation of phosphopeptides and de-phosphopeptides. This column has been used in the past to separate tryptic phosphopeptides of phosphorylated proteins from the bulk of non-phosphorylated tryptic peptides (10, 11). We have changed the loading and elution conditions from published procedures, to avoid the loss of a sub-fraction of phosphopeptides, while still achieving quantitative removal of the non-phosphorylated peptides. A typical column profile is shown in *Figure 1*.

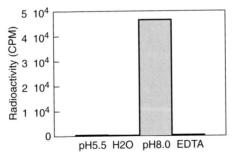

Figure 1. Separation of phosphopeptides on a ferric ion_iminodiacetic acid (Fe^{3+}_IDA) column. A tyrosine-containing peptide library mixture was phosphorylated using c-Src kinase and [γ-^{32}P]ATP. After removing the ATP from the peptide mixture, phosphopeptides were separated from unphosphorylated peptides on the Fe^{3+}–IDA column. The column was washed sequentially with 3 ml of buffer B (pH 5.5) and distilled water and phosphopeptides were eluted with buffer A (pH 8.0). As shown here, elution of phosphopeptides can be followed by monitoring radioactivity.

Protocol 4. Phosphopeptide separation on a ferric
ion–iminodiacetic acid column

Equipment and reagents
- iminodiacetic acid (IDA)-coupled agarose
  beads (Pierce, Rockford, IL)
- $FeCl_3$ (20 mM)
- buffer A (500 mM NH_4HCO_3, pH8.0)
- buffer B (50 mM MES, 1M NaCl, pH 5.5)
- EDTA (100 mM, pH 8.0)
- peristaltic pumps
- ABI peptide sequencer (Perkin–Elmer)

Method

1. Charge a 0.3 ml column of IDA beads using 2 ml of $FeCl_3$, at 0.5 ml/ min.

2. Wash with 3 ml of water at 1 ml/min.

3. Wash with 3 ml of buffer A at 1 ml/min.

4. Wash again with 3 ml of water.

5. Equilibrate with 3 ml of buffer B.

6. Dissolve the dried sample of peptide/phosphopeptide mixture in 200 μl of buffer B and load carefully on to the ferric column.

7. Elute the column with 3 ml of buffer B followed by 3 ml of distilled water at 0.2 ml/min.

8. Elute the phosphopeptides with 3 ml of buffer A.

9. Elute the Fe^{3+} with 100 mM EDTA.

10. Collect the eluate from step 8 and lyophilize and redissolve in water several times to remove the ammonium bicarbonate.

11. Resuspend the phosphopeptide mixture in 40 μl of water and adjust to neutral pH.

12. Sequence 20 μl of the phosphopeptide mixture on the protein sequencer.

The Fe^{3+}–IDA column can efficiently separate phosphorylated peptides from unphosphorylated species. However, a small percentage (~0.1%) of the degenerate unphosphorylated peptides, which are rich in acidic amino acids, co-purify with the phosphopeptides, because they bind weakly to the ferric column. This can be very problematic when using peptide libraries in which acidic residues (Asp and Glu) are fixed, and these should be avoided if possible. If acidic amino acids do have to be included in the fixed positions, the pH value of buffer A should be increased to 6.5 and the volume of IDA beads should be reduced.

Enough phosphopeptides (e.g. up to 1% of the input peptide library) must be purified for sequencing so that the phosphorylated peptides are in a large excess over the contaminating unphosphorylated peptides. In a reaction in which 1% of the peptide mixture is phosphorylated, the total quantity of phosphopeptides is $(1.8 \text{ mM}) \times (0.3 \text{ ml}) \times (0.01) = 5.4$ nmoles. Roughly 1–2 nmoles of phosphopeptide mixture is typically added to the sequencer. This means that in a cycle in which all 15 residues are equally abundant, the yield of each amino acid is approximately 60 pmoles.

4.4. Analysis of results

Sequencing of the phosphopeptide mixture reveals the abundance of a particular amino acid at each degenerate position. In order to determine the substrate preference of a kinase, it is necessary to obtain sequence data for three mixtures: the original peptide library, the phosphopeptides purified from the library, and the peptides purified from the mock experiment in which the protein kinase was omitted (see *Protocol 2*).

4.4.1 Interpretation of results

In theory, at a given cycle of sequencing, the abundance of each amino acid in the phosphopeptide mixture could be divided by its abundance in the library, to determine the preference of the protein kinase at a particular position. Using this method, variations in the abundance of amino acids at a particular position in the library, or variations in yield for that amino acid from the sequencer, will cancel out. If the kinase is insensitive to the amino acid at a particular position, the relative abundance of all amino acids at this cycle in the phosphopeptide mixture will be the same as in the library and every amino acid will give a value of one.

If much less than 1% of the total mixture is phosphorylated, a correction must be made for contamination with unphosphorylated peptides. The contamination can be estimated from the quantity of Ser (or Tyr) at residue 7, since phosphorylated serine or threonine will not be detected at this position. In control reactions in which a mock phosphorylation has been carried out without the protein kinase, the peptides purified on the Fe^{3+}–IDA column are usually rich in Asp and Glu at every degenerate cycle, presumably due to the weak interactions between acidic side chains and the ferric ion. The abundance of each amino acid at each cycle from this control is subtracted to correct for the background.

To calculate the relative preference of amino acids at each degenerate position, the corrected data are compared with the starting mixture to generate the ratios of amino acid abundance. The sum of the abundance of each amino acid at a given cycle is normalized to the number of amino acids present. Each amino acid at a particular position thus has a value of 1 in the absence of selectivity. Theoretically, any value greater than 1 should indicate a preference for that particular amino acid. However, values higher than 1.5 are generally considered significant because of the complexity of the data and the subsequent calculations. The entire process is summarized in *Protocol 5*.

Protocol 5.   Data analysis to determine preferred sequence

Method

1. Normalize the amount of each amino acid at each degenerate positions. To correct for losses during sequencing, we routinely normalize the total amount of amino acids (in pmoles) at each degenerate position to that of the same amino acid at the first degenerate position, P_1:

   $P(ij)$ is the amount of amino acid j at position i for the kinase experiment;

   $P_n(ij)$ is the normalized amount of amino acid j at position i for the kinase experiment;

   $P_n(ij) = P(ij) \times \text{Sum}(P_1)/\text{Sum}(P_i)$.

2. Repeat the normalization in step 1 for the control experiment without kinase, and for the sequencing of the original library.

   $C_n(ij)$ indicates the normalized amount for amino acid j at position i in the control experiment;

   $R_n(ij)$ indicates the normalized amount for amino acid j at position i in the original peptide library.

3. Subtract the control experimental values from the kinase experimental values: $P_n(ij) - K \times C_n(ij)$; K can be calculated by the relative amount of fixed Ser or Tyr, $K = P(\text{Ser})/C(\text{Ser})$.

4. Calculate the relative abundance $A(ij) = [P_n(ij) - K \times C_n(ij)]/R_n(ij)$.

5. Normalize the relative abundance for total number of amino acids included at each degenerate position. If this was 18, then $A_n(ij) = A(ij) \times 18/\text{Sum}(A_{ij})$, where $A_n(ij)$ represents the enrichment value for amino acid j at position I.

   Graphic plots showing the enrichment values of amino acids at all degenerate positions can be generated after the calculation. They reveal the substrate preference for individual protein kinases. In *Figure 2*, the specificity of Cdc2/ cyclin A on the secondary Ser–Pro degenerate peptide library (MAXXXXSP-XXXAKKK) is plotted. The value and significance of this approach is evident from the reproducibility of the results obtained with a given protein kinase, and by the agreement between the recognition motifs obtained with this approach and those obtained by the more traditional approaches discussed in Section 2.2.

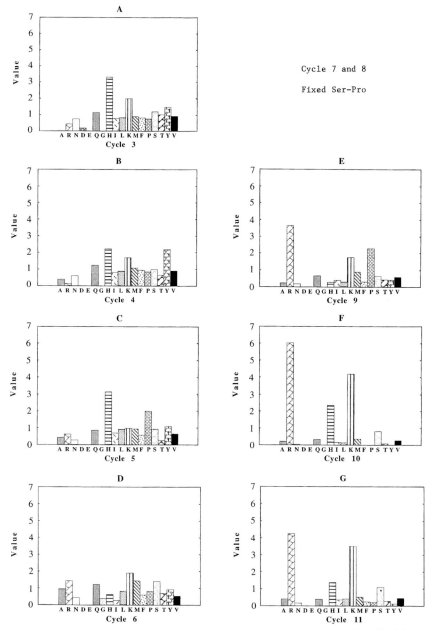

Cycle 7 and 8

Fixed Ser-Pro

Figure 2. Substrate specificity of the cyclin A/Cdc2 protein kinase is identified by the degenerate peptide library (MAXXXXSPXXXAKKK). Each box indicates the relative abundance of 18 amino acids at a given sequencing cycle. The first degenerate position (cycle 3) of the library is indicated in box A. The fixed postions (Ser_Pro) (cycles 7 and 8) are not shown. Therefore, boxes A, B, C, and D indicate amino acid preferences at the -4, -3, -2, and -1 positions on the N-terminal side of the phosphorylation site, and boxes E, F, and G represent the preferences at the +2, +3, and +4 positions on the C-terminal side of the phosphorylation site.

5. Use of the recognition motif obtained to further study the protein kinase

5.1 Prediction of physiological substrates for the protein kinase

This method should be extremely helpful in predicting the optimal substrates and *in vivo* targets for various protein kinases. It will be particularly useful in studying substrates of protein kinases in an organism whose genome has been completely sequenced. Using programmes such as BLAST, FASTA, or FINDPATTERNS in the GCG suite, the recognition motifs can be used to search protein databases. Alternatively, new search programmes can be written using the peptide library selection profile as matrices. Any proteins that match are potential *in vivo* substrates of the kinase under study. The sequence of a particular protein can also be scanned for potential phosphorylation sites for a given protein kinase. Both approaches provide short-cuts in the study of signalling pathways regulated by the protein kinase.

5.2 Development of peptide inhibitors for protein kinases

The recognition motif identified by this approach may also facilitate the design of inhibitors for protein kinases. First, peptides or peptide mimetics based on the optimal substrate sequence identified may themselves be developed as inhibitors. Secondly, the optimal sequences identified by this approach can be used to develop peptide kinase assays (see Chapters 9 and 10) suitable for screening of chemical libraries for potential inhibitors. Finally, structural analysis of protein kinases in complex with peptide substrates identified by this approach may provide a basis for the modelling and design of drugs that specifically inhibit the protein kinase.

References

1. Hanks, S. K., Quinn, A. M., and Hunter, T. (1988). *Science*, **241**, 42.
2. Glass, D. B., Cheng, H. C., Mueller, L. M., Reed, J., and Walsh, D. A. (1989). *J. Biol. Chem.*, **264**, 8802.
3. Knighton, D. R., Zheng, J. H., Ten Eyck, L. F., Xuong, N. H., Taylor, S. S., and Sowadski, J. M. (1991). *Science*, **253**, 414.
4. Geysen, H. M., Meloen, R. H., and Barteling, S. J. (1984). *Proc. Natl. Acad. Sci. USA*, **81**, 3998.
5. Wu, J., Ma, Q., N., and Lam, K. S. (1994). *Biochemistry*, **33**, 14825.
6. Songyang, Z., Shoelson, S. E., Chaudhuri, M., Gish, G., Pawson, T., Haser, W. G., King, F., Roberts, T., Ratnofsky, S., Lechleider, R. J., and B. G. Neel, R. B. B., J. E. Fajardo, M. M. Chou, H. Hanafusa, B. Schaffhausen, and L. C. Cantley (1993). *Cell*, **72**, 767.
7. Songyang, Z., Blechner, S., Hoagland, N., Hoekstra, M. F., Piwnica, W. H., and Cantley, L. C. (1994). *Curr. Biol.*, 4, 973.

8. Songyang, Z., Carraway III, K. L., Eck, M. J., Harrison, S. C., Feldman, R. A., Mohammadi, M., Schlessinger, J., Hubbard, S. R., Smith, D. P., Eng, E., Lorenzo, M. J., Ponder, B. A. J., Mayer, B. J., and Cantley, L. C. (1995). *Nature*, **373**, 536.
9. Hunter T. and Sefton B. M. (1990). In *Methods in enzymology*, Vol. 200. Academic Press, London.
10. Muszynska, G., Andersson, L., and Porath, J. (1986). *Biochemistry*, **25**, 6850.
11. Muszynska, G., Dobrowolska, G., Medin, A., Ekman, P., and Porath, J. O. (1992). *J Chromatogr.*, **604**, 19.
12. Kemp, B. E., Benjamini, E., and Krebs, E. G. (1976). *Proc. Natl. Acad. Sci. USA*, **73**, 1038.

17

Crystallization of protein kinases and phosphatases

BOSTJAN KOBE, THOMAS GLEICHMANN, TRAZEL TEH,
JÖRG HEIERHORST, and BRUCE E. KEMP

1. Introduction

X-ray crystallography is the most powerful method for determination of macromolecular three-dimensional structures, and it has made enduring contributions to the understanding of the function of protein kinases and phosphatases. Determination of protein crystal structures requires single crystals that are suitable for high resolution X-ray diffraction analysis (*Figure 1*). Crystallization will only occur within a limited range of conditions that need to be empirically determined. Despite numerous successful crystallization experiments, our understanding of protein crystallogenesis and macromolecular crystal growth processes remains limited. Crystallization of proteins can be broken down into two aspects: screening (based mainly on trial and error), where one attempts to obtain crystals; and optimization, where one strives to improve the diffraction quality of the crystals.

A comprehensive discussion of protein crystallization is beyond the scope of this chapter. Instead, some general suggestions will be provided, as well as specific practical guides. Readers are encouraged to refer to the primary literature on the topic to obtain further details, for example, *Crystallization of nucleic acids and proteins* in the Practical Approach Series (1). As crystallization of protein kinases and phosphatases follows the same guidelines as that of other classes of proteins, the general discussions of crystallization outlined in this chapter are equally applicable to proteins other than protein kinases and phosphatases.

Although often the rate-limiting bottleneck, crystallization is only one of the many steps in the process of structure determination of a protein by X-ray crystallography. The various steps involved are:

- protein purification;
- crystallization;

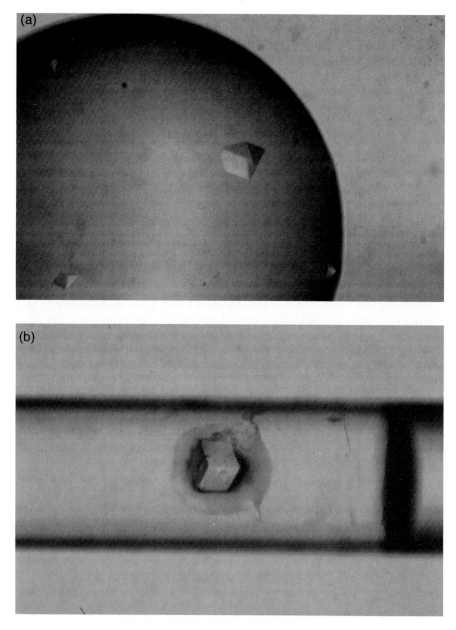

Figure 1. Examples of protein kinase crystals. (a) Crystals of *Aplysia* twitchin kinase fragment in a hanging drop. (b) A crystal of *C. elegans* twitchin kinase (5890–6380) mounted in an X-ray capillary.

- measurement of diffraction data:
  - (i) preliminary characterization (determination of crystal symmetry and unit cell parameters);
  - (ii) diffraction data collection;
  - (iii) diffraction data processing (integration and scaling);
- phase determination [multiple isomorphous replacement (MIR), anomalous scattering (AS) and multiwavelength anomalous dispersion (MAD), molecular replacement, density modification, direct methods, and combinations of these methods];
- electron density map interpretation and model building;
- crystallographic refinement.

The steps preceding and following crystallization will also be briefly described in this chapter.

2. General principles of crystallization

To design crystallization experiments rationally, an understanding of the physical chemistry behind protein crystallization is crucial. Crystals represent a solid phase of a protein that can form when the solubility of the protein in a given solution is exceeded. Crystals therefore grow from *supersaturated* solutions; such solutions are thermodynamically unstable, and will return to equilibrium. Supersaturation can be achieved by varying any parameter that affects the chemical potential of the solution. Such parameters include protein concentration, salt concentration, temperature, and many others. Because supersaturation is the driving force for crystallization, it is the parameter that one has to control in order to optimize the crystals obtained.

Crystal growth consists of three steps: nucleation, growth, and cessation of growth. Cessation of growth will usually result from the depletion of the protein from solution, but may also arise for other reasons, such as growth defects. Crystals will not spontaneously grow out of all supersaturated solutions. If the degree of supersaturation is low, the energy barrier required for nucleation may not be overcome, and crystals will not nucleate; existing crystals, however, may grow. Supersaturation can therefore be divided into a metastable region, where crystals grow but do not nucleate, and a labile region, where crystals nucleate *and* grow. This is indicated in the two-dimensional solubility diagram in *Figure 2*. In this diagram, protein solubility is shown as a function of one parameter (the crystallizing agent concentration), all other parameters being constant. The solubility curve divides the undersaturated and supersaturated zones. Under the solubility curve, the solution is undersaturated and the protein will never crystallize. Above the solubility curve, the concentration of the protein is higher than the concentration at equilibrium for a given crystallizing agent concentration. The level of supersaturation is the ratio of the protein concentration to the solubility value.

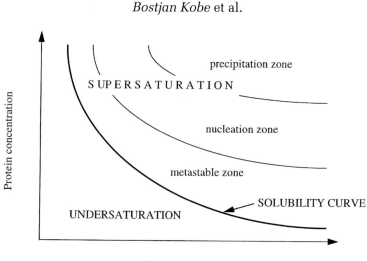

Figure 2. An example of a schematic two-dimensional solubility diagram. Protein concentration is shown as a function of crystallizing agent concentration. The solubility curve and different zones are indicated.

Where the degree of supersaturation is very high, excess protein immediately separates from the solution and forms an amorphous precipitate (the precipitation zone). This is because, in crystals, specific contacts must be formed between individual protein molecules. When the degree of supersaturation is high, formation of these contacts becomes rate limiting, resulting in random associations of protein molecules instead, and formation of an amorphous solid phase instead of crystals.

In the process of attempting protein crystallization, one is searching for conditions of supersaturation that will allow growth of crystals instead of an amorphous precipitate. Such conditions turn out to be scarce for most proteins, mainly because the forces that hold together the protein molecules in crystals are weak. This is the reason for the intrinsic difficulty of protein crystallization. Because of the large number of parameters that affect supersaturation, searching for crystallization conditions becomes a multiparametric process. The major parameters affecting crystallization are:

- the protein sample: (i) purity; (ii) concentration; (iii) presence of ligands;
- the crystallization solution: (i) type of precipitant; (ii) concentration of precipitant; (iii) pH; (iv) ionic strength; (iv) additives and other constituents of crystallization solution;
- environmental parameters: (i) temperature (and temperature fluctuations); (ii) pressure; (iii) (micro)gravity;
- the crystallization method: (i) rate of equilibration; (ii) volume of crystallization sample; (iii) surfaces on crystallization vessels.

3. Expression and purification of proteins

Unfortunately, milligram quantities of very pure protein are usually required for successful crystal growth and crystal structure determination. Over-expression in bacteria or another suitable system is consequently a necessity in most cases. One of the most important parameters influencing crystal growth is protein purity. Although the impact of contaminants will vary for different proteins, it is a general rule that better crystals are obtained with the purest samples. In general, lower concentrations of contaminants affect nucleation, whereas higher concentrations influence crystal growth morphology. As a rule, the greatest success is achieved if the protein sample is:

- >95% pure by Coomassie Blue-stained SDS–PAGE;
- substantially pure by isoelectric focusing;
- monodisperse by dynamic light scattering (2);
- conformationally homogeneous.

However, sometimes even heterogeneous samples such as the *Aplysia* twitchin kinase, which autophosphorylates at several sites during bacterial expression (unpublished results), can be successfully crystallized.

3.1 Production of fragments and limited proteolysis

Flexible proteins exist in a variety of conformational states at equilibrium, and this increases the difficulty in developing an ordered three-dimensional array. This problem tends to be greater with multidomain proteins and, accordingly, it may be advantageous to express well-defined domains separately. In protein kinases and phosphatases, these domains are usually evident from sequence comparisons. An alternative experimental approach to examine the structural domains of multidomain proteins is to utilize limited proteolysis with one or more proteases, where residual fragments are then identified by N-terminal sequencing and/or mass spectrometry. Proteolysis of a native protein can be limited by performing the digestion under incubation conditions that prevent the proteolytic reaction from reaching completion, thus generating fewer and larger protein fragments. Sites susceptible to cleavage in native proteins include stretches of polypeptide chain which are either in exposed loops within compact domains, or between domains.

An example of such an approach, utilizing thermolysin as the protease, is presented in *Protocol 1*. It may be advantageous to subsequently produce the fragments identified by this approach by the expression of recombinant DNA encoding the fragment, rather than attempt to purify the fragment from digests. Care should be taken that the truncation does not alter the biochemical properties of the protein, and hence give rise to misleading conclusions about the structure and function of the intact protein. For this reason, the biochemical function of the fragments should be thoroughly characterized.

Protocol 1. Identification of stable domains in native proteins by limited proteolysis

Equipment and reagents

- Protein sample, >5μg
- Protease

- SDS sample buffer and/or protease inhibitor heating block
- SDS–PAGE apparatus

Method[a]

1. Preparation of protein sample. The protein sample should be >90% pure by SDS–PAGE, and in a buffer which provides protein stability, and is compatible with the pH required for optimum activity of the protease. For preliminary experiments, 5 μg of protein is sufficient for digestion, as this can be readily analysed by SDS–PAGE. The amount of protein can be scaled up for subsequent sequencing and mass spectrometry.

2. Preparation of protease. In preliminary experiments, screen a number of different proteases to see which gives the most satisfactory results. Use proteases with relatively broad specificities and of reasonable purity. For digestion using thermolysin, prepare a 1 mg/ml solution in distilled water containing 2 mM Ca^{2+}.

3. Digestion reaction. To establish optimal conditions for the digestion, use a range of protease concentrations and reaction times. Add the protease to the protein sample at an enzyme/protein ratio ranging from 1:1 to 1:100. Include appropriate controls comprising protein and enzyme alone. For thermolysin digestion, incubate at 37°C or at room temperature, for times ranging from seconds to hours.

4. Termination of reaction. For SDS–PAGE analysis of proteolytic fragments, remove 5 μg aliquots after each incubation time-point into SDS sample buffer, with the addition of 2–5 mM EDTA to prevent further action of the thermolysin.

[a] For a more detailed description of proteolytic procedures, please refer to *Proteolytic enzymes* from the Practical Approach Series (3).

3.2 Preparation before and during crystallization experiments

Setting up a crystallization screen is neither time consuming nor difficult, but it is often tempting to proceed hastily without making some indispensable checks that will ultimately save time. On the other hand, it is probably best to commence crystallization trials early, before spending a considerable period ensuring that one has the best possible sample in hand. The list below suggests some basic tasks to perform before commencing crystallization, so that the best strategy can be chosen. Some of these can also be conducted in parallel with initial crystallization trials, to help explain the results and aid the design of further trials. The first two questions can be addressed by examining the published literature, while the remaining tasks may require some experimentation:

- ligands and effectors: should they be included in crystallization solutions?
- what crystallization conditions were used for related proteins?
- SDS–PAGE to determine purity, presence of contaminants
- biochemical assays to determine the functional integrity of the protein
- protein concentration assays to choose an appropriate concentration for crystallization
- determination of the total quantity of protein: this will influence the screening strategy to determine crystallization conditions
- mass spectrometry to determine the integrity of the preparation, and the possible presence of post-translational modifications
- native gel electrophoresis and size exclusion chromatography to determine the homogeneity with respect to oligomeric species
- non-reducing SDS–PAGE to assess the possibility of aberrant disulfide cross-linking
- dynamic light scattering to determine if the sample is monodisperse
- isoelectric focusing to determine possible heterogeneity of isoelectric properties and deamidation (which may not be evident from mass spectrometry)
- circular dichroism to assess if the protein is folded
- examination of storage conditions to provide a regime that prevents the sample changing with time (e.g. freezing in small aliquots)

4. Crystallization

Many parameters affect crystallization (see Section 2). When searching for crystallization conditions, one needs to choose a strategy that will sample many conditions using as little protein as possible. Unfortunately, there is no simple and straightforward path to fulfil these requirements. There are usually several parallel pathways that may be similarly efficient, and all will encounter multiple-choice crossroads along the way. Lateral thinking can be crucial, and it is important to have one's eyes open so as not to miss any leads that might point in the right direction. In this section, different practical aspects of crystallization will be described, culminating in a flow diagram (Section 4.14, *Figure 4*) that should prevent disorientation in the 'crystallization maze'.

4.1 Crystallization methods

The aim of any crystallization method is to bring the protein solution into a supersaturated state. The best methods achieve this slowly, allowing for the formation of numerous weak but specific contacts between the protein molecules, these being required for crystal growth. *Table 1* lists some of the commonly used crystallization methods. The most popular method is the hanging drop vapour diffusion technique, because it uses small amounts of protein (less than 1 μl protein solution per trial), therefore permitting more comprehensive screens of initial crystallization conditions. The processes

Table 1. Commonly used crystallization methods

Method	Principle	Microtechnique
Vapour diffusion	Droplet with protein solution equilibrated against reservoir containing crystallization agent at higher concentration, in a sealed container. Equilibration proceeds through vapour diffusion of water (and other volatile species).	Hanging drop Sitting drop
Dialysis	Protein solution, separated by a semi-permeable membrane, equilibrated against a solution with crystallizing agent that can pass the membrane.	Dialysis button
Free interface diffusion	Less dense solution poured gently over more dense solution (or protein solution over frozen crystallizing agent), left to equilibrate.	Small diameter tubes
Batch method	Protein solution is mixed with crystallizing agent; supersaturation is reached instantaneously.	Microbatch

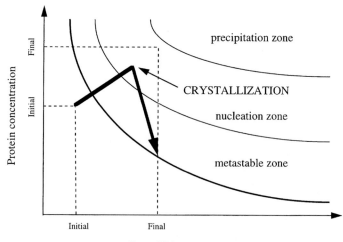

Figure 3. Solubility diagram in a vapour diffusion experiment, in the case when crystallization occurs.

394

occurring during a vapour diffusion experiment are presented in a solubility diagram in *Figure 3*.

Protocol 2.   Crystallization by hanging drop vapour diffusion

Equipment and reagents
- Crystallization plate (e.g. Linbro)
- Siliconized cover-slips
- Sealant (e.g. vacuum grease or Vaseline)

Method

1. Grease the rims of individual compartments of a crystallization plate with vacuum grease, or other sealant, dispensed from a syringe.

2. Fill up the reservoir with 1 ml of crystallizing agent solution.

3. Mix 1 µl of protein solution with 1 µl of reservoir solution on a cover-slip. Keep the droplet hemispherical and try not to introduce bubbles, as this will give a non-reproducible surface area and make it harder to see crystals.

4. Invert the cover-slip with your fingers (or tweezers) and gently press it on to the greased rim. Make sure it is completely sealed.

5. When the entire plate is set up, put plasticine in the corners to raise the plate cover above the cover-slips, thus avoiding contact between the cover plate and the cover-slips. Cover the plate and leave it in an undisturbed, temperature-controlled location (e.g. a cabinet or an incubator for trials at around room temperature, or a fridge or cold room for lower temperatures.)

4.2 Crystallizing agents

While many properties of the protein solution (e.g. temperature, pH) can be manipulated to bring it into a supersaturated state and induce crystallization, precipitating agents (precipitants) are the most commonly used. These can be grouped into three categories: *salts* (e.g. ammonium sulfate), *polymers* [e.g. polyethylene glycol (PEG)], and *organic solvents* [e.g. 2-methyl-2,4-pentanediol (MPD)], see *Table 2*. The three groups act by slightly different mechanisms, but share the common property of inducing supersaturation by competing with the protein molecules for available water molecules. Protein molecules must bind water to remain solvated, and when they are deprived of it, they are compelled to associate with other protein molecules, leading to the formation of crystals or precipitate. PEG additionally modifies the natural structure of the solvent itself, creating a complex network that excludes macromolecules. At low ionic strength, salt can have an inverse effect, increasing protein solubility with increasing salt concentration ('salting-in').

Table 2. Commonly used precipitants

Salts	Polymers	Organic solvents
Sulfate	PEG	MPD
Ammonium	4000	Ethylene glycol
Lithium	8000	Ethanol
Sodium	400	*t*-Butanol
Magnesium	10 000	Isopropanol
Nickel	20 000	Dioxane
Formate	PEG monomethyl ether	1,6-Hexanediol
Sodium	2000	2,5-Hexanediol
Magnesium	5000	Methanol
Citrate	550	1,3-Propanediol
Sodium	Polypropylene glycol P400	Acetone
Ammonium	Polyethylenimine	Acetonitrile
Chloride	Jeffamine M-600	Dimethyl sulfoxide
Sodium	Polyamine	1,3-Butyrolactone
Ammonium		
Potassium		
Acetate		
Sodium		
Ammonium		
Potassium, sodium tartrate		
Sodium sulfite		
Phosphate		
Sodium, monobasic		
Potassium, monobasic		
Ammonium dihydrogen		
Sodium bromide		
Ammonium nitrate		
Cetyltrimethyl ammonium salts		

The only essential requirement of a precipitant is that it does not perturb the structural integrity of the protein.

4.3 Initial screening

In the initial crystallization trials, there is a plethora of possible choices of conditions. The most popular method for initial screening has become the *sparse matrix sampling* technique, which uses a set of conditions based on successful crystallization of other proteins to sample a large range of buffer, pH, precipitant, and additive variables (4). The ease, speed, availability of commercial kits (Crystal Screen, Hampton Research), and high success rates account for the popularity of this technique. The original 50 conditions have been extended to more general conditions and more specialized screens (e.g. low ionic strength screen, membrane protein screens), and many variations have been developed in different laboratories.

A more rational statistical approach to reducing the number of variables, while thoroughly sampling the multidimensional space, is the *incomplete factorial search* (5). The advantage of statistical approaches over sparse matrix screens is that experiments can be quantitatively assessed, identifying variables that are favourable or unfavourable for crystallization, and even positive interactions between different variables (6). Another method derived from the incomplete factorial search idea uses a combination of random sampling, quantitative evaluation, and weighted sampling (7). Alternative sampling methods include orthogonal arrays (8).

Reverse screening (9) is a screening procedure that attempts to minimize the amount of protein used, and is particularly useful for situations where only a limited amount is available. This is required by initially determining the solubility of the protein using a *footprint screen* (10), then establishing conditions under which the protein is highly supersaturated and hence conducive to nucleation, and finally optimizing the crystals by modifying solubility, introducing ligands, and seeding.

4.4 Crystal optimization

Crystals obtained in the initial screens will usually need optimization. This is done by systematically varying individual parameters in small increments while keeping others constant, centred around the conditions where the initial crystals were obtained. This is in fact a *full factorial screen*, also known as a *grid screen*. *Incomplete factorial experiments* can be used instead to reduce the number of experiments. It is best to set a hierarchy of parameters to be evaluated, initially evaluating the effect of changing one parameter and, based on these results, proceeding to the next in an iterative manner. Precipitant concentration, pH, and temperature are the most important parameters to be optimized at the outset. The goal is to obtain single crystals of sufficient size (>0.1 mm in all dimensions) for a diffraction experiment.

If the optimization of the initial parameters does not produce suitable crystals, there are once again multiple choices of how to proceed, and a hierarchical list should be compiled (see Section 4.14). At this point, it is desirable to check that the protein integrity has been maintained throughout the trial. One can:

- change the protein concentration
- substitute the original crystallization agents with similar ones (e.g. ammonium sulfate with lithium sulfate, PEG 4000 with PEG 8000, one buffer for another at the same pH)
- add new crystallization agents (see Section 4.5)
- change the crystallization method [change the initial ratio of protein solution to reservoir solution in the drop, change the volume of the drop, try a sitting drop, try a different crystallization method (Section 4.1), try seeding (Section 4.6), try crystallization in gels or oils (Section 4.7), or filter the protein solution to reduce nucleation (11, 12)].

Another way to improve crystallization that should be considered is to alter the protein sample itself (see Section 4.9).

4.5 Additives

An additive is any chemical that modifies the original set of crystallizing agents in order to attempt to obtain and/or improve crystals. An additive can be:

- a structurally required metal ion (often Mg^{2+} or Ca^{2+})
- a reducing agent (e.g. DTT)
- a co-factor such as ATP (or its non-hydrolysable analogues, AMPPNP and AMPPCP, used for crystallization of p38 MAP kinase and phosphorylase kinase)
- a substrate peptide
- a heavy atom compound
- a phosphate analogue (e.g. tungstate, molybdate, vanadate)

Once crystals have been obtained and flash-freezing is attempted (see Section 5), cryoprotectants could be required as additives for crystal growth (13), especially if crystals crack when transferred to the solution containing cryoprotectant. In general, additives will be present in small amounts, and will subtly affect crystal growth rather than acting as a precipitant. Sometimes, additives can have profound effects, and may even prevent crystal growth, or denature the protein (especially heavy atom compounds).

The presence of flexibly linked domains may result in conformational heterogeneity and prevent crystallization, or result in poorly diffracting crystals. One approach to circumvent this problem is to identify individual domains or sets of domains that may represent more rigid entities, and express these for crystallization (Section 3.1). However, the disadvantage is that information on the domain arrangements is lost, and this may be important for understanding the function of modular proteins. As studies of modular arrangements will undoubtedly be an important topic in future years, methods that stabilize flexible proteins will be increasingly needed. Cosmotropic (i.e. stabilizing) solute additives, such as polyhydric alcohols, sugars, and amino and methylamino acid, have been used to preserve protein structures (14–16).

An efficient way to screen the effects of diverse additives is to use the commercially available additive screens (Hampton Research). Some common additives are listed below, grouped according to their effects:

- protein stability enhancers: glycerol, polyethylene glycols (e.g. PEG 4000), sucrose, glucose, sorbitol, sarcosine, *N*-acetyl glucosamine, ammonium sulfate
- reducing agents: DTT, 2-mercaptoethanol, glutathione, cysteine
- protein solubility enhancers: detergents, salts

- crystal contact mediators: heptane triol, negative and positive ions, polyamines (spermine, spermidine), polyanions, poly(lysine), hexadecanoic acid, 1,7-diaminoheptane, dimethyl ethylammonium propane sulfonate, amino acids, betaine, taurine
- metal chelators: EDTA, EGTA
- crystal growth rate reducers: dioxane, dimethyl sulfoxide, PEG 200, ethanol, DMF, *t*-butanol, NaCl
- protease inhibitors: PMSF, leupeptin, pepstatin, benzamidine, aprotinin, antipain
- chaotropes: guanidine hydrochloride, urea, trimethylammonium hydrochloride, phenol
- co-precipitants: PEG + salt (LiCl), PEG + isopropanol, ammonium sulfate + thiocyanate, others + MPD

4.6 Seeding

Seeding is a technique where crystal 'seeds' are introduced into a fresh protein solution under conditions of supersaturation, where crystal growth but not nucleation can occur (the metastable region). This technique is used to optimize the crystals obtained by spontaneous nucleation, or in situations where further attempts at obtaining crystals with a new protein batch have failed. It also has other uses (17); for example, to obtain crystals of a complex where crystals of a native protein are available (cross-seeding). There are two major seeding techniques: macroseeding (*Protocol 3*), where individual crystals are washed and introduced as seeds, and microseeding (*Protocol 4*), where microscopic crystal seeds are introduced. Streak seeding (see *Protocol 5*) is a variant of microseeding, where an animal whisker is used for seed transfer. The solution in which the seeds are introduced should normally be pre-equilibrated, with the precipitant and/or protein concentrations lower than in the conditions where crystallization spontaneously occurs, but high enough to prevent the seeds from dissolving. Finding the optimum conditions will require screening.

Protocol 3. Macroseeding

Equipment and reagents
- Multiwell sitting drop plate, or equivalent
- Pre-equilibrated crystallization drop

Method
1. Under a microscope, pick up a crystal from a drop using a capillary attached with rubber tubing to a mouthpiece, syringe, or variable pipette (e.g. Gilson Pipetman).

Protocol 3. *Continued*

2. Wash the crystal repeatedly in stabilizing solutions (e.g. sequentially in at least four adjacent depressions of a multiwell sitting drop plate, each containing 0.1 ml stabilizing solution).

3. Transfer the crystal into a pre-equilibrated drop.

Protocol 4. Microseeding

Equipment and reagents

• Pre-equilibrated crystallization drop   • Seed beads or equivalent

Method

1. Prepare a seed stock by crushing a few crystals [e.g. in a glass homogenizer or using seed beads (Hampton Research)]. The 'seed stock' can be stored, in most cases, for future use.

2. Make several dilutions of the seed stock (usually in the 10^{-3}–10^{-7} range) in a precipitant solution that will not dissolve the seeds.

3. Introduce a small amount of seed solution into a pre-equilibrated drop.

Protocol 5. Streak seeding

Equipment and reagents

• Animal whisker   • Wax
• Thick-walled capillary   • Pre-equilibrated crystallization drop

Method

1. Make the probe by attaching an animal (e.g. cat) whisker to a thick-walled capillary using wax.

2. Wash the probe in ethanol and water, and wipe dry.

3. Touch an existing crystal (a gentle touch is usually sufficient) to dislodge some seeds.

4. Run the probe through the pre-equilibrated drop in a straight line. Crystals will grow along the streak line.

5. You can streak subsequent drops without loading new seeds. This will achieve seed dilution.

6. Whiskers can be reused, but will transfer fewer and fewer seeds with time.

4.7 Special methods: crystallization in gels, oils, and using robotics

Crystal growth in gels sometimes produces better quality crystals, especially in the case of fragile crystals. Silica gels and agarose gels have been used with success (18). Both are normally used in a free-interface diffusion set-up, where the precipitant solution is layered on top of a gelled protein solution, but the agarose gels can also be used in a vapour diffusion set-up. Oils have been used for 'containerless' crystallization, leaving the crystallization droplet suspended between two immiscible oils. Oils are also used for sealing sitting drops in automated screens using robotics (1).

4.8 Crystallization of membrane proteins

Crystallization of membrane proteins follows procedures similar to those for soluble proteins, once the membrane lipids have been replaced by mild, non-denaturing detergents to render these proteins soluble. The major difficulty with respect to crystallization appears to be that the surfaces covered by detergent are unable to form crystal contacts, leaving only the remaining polar surfaces of the protein. Recent novel approaches include co-crystallization with antibody (Fab or ScFv) fragments to increase the polar surface area (19), and the introduction of lipid cubic phases to nucleate and support crystal growth (20). For most membrane-bound protein kinases and phosphatases, the best approach may be to circumvent the problem of membrane protein crystallization by removing the membrane spanning region(s) via proteolytic cleavage or recombinant expression, unless the region of interest includes the membrane-spanning portion. As most membrane-anchored protein kinases and phosphatases contain only single membrane-spanning segments, and the extra-membrane domains are soluble if expressed on their own, this suggests that there will not generally be a rigid association with the membrane-spanning regions; consequently, crystals of full-length protein may be extremely difficult to obtain. However, crystallographic studies on the truncated extra-cellular and intracellular regions of receptors have already provided substantial insights into the transmembrane signal transduction process [e.g. the insulin receptor (21)].

4.9 Drastic actions

If after extensive screening no crystals are obtained, or the available crystals remain unsuitable for diffraction experiments, one has to return to the protein itself. Further inspection of the protein quality may reveal possible reasons for crystallization problems, e.g. insufficient purity or heterogeneity of oligomeric species (Section 3.2). In such cases attempts should be made to rectify these problems, and crystallization repeated with the improved protein sample. If conformational heterogeneity is thought to be the problem, more rigid fragments may be identified by limited proteolysis (Section 3.1). Further possibilities are to change the source of the protein (organism or tissue), the protein expression system, or the protein family member. For example, myosin light chain kinases have proved extremely difficult to crystallize, but crystals of the related twitchin kinase were readily obtained. Introducing mutations may have similar effects, by increasing protein stability or solubility, or by aiding the formation of crystal contacts [e.g. (22, 23)]. The presence of glycosylation used to be thought of as unfavourable for crystallization. Recent results, however, suggest that the heterogeneity of glycosylation may be more of a problem than the glycosylation itself, and in some cases sugar chains can mediate important crystal contacts. Glycosylation can be manipulated by enzymatic treatment or expression in appropriate hosts, and will not usually be an issue for protein kinases and phosphatases. Proteins can also be chemically modified to facilitate crystallization [e.g. by alkylation to prevent cysteine oxidation (24), or by methylation of lysine residues (25)]. Other alternatives are co-crystallization with antibody fragments (19) and physiological binding partners (26), or crystallization as fusion proteins (27). Finally, heterologous nucleation catalysts may be introduced to facilitate crystallization (28, 29).

4.10 Evaluation of crystallization experiments

The results of crystallization trials are examined using a light microscope, ideally a stereo-zoom dissecting microscope with a crossed polarizing attachment, and at least $10 \times$ magnification. With a higher magnification, less obvious crystalline specimens will be more readily detected. Normally, crystallization experiments should be examined immediately after set-up (to find any samples that precipitated immediately after mixing, and to spot any objects that may later be mistaken for crystals), every day for the next few days, and less frequently thereafter. In vapour diffusion experiments, equilibrium will be reached in days to weeks, depending on the crystallization mixture. Crystals have, however, been reported to appear months or years after set-up! Unexpected processes, such as proteolysis, oxidation, or improperly sealed cover-slips, are usually responsible for crystal growth in such cases. It is therefore advisable that experiments be occasionally checked long after set-up, and not discarded for a good length of time.

There are usually three easily distinguishable outcomes of a crystallization experiment:

- *a clear drop*, which suggests that nucleation or precipitation zones are not reached in the phase diagram; the precipitant (and/or protein) concentration therefore needs to be increased

- *an amorphous precipitate*, which suggests that the equilibrium is too far beyond the crystal nucleation zone; the precipitant (and/or protein) concentration needs to be decreased (this does not, however, guarantee that crystals can actually be obtained)

402

- *crystals*, the quality of which can be improved by using grid screens and additives

It is not always straightforward to distinguish between amorphous and crystalline material. Microcrystals can easily be overlooked. A good stereo-microscope helps, and one should look for defined edges in crystals, and optical properties such as birefringence (refraction of polarized light). Birefringence depends, however, on the crystal order and symmetry (e.g. it is absent in crystals with cubic symmetry). Streak seeding can be used to prove if a precipitate is crystalline, but if it does not yield crystals, the results will be inconclusive.

Even if no crystals have been produced, examination of crystallization trials can provide valuable information. Certain types of precipitation are considered to be more promising than others. A fine, often yellowish or brownish precipitate is generally considered a bad sign, indicating denatured protein. Granular or oily precipitates are a more positive sign, and suggest further optimization using similar crystallizing agents. Examination also reveals information about protein stability versus pH, precipitant type, and time. For example, if all the drops at pH 5 or lower contain fine brownish precipitate, one can conclude that the protein is unstable under such conditions and that low pH should be avoided in further experiments.

Factorial experiments allow statistical evaluation of crystallization results, but the main problem is how to apply the appropriate scoring scale to quantify the results. Scoring scales containing from 3 to 9 values have been suggested (6, 7, 30) (*Table 3*).

Table 3. An example of a crystallization experiment evaluation scale

Score	Observation
1	Cloudy/yellowish precipitate
2	Particulate or gelatinous precipitate
3	Spherulites (radial aggregates of microscopic needles)
4	Crystal needles
5	Crystal plates
6	Crystal prisms

4.11 Crystal analyses

If crystals are obtained, some key initial questions must be addressed. Are the crystals protein crystals, or crystals of some small molecule ingredient of the crystallization solution? Is it the protein of interest that has crystallized, or an impurity, or a proteolytic product? Some common tests (and caveats) to answer these questions are listed in *Table 4*.

Table 4. Crystal analyses

Test method	Resulting information	Caveat
Examination of reservoir solution	If similar crystals form in the reservoir solution, they are not protein crystals	Crystallization drop ingredients not present in the reservoir may form crystals
Control crystallization experiment with protein absent	If similar crystals form in the control experiment, they are not protein crystals	Protein may induce small molecule crystal growth
Click-test: touch the crystal with a microtool	Protein crystals are soft. Small molecule crystals are hard and 'click' upon breaking	Need good size crystals
Staining	Different stains work through different principles, but will differentially stain protein (or macromolecule) and small molecule crystals	Need good size crystals
X-ray diffraction	If diffraction observed, unit cell dimensions will discriminate between macromolecule and small molecule crystals	Need diffracting crystals
Dissolve crystals	Analyze solution with SDS–PAGE, immunoblot, sequencing, mass spectrometry	Need to wash crystals well to prevent transferring protein from crystallization solution
Crystal density measurements (e.g. Ficoll gradient)	Will discriminate macromolecule from small molecule crystals, and give further information on number of molecules in asymmetric unit	Need good size crystals

4.12 Co-crystallization and soaking

Co-crystallization and soaking are used to obtain crystals of protein–ligand complexes. Co-crystallization is identical to a regular crystallization experiment, but with the ligand added to the protein solution. For small molecule ligands, an excess of the ligand often produces the best results. Soaking, on the other hand, uses crystals grown in the absence of ligand; these are transferred (directly or gradually) into a ligand-containing solution and left soaking for hours, days, or sometimes weeks. Protein crystals are typically 30–80% solvent, and contain large solvent channels, allowing the ligand to diffuse into the crystal and bind to the protein. The advantages of soaking are that crystallization screening is not required, and that the complex crystals will be isomorphous with the native ones (i.e. have similar unit cell dimensions), facilitating structure determination. The disadvantage is that the binding site may be blocked by existing crystal contacts, resulting in either no binding, or crystal break-up; also, larger ligands may not be able to diffuse through the solvent channels. Crystal cross-linking by glutaraldehyde has been used to prevent fragile crystals from cracking during soaking procedures (31).

4.13 Protein kinases and phosphatases already crystallized

Crystallization is governed by the formation of specific contacts between protein molecules, and this may alter even if only slight changes are made to the protein. However, knowledge of successful crystallization conditions for a protein has proved to be useful in the past for the crystallization of mutants, ligand complexes, and even homologous proteins. For these reasons, we have assembled information on protein kinases and phosphatases that have been crystallized in *Table 5*. This information can be used as a starting point to reproduce crystallization of a previously crystallized molecule, or of variants and ligand complexes, and to design factorial screens for related proteins. Differences in the protein purification procedure and purity, in the reagents used, and the crystallization methodology can influence the outcome of a crystallization experiment, so the published conditions should be used only as a starting point, and conditions should be reoptimized.

An examination of the crystallization conditions used for Fab fragments of immunoglobulins, for example, shows a narrow set of conditions where these crystals grow (32). Although this is not clearly evident for protein kinases or phosphatases, examination of crystallization conditions for a closely related kinase or phosphatase class may suggest a starting set of conditions for screening. A few general observations can be made based on inspection of *Table 5*:

- the hanging drop is clearly the most common crystallization method;
- the pH range used is relatively narrow (pH 5 to 8.8, even narrower for protein kinases);
- polyethylene glycols and ammonium sulfate are the most successful precipitants;
- a reducing agent is usually a required additive.

A further source of useful crystallization conditions is the biological macromolecule crystallization database (BMCD, http://ibm4.carb.nist.gov: 4400/bmcd.html) (32, 33).

4.14 A flow diagram for crystallization

A flow diagram to guide the investigator through the labyrinth of possible routes to crystallization is given in *Figure 4*. Crystallization is not a uni-directional protocol, but an iterative procedure where many options are possible at any one stage. The best pathway through the scheme will be different in any specific case.

Table 5. Crystallized protein kinases and phosphatases

Protein kinases

Protein	Crystallization (reservoir) solution	T (°C)	Protein concentration	Ligand	Crystallization method	Comments	Ref
cAMP-dependent protein kinase, catalytic subunit, mouse	10 mM DTT, 8% PEG 400	4	10 mg/ml in: 150 mM NH_4(OAc), 50 mM Bicine (pH 8)	PKI(5-24) (3 × molar excess) in 10 mM DTT, ± $MgCl_2$ (5 × molar excess), ATP (20 × molar excess)	Hanging drop (equal volumes of protein solution, reservoir solution, PKI solution)	15% methanol added to reservoir	43
cAMP-dependent protein kinase, catalytic subunit, mouse	15% MPD, M $(NH_4)_2SO_4$, 0.1 M Bicine (pH 8)	4	0.5 mM in: 4% MPD, 0.1 M Bicine (pH 8)	3 mM adenosine	Hanging drop		44
cAMP-dependent protein kinase, catalytic subunit, porcine heart	15% methanol (pH 5.6)	5	18 mg/ml in: 75 mM LiCl, 0.1 mM EDTA, 1 mM DTT, 1.5 mM octanoyl-N-methylglucamide, 25 mM MES–Bis–Tris (pH 6.4)	PKI(5-24) (molar excess), staurosporine (molar excess)	Hanging drop		45, 46
cAMP-dependent protein kinase, catalytic subunit, bovine, recombinant	15% methanol	5	20 mg/ml in: 75 mM LiCl, 0.1 mM EDTA, 1 mM DTT, 1.5 mM octanoyl-N-methylglucamide, 25 mM MES–Bis–Tris (pH 6.3-6.6)	1 mM PKI (5-24), 1.5 mM inhibitor (H7, H8 or H89)	Hanging drop		47

Protein	Precipitant/reservoir	Temperature (°C)	Protein solution	Additive	Method	Reference
cAMP-dependent protein kinase, catalytic subunit, mouse, polyhistidine-tagged	20% MPD, 0.1 M Bicine (pH 8)	4	0.25 mM in: 4% MPD, 0.1 M Bicine (pH 8)	0.75 mM PKI(5–24), 4 mM adenosine	Hanging drop (20 μl)	48
MAP kinase ERK2	18% PEG 8000, 0.2 M (NH$_4$)$_2$SO$_4$, 5 mM DTT, 0.1 mM NaN$_3$, 50 mM MES (pH 5.9)	21	6 mg/ml in: 0.2 M NaCl, 5 mM DTT, 1 μg/ml aprotinin, 2 μg/ml leupeptin, 120 μg/ml benzamidine, 2 μg/ml antipain, 2 μg/ml pepstatin, 20 mM Tris–HCl (pH 7.5)		Hanging drop (5 μl protein + 5 μl reservoir solution)	49
MAP kinase ERK2, phosphorylated	20% PEG 4000, 0.2 M (NH$_4$)$_2$SO$_4$, 0.1 M MES (pH 6.5)		8 mg/ml in: 0.15 M NaCl, 5 mM DTT, 0.1 mM EDTA, 0.05% n-octyl glucoside, 10 mM Hepes (pH 7.5)		Hanging drop (equal volumes of protein and reservoir solutions)	50
MAP kinase p38	0.56 M sodium citrate, 130 mM (NH$_4$)$_2$SO$_4$, 40 mM Hepes (pH 7.5)	RT	40 mg/ml in: 0.1 M NaCl, 10 mM DTT, 5% glycerol, 50 mM Hepes (pH 7.5)		Hanging drop (2 μl protein + 3 μl reservoir solution)	51
MAP kinase p38	18% PEG 8000, 0.2 M Mg(OAc)$_2$, 0.1 M Hepes (pH 7.4)		10 mg/ml in: 50 mM NaCl, 1 mM EDTA, 10 mM DTT, 10 μg/ml leupeptin, 1 mM benzamidine, 2 μM pepstatin, 25 mM Hepes (pH 7.4)		Microseeding	52

Table 5. *Continued*

Protein	Crystallization (reservoir) solution	T (°C)	Protein concentration	Ligand	Crystallization method	Comments	Ref
MAP kinase p38	28–32% PEG 1500, 5 mM DTT	RT	6 mg/ml	1.4 mM inhibitor ± 1.4 mM AMPPNP or AMPPCP	Hanging drop (1.4 μl protein + 1.4 μl reservoir solution)	Seeding accelerates nucleation	53
Phosphorylase kinase (1-298)	5% PEG 8000, 10% glycerol, 10 mM DTT, 0.02% NaN₃, 50 mM Hepes/NaOH (pH 6.9)	14 or 16 (with ATP)	10–12 mg/ml, 2-3 mg/ml with substrate peptide in: 2% glycerol, 10 mM DTT, 0.02% NaN₃, 0.1 mM EDTA, 10 mM Hepes/NaOH (pH 8.2)	3 mM ATP, 10 mM Mg(OAc)₂; or 3 mM AMPPNP, 10 mM MnCl₂; or 10 mM substrate peptide, 3 mM AMPPNP, 10 mM MnCl₂	Hanging drop (3 μl protein + 3 μl reservoir solution; 1 μl protein + 1 μl reservoir solution for substrate peptide)		54, 55
Cyclin-dependent kinase 2	200 mM Hepes (pH 7.4)	4	10 mg/ml in: 1 mM EDTA 20 mM Hepes (pH 7.4)		Sitting drop (equal volumes of protein and reservoir solutions)	Gradually increase Hepes concentration in reservoir solution to 0.8 M	56
Cyclin-dependent kinase 2	50 mM NH₄(OAc), 10% PEG 4000, 0.1 M Hepes (pH 7.4)	20	10 mg/ml in: 15 mM NaCl, 10 mM Hepes (pH 7.4)	± staurosporine	Hanging drop (1 μl protein + 1 μl reservoir solution)		57
Cyclin-dependent kinase 2	28% saturated (NH₄)₂SO₄ (35% with p27), 1 M KCl, 5 mM DTT, 10 mM ATP, 40 mM Hepes (pH 7)	4	25 mg/ml, 30 mg/ml for p27, in: 0.2 M NaCl, 5 mM DTT, 40 mM Hepes (pH 7)	Cyclin A (173-432) or cyclin A (173-432) + p27 (22-106)	Hanging drop (equal volumes of protein and reservoir solutions)	Unphosphorylated and phosphorylated grow under same conditions	26, 58

Kinase	Crystallization conditions		Protein solution	Additives		Crystallization method	Ref
Cyclin-dependent kinase 2	17-19% PEG monomethyl ester 5000, 0.1 M KCl, 0.1 M Tris (pH 7.5)	20	15-20 mg/ml in: 150 mM NaCl, 5% glycerol, 1 mM EDTA, 1 mM DTT, 10 mM Hepes (pH 7.4)			Hanging drop (equal volumes of protein and reservoir solutions)	59
Casein kinase-1, S. pombe (1-298)	1.55 M (NH$_4$)$_2$SO$_4$, 50 mM sodium citrate (pH 5.6)	16	0.3 mM in: 0.1 mM EDTA, 0.1% 2-mercaptoethanol, 1 mM DTT, 0.2 M NaCl, 10 mM MOPS (pH 7.0)	6 mM ATP, 1.5 mM MgCl$_2$	CksHs1	Hanging drop (5 μl protein + 5 μl reservoir solution)	60
Casein kinase-1, S. pombe (1-298)	1.4-1.7 M (NH$_4$)$_2$SO$_4$, 2% methanol, 2.5-25 mM sodium citrate (pH 4.6)	16	10 mg/ml in: 0.1 mM EGTA, 0.1% 2-mercaptoethanol, 1 mM DTT, 0.05 M NaCl, 10 mM MOPS (pH 7.0)	2 mM CKI7 in 2% dimethyl sulfoxide		Hanging drop (mix protein and inhibitor, equilibrate against reservoir)	61
Casein kinase I δ, rat (1-317)	14-16% PEG 3400, 50 mM sodium citrate, 50 mM potassium phosphate (pH 6.8)		12-15 mg/ml in: 1 mM EDTA, 1 mM DTT, 0.2 M NaCl, 5 mM β-octyl glucoside, 50 mM Tris-HCl (pH 7.5)			Sitting drop (3 μl protein + 3 μl reservoir solution)	62
Protein kinase CK2, Zea mays, catalytic subunit	25% PEG 3400, 200 mM sodium acetate, 0.1 M Tris-HCl (pH 8.0)		8 mg/ml in: 0.5 M NaCl, 7 mM 2-mercaptoethanol, 25 mM Tris-HCl (pH 8.5)			Vapor diffusion (3 μl protein + 3 μl reservoir + 3 μl 6 mM ATP + 3 μl 1.5 mM MgCl$_2$ solutions)	63
Twitchin kinase, C. elegans (5890-6262)	20-22% PEG 4000, 0.2 M MgSO$_4$, 10-15% glycerol,	4	10 mg/ml in: 50 mM NaCl, 1 mM DTT,			Hanging drop (2-3 μl protein + equal volume	64

Table 5. *Continued*

Protein	Crystallization (reservoir) solution	T (°C)	Protein concentration	Ligand	Crystallization method	Comments	Ref
	0.1 M Hepes (pH 7.25–7.5)		1 mM EDTA, 1mM PMSF, 25 mM Hepes (pH 7.5)		reservoir solutions)		
Twitchin kinase, *C. elegans* (5890–6380)	4 M sodium formate	21	5 mg/ml in: 20 mM Tris (pH 7.4), 200 mM NaCl, 2 mM DTT		Hanging drop (2 μl protein + 2 μl reservoir solutions)		65
Twitchin kinase, 387-residue fragment, *Aplysia*	15% PEG 4000, 1 mM DTT, 0.1 M sodium citrate (pH 5)	21	2.4 mg/ml in: 50 mM NaCl, 1 mM DTT, 1 mM EDTA, 1 mM PMSF, 25 mM Hepes (pH 7.5)		Hanging drop (4 μl protein + 4 μl reservoir solutions)		65
Calcium/calmodulin-dependent protein kinase I (1–320)	2 M (NH₄)₂SO₄, 2% PEG 400, 0.1 M Hepes (pH 7.5)	21	25 mg/ml in: 20 mM Tris–HCl (pH 8.0), 1 mM DTT, 0.01 mg/ml leupeptin, 0.02% NaN₃, 70 mM MgSO₄		Hanging drop (1 μl protein + 1 μl reservoir solution)		66
Insulin receptor tyrosine kinase domain (978–1283)	20% PEG 6000, 0.2 M malate-imidazole (pH 7.5)	21	10 mg/ml		Hanging drop (equal volumes of protein and reservoir solutions)	Macroseeding	21
Insulin receptor tyrosine kinase domain (978–1283), phosphorylated	22% PEG 8000, 2% ethylene glycol, 0.1 M Tris–HCl (pH 7.5)	4	10 mg/ml in: 200 mM NaCl, 0.2 mM Na₃VO₄, 20 mM Tris–HCl (pH 7.5)	2 mM AMPPNP, 6 mM MgCl₂, 0.5 mM peptide substrate	Hanging drop (2 μl protein + 2 μl reservoir solution)		67
FGF receptor tyrosine kinase	16% PEG 10000, 0.3 M (NH₄)₂SO₄,	4	10 mg/ml in: 10 mM NaCl,	± 10 mM AMP-PCP,	Hanging drop (2 μl protein +		68

Protein	Crystallization conditions	Temperature (°C)	Protein solution	Ligands	Method	Reference
domain (456-765)	5% ethylene glycol, 0.1 M bis-Tris (pH 6.5)				(2 µl protein + 2 µl reservoir solution)	
Lck tyrosine kinase domain (225-509)	2 M $(NH_4)_2SO_4$, 0.1 M bis-Tris (pH 6.5)	4	2 mM DTT, 10 mM Tris–HCl (pH 8) 0.64 mg/ml	20 mM $MgCl_2$, 0.1 mM ATP, 0.1 mM substrate peptide	Hanging drop (equal volumes protein and reservoir solutions); Macroseeding in 0.51 mg/ml protein, 0.08 mM ATP, 0.08 mM substrate peptide, 0.6 M $(NH_4)_2SO_4$, 50 mM bis-Tris (pH 6.5), reservoir 1.6 M $(NH_4)_2SO_4$, 0.1 M bis-Tris (pH 6.5)	69
c-Src tyrosine kinase (human, 86-536)	0.8 M sodium tartrate, 20 mM DTT, 50 mM Pipes (pH 6.5)	RT	15 mg/ml in: 0.1 M NaCl, 5 mM DTT, 20 mM Hepes (pH 7.6)	10 mM quercetin or 2AMP-PNP	Hanging drop (1 µl protein + 1 µl reservoir solution); Streak seeding	70
c-Src tyrosine kinase (chicken, 81-533)	16–18% PEG 2000, 10% PEG 400, 0.3 M NaCl, 1 mM NaN_3, 1 mM EDTA, 1 mM DTT, 0.1 M Tris (pH 8-8.2)	RT		0.1 mM ATP, 0.1 mM substrate peptide	Macroseeding	71
Hck tyrosine kinase	150 mM calcium acetate, 7% PEG 8000, 16% ethylene glycol, 0.1 M cacodylate (pH 6.5),		50 mg/ml in: 1mM DTT, 10 mM Hepes (pH 7.5)	10 mM quercetin or 2AMP-PNP	Hanging drop (1 µl protein + 1 µl reservoir solution); Streak seeding	72
Protein phosphatases						
Protein phosphatase-1	0.2 M $MgCl_2$, 32% PEG 4000, 65 mM Tris (pH 8.8)		8 mg/ml in; 50 mM Tris, 500 mM NaCl, 10 mM 2-mercaptoethanol, 0.9 mM $MnCl_2$, 0.1 mM EGTA	Microcystin (4× molar excess)	Hanging drop (1 µl protein + 0.8 µl reservoir); 0.2 µl 1 M 1,7-diaminoheptane improves morphology	73

Table 5. *Continued*

Protein	Crystallization (reservoir) solution	T (°C)	Protein concentration	Ligand	Crystallization method	Comments	Ref
Protein phosphatase-1	50 mM Tris, 3% monomethyl PEG 2000, 5 mM DTT		5 mg/ml in: 300 mM NaCl, 2 mM DTT, 0.4 mM MnCl$_2$, 10 mM Tris (pH 7.8)	Microcystin (4× molar excess)	Hanging drop (mix equal volumes protein and 0.1 M Tris (pH 8.5), 12.5 mM sodium tugstate, 6% monomethyl PEG 2000, 5 mM DTT, invert 8 µl over reservoir	Microseeding	74
Protein phosphatase-1	2 M (NH$_4$)$_2$SO$_4$, 2% PEG 400, 2 mM DTT, 0.1 M Hepes (pH 7.5)	20	8 mg/ml in: 300 mM NaCl, 2 mM DTT, 0.4 mM MnCl$_2$, 10 mM Tris-HCl (pH 7.8)	G-subunit (63-75) (3× molar excess)	Hanging drop (2 µl protein + 2 µl reservoir solution)		75
Calcineurin A (17–392)-calcineurin B	8% PEG 8000, 0.1 M potassium phosphate, 20 mM 2-mercaptoethanol	4	25–55 mg/ml in: 0.1 M MnCl$_2$, 0.1 M CaCl$_2$, 2 mM 2-mercaptoethanol, 25mM Tris-HCl (pH 8.0)	FKBP12-FK506	Hanging drop (6 µl protein + 2 µl reservoir solution)		76
Calcineurin A-calcineurin B	PEG 6000, CaCl$_2$	RT		FKBP12-FK506			77
Calcineurin A-calcineurin B	6% PEG 6000, 10 mM HoCl$_3$, 0.1 M bis-Tris, 1mM DTT						77
Protein serine/threonine phosphatase 2C	8-12% PEG 8000, 15% glycerol, 2 mM DTT, 50 mM potassium phosphate (pH 5.0)	4	15 mg/ml in: 50 mM NaCl, 2 mM DTT, 1 mM MnCl$_2$, 10 mM Tris-HCl (pH 7.5)		Hanging drop (equal volumes protein and reservoir solutions)		78

Protein	Reservoir/precipitant solution	Temperature (°C)	Protein concentration	Method	Notes	Reference
Protein–tyrosine phosphatase 1B	0.2 M Mg(OAc)$_2$, 14% PEG 8000, 0.1 M Hepes (pH 7.5)	4	10 mg/ml		Conditions same for C215S mutant + substrate peptide (3-fold excess), mixed before crystallization (79)	80
Receptor protein tyrosine phosphatase α, domain 1 (202-503)	0.3 M NH$_4$(OAc), 12% PEG 8000, 50 mM NaOAc (pH 5.0)		15–20 mg/ml	Hanging drop		81
Receptor protein tyrosine phosphatase μ, domain 1 (874-1168)	2 mM DTT, 0.5 mM EDTA. 0.75 M sodium citrate (pH 5.5)	4	7.5 mg/ml			82
SHP-2 tyrosine phosphatase (1-527)	13% PEG 4000, 20 mM DTT, 0.1 M Tris (pH 8.5)	RT	15 mg/ml in: 0.1 M NaCl, 20 mM DTT, 0.1 M Tris (pH 8.5)	Sitting drop (equal volumes protein and reservoir solutions)	5 mM trimethyl lead acetate and 1.2 mM cetyltrimethyl ammonium bromide in drops improves morphology; macroseeding	83
Protein tyrosine phosphatase, Yersinia, (163-468)	10-15% PEG 1500, small amounts MPD	22	12 mg/ml in: 0.1% 2-mercaptoethanol, 1 mM imidazole (pH 7.2)	Hanging drop (equal volumes of protein and reservoir solutions)	Optional macroseeding	84
Protein tyrosine phosphatase, Yersinia, (163-468)	22% PEG 4000, 200 mM Li$_2$SO$_4$, 10% isopropanol, 0.1% 2-mercaptoethanol, 0.1 M Tris–HCl (pH 8.5)		20 mg/ml	Hanging drop (equal volumes of protein and reservoir solutions)	1 mM sodium tungstate	85

Table 5. *Continued*

Protein	Crystallization (reservoir) solution	T (°C)	Protein concentration	Ligand	Crystallization method	Comments	Ref
Protein tyrosine phosphatase, *Yersinia*, (163-468)	220 mM NaNO$_3$, 27% PEG 1500, 10% MPD, 0.1% 2-mercaptoethanol, 10 mM imidazole (pH 7.2)		23 mg/ml	sodium nitrate	Hanging drop (equal volumes of protein and reservoir solutions)		85
VHR dual specificity protein phosphatase	20-30% PEG 4000, 44 mM Li$_2$SO$_4$, 0.2% 2-mercaptoethanol, 0.1 M Hepes (pH 7)	22	5 mg/ml in 10-15% PEG 4000, 22 mM Li$_2$SO$_4$, 0.1% 2-mercaptoethanol, 50 mM Hepes (pH 7)		Hanging drop (20 μl)	Optional microseeding	86
Cdc25A phosphatase, catalytic domain (336-523)	18-20% PEG 3350, 0.025% β-octylglucoside, 0.11 M sodium citrate (pH 5.8)	4	10 mg/ml in: 200 mM NaCl, 1 mM DTT, 20 mM Hepes (pH 7.4)		Hanging drop (4 μl protein + 4 μl reservoir solution)	Optional microseeding	87
Low-molecular-weight phosphotyrosine protein phosphatase	40% (NH$_4$)$_2$SO$_4$, 10 mM DTT, 0.1 M acetate (pH 5.5)	20	15 mg/ml in: 50 mM NaCl, 2 mM DTT, 1 mM MnCl$_2$,			Macroseeding	88
Low-molecular-weight phosphotyrosine protein phosphatase	20% PEG 4000, 10% 2-propanol, 0.1 M Hepes (pH 7.5)		10 mM Tris–HCl (pH 7.5) 10 mg/ml in water	Hanging drop	Macroseeding Sitting drop	Microseeding, macroseeding	89, 90

Crystallizations of fragments of protein kinases and phosphatases that do not include their catalytic domains are not shown. Only crystallizations that have led to successful structure determination are entered. We apologize for any possible omissions from this table. RT, room temperature.

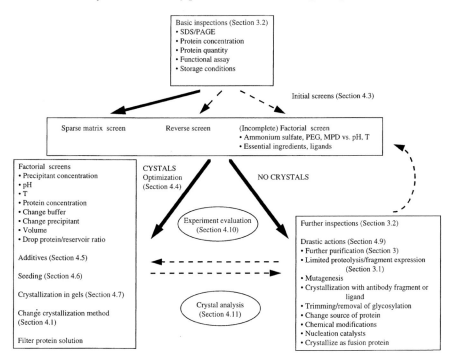

Figure 4. Flow diagram for crystallization. Refer to the text for details on any individual procedure.

5. Preparation of crystals for measurement of diffraction data

Once crystals are obtained, they are used to measure the diffraction data. Protein crystals contain solvent, and therefore dehydrate if exposed to the air. For diffraction experiments, they must either be transferred into sealed capillaries (see *Protocol 6, Figure 1*), or flash-cooled at temperatures around 100 K (see *Protocol 7*). The latter will usually require supplementation of the stabilizing solution with a cryoprotectant, to prevent the solvent from forming ice crystals and destroying the crystal, rather than forming a glass-like phase. Common cryoprotectants are 13–30% glycerol, 11–30% ethylene glycol, 25–35% PEG 400, 20–30% MPD, and others such as xylitol, erythritol, and glucose (34). The advantage of cryogenic temperatures is that they generally eliminate crystal decay in the X-ray beam. The crystals should initially be characterized at ambient temperatures, and then collection of a higher quality diffraction data set should be attempted at cryogenic temperatures. This way, the effect of flash-freezing on the diffraction qualities can be assessed.

X-rays are generated using laboratory X-ray generators (usually using a copper rotating anode), or synchrotron radiation. Most standard detectors use phosphorous image plates. Certain popular data processing programs can be used with diverse detector set-ups [e.g. the HKL suite (35)].

Protocol 6. Mounting crystals in capillaries

Equipment and reagents

- Microscope
- Thin-walled capillary
- Rubber tubing
- Mouthpiece or equivalent
- Wax
- Filter paper wick or equivalent

Method

1. Cleanly break a thin-walled, glass or quartz capillary (0.5–1.5 mm diameter depending on crystal size; to break a quartz capillary, a diamond knife is required) at the thin end and attach with rubber tubing to a mouthpiece, syringe, or variable pipette (e.g. Gilson Pipetman).

2. Under the microscope, draw the crystal into the capillary.

3. Position the plug of liquid containing the crystal ~1 cm from the cut end, and seal with wax.

4. Allow the crystal to sink to the meniscus closest to the sealed end. Remove the excess liquid from the capillary with a fine pipette (e.g. a Pasteur pipette drawn out to a capillary in a flame, or a capillary with a smaller diameter), or a filter-paper strip or wick, leaving only a small amount surrounding the crystal.

5. Cut the other end of the capillary at a desired distance (depending on the camera set-up) and seal with wax.

6. Attach the capillary to the goniometer head with plasticine.

Protocol 7. Flash-freezing

Equipment and reagents

- Microscope
- Cryoloop
- Stream diverter
- Cryostream cooler
- Solution with cryoprotectant

Method

1. Transfer the crystal into a solution supplemented with cryoprotectant. The transfer time (e.g. direct or gradual) and the time of soaking should be determined by trials.

---

2. Pick up the crystal using a cryoloop (a thin nylon or similar material loop attached to a cap that fits on to the goniometer head; the diameter of the loop should be just larger than the crystal). The entire set-up, including loops of different sizes and a magnetic attachment system, is available from Hampton Research. Try to keep the amount of liquid in the loop to a minimum.

3. Transiently divert the cryo stream with a piece of cardboard and attach the loop to the goniostat.

4. Quickly remove the stream diverter; the cryo stream will rapidly cool the crystal and the mother liquor should form clear, vitreous ice. Alternatively, better results are sometimes obtained by initially freezing the crystal in liquid nitrogen or liquid propane, before transferring it into the cryo stream.

6. Phase determination, electron density map interpretation, and refinement

Diffraction data are used to calculate an electron density map, which describes the distribution of electrons in the crystal. This distribution can then be interpreted as an atomic model of the molecules constituting the crystal, and the model refined for best fit with the experimental data. Unfortunately, a diffraction experiment provides only one part of the data required for the calculation of an electron density map, the so-called structure factors. The other part, the phases, has to be obtained in other ways. The commonly used methods include multiple isomorphous replacement (MIR), anomalous scattering (AS) and multi-wavelength anomalous dispersion (MAD), molecular replacement, density modification, direct methods, and combinations of these. In the case of molecular replacement, a structurally similar molecular model is required, and is used to provide the initial phases for the map calculation. In the case of protein kinases and phosphatases, it will increasingly be the case that such a model is accessible; an available structure of the molecule [Protein Data Bank (36)] most similar in sequence will normally be used. If molecular replacement is an option, it is the method of choice. Other methods such as MIR, AS, and MAD require further experimental work, such as modification of the crystals by binding heavy atom compounds (e.g. soaking of crystals in the presence of these compounds) or anomalous scatterers (e.g. substitution of methionines in the protein with selenomethionine, followed by crystallization of the modified protein), and collection of diffraction data from these modified crystals.

All the structure determination methods will require extensive computation, and diverse software is available for these calculations. Popular examples (37–39) are the CCP4 suite which contains programs for most structure determination requirements (40), the program 'O' for electron density interpretation and model building using a graphics workstation (41), and the program X-PLOR for crystallographic refinement (42).

7. Conclusions

We have attempted to provide both an overview of the structure determination process by X-ray crystallography, and a more detailed description of crystallization procedures with hands-on protocols to get newcomers to the field started. Nevertheless, further reading is encouraged using the provided reference list, as well as extra literature searching. It should be emphasized that although crystallization may seem initially confusing, it is neither a difficult nor an expensive endeavour, and is well suited to collaboration between a critical molecular biologist, having complete control over the protein sample to be crystallized, and a crystallographer. The equipment required for the structure determination steps subsequent to crystallization is unfortunately expensive. However, with the explosive growth of macro-molecular crystallography, there is almost certainly a crystallographer near you who would be more than happy to help out after being shown the protein kinase or phosphatase crystals that you have grown!

References

1. Ducruix, A. and Giege, R. (ed.) (1992). Crystallization of nucleic acids and proteins. IRL Press, Oxford.
2. Ferre-D'Amare, A.R. and Burley, S.K. (1994). Structure **2**, 357.
3. Beynon, S. and Bond, S. (ed.) (1989). Proteolytic enzymes. IRL Press, Oxford.
4. Jancarik, J. and Kim, S.-H. (1991). J. Appl. Cryst. **24**, 409.
5. Carter, C.W., Jr and Carter, C.W. (1979). J. Biol. Chem. **254**, 12219.
6. Carter, C.W., Jr (1992). In Crystallization of nucleic acids and proteins: a practical approach (ed. Ducruix, A. and Giegé, R.), pp. 47–72. Oxford University Press, Oxford.
7. Shieh, H.-S., Stallings, W.C., Stevens, A.M., and Stegeman, R.A. (1995). Acta Crystallogr. **D51**, 305.
8. Kingston, R.L., Baker, H.M., and Baker, E.N. (1994). Acta Crystallogr. **D50**, 429.
9. Stura, E.A., Satterthwait, A.C., Calvo, J.C., Kaslow, D.C., and Wilson, I.A. (1994). Acta Crystallogr. **D50**, 448.
10. Stura, E.A., Nemerow, G.R., and Wilson, I.A. (1992). J. Crystal Growth **122**, 273.
11. Blow, D.M., Chayen, N.E., Lloyd, L.F., and Saridakis, E. (1994). Protein Sci. **3**, 1638.
12. Hirschler, J., Charon, M.-H., and Fontecilla-Camps, J.C. (1995). Protein Sci. **4**, 2573.
13. Chayen, N.E.,, Boggon, T.J., Cassetta, A., Deacon, A., Gleichmann, T., Habash, J., Harrop, S.J., Helliwell, J.R., Nieh, Y.P., Peterson, M.R., Raftery, J., Snell, E.H., Haedener, A., Niemann, A.C., Siddons, D.P., Stojanoff, V., Thompson, A.W., Ursby, T., and Wulff, M. (1996). Quart. Rev. Biophys. **29**(3), 227.

14. Sousa, R. and Lafer, E.M. (1990). In Methods: a companion to Methods in enzymology (ed. Carter, C.W., Jr.), Vol. 1, p. 50. Academic Press, New York.
15. Jeruzalmi, D. and Steitz, T.A. (1997). J. Mol. Biol. **274**, 748.
16. Sousa, R. (1995). Acta Crystallogr. **D51**, 271.
17. Stura, E.A. and Wilson, I.A. (1992). In Crystallization of nucleic acids and proteins: a practical approach (ed. Ducruix, A. and Giegé, R.), p. 99. Oxford University Press, Oxford.
18. Robert, M.C., Provost, K., and Lefaucheux, F. (1992). In Crystallization of nucleic acids and proteins: a practical approach (ed. Ducruix, A. and Giegé, R.), p. 127. Oxford University Press, Oxford.
19. Ostermeier, C., Iwata, S., Ludwig, B., and Michel, H. (1995). Nature Struct. Biol. **2**, 842.
20. Landau, E.M. and Rosenbusch, J.P. (1996). Proc. Natl. Acad. Sci. USA **93**, 14532.
21. Hubbard, S.R., Wei, L., Ellis, L., and Hendrickson, W.A. (1994). Nature **372**, 746.
22. Dyda, F., Hickman, A.B., Jenkins, T.M., Engelman, A., Craigie, R., and Davies, D.R. (1994). Science **266**, 1981.
23. Zhang, F., Basinski, M.B., Beals, J.M., Briggs, S.L., Churgay, L.M., Clawson, D.K., DiMarchi, R.D., Furman, T.C., Hale, J.E., Hsiung, H.M., Schoner, B.E., Smith, D.P., Zhang, X.Y., Wery, J.P., and Schevitz, R.W. (1997). Nature **387**, 206.
24. Xiao, B., Jones, D., Madrazo, J., Soneji, Y., Aitken, A., and Gamblin, S. (1996). Acta Crystallogr. **D52**, 203.
25. Rayment, I., Rypniewski, W.R., Schmidt-Bäse, K., Smith, R., Tomchick, D.R., Benning, M.M., Winkelmann, D.A., Wesenberg, G., and Holden, H.M. (1993). Science **261**, 50.
26. Jeffrey, P.D., Russo, A.A., Polyak, K., Gibbs, E., Hurwitz, G., Massague, J., and Pavletich, N.P. (1995). Nature **376**, 313.
27. Center, R.J., Kobe, B., Wilson, K.A., Teh, T., Howlett, G.J., Kemp, B.E., and Poumbourios, P. (1998). Protein Sci. **7**, 1612.
28. McPherson, A. and Schlichta, P. (1988). Science **239**, 385.
29. Hemming, S.A., Bochkarev, A., Darst, S.A., Kornberg, R.D., Ala, P., Yang, D.S.C., and Edwards, A.M. (1995). J. Mol. Biol. **246**, 308.
30. Abergel, C., Moulard, M., Moreau, H., Loret, E., Cambillau, C., and Fontecilla-Camps, J.C. (1991). J. Biol. Chem. **266**, 20131.
31. Stura, E.A. and Chen, P. (1992). In Crystallization of nucleic acids and proteins: a practical approach (ed. Ducruix, A. and Giegé, R.), p. 241. Oxford University Press, Oxford.
32. Gilliland, G.L. and Bickham, D. (1990). Methods: a companion to Methods enzymology (ed. Carter, C.W., Jr.), Vol. 1, p. 6. Academic Press, New York.
33. Gilliland, G.L. (1997). Methods in enzymology (ed. Carter, C.W., Jr. and Sweet, R.M.), Vol. 277, p. 546. Academic Press, New York.
34. Rodgers, D.W. (1994). Structure **2**, 1135.
35. Otwinowski, Z. (1993). In Proceedings of the CCP4 study weekend. Data collection and processing (ed. Sawyer, L., Isaacs, N., and Bailey, S.), p. 56. SERC, Daresbury Laboratory, Warrington, UK.
36. Bernstein, F.C., Koetzle, K.F., Williams, G.J.B., Meyer, E.F., Jr, Brice, M.D., Rodgers, J.R., Kennard, O., Shimanouchi, T., and Tasumi, M. (1977). J. Mol. Biol. **112**, 535.
37. Finzel, B.C. (1993). Curr. Opin. Struct. Biol. **3**, 741.

38. Carter, C.W. and Sweet, R.M. (ed.) (1997). Methods in enzymology, Vol. 276, Macromolecular Crystallography, Part A. Academic Press, London.
39. Carter, C.W. and Sweet, R.M. (ed.) (1997). Methods in enzymology, Vol. 277, Macromolecular Crystallography, Part B. Academic Press, London.
40. CCP4, (1994). Acta Crystallogr. **D50**, 760.
41. Jones, T.A., Zou, J.-Y., Cowan, S.W., and Kjeldgaard, M. (1991). Acta Crystallogr. **A47**, 110.
42. Brünger, A.T., Kuriyan, J., and Karplus, M. (1987). Science **235**, 458.
43. Knighton, D.R., Zheng, J., Ten Eyck, L.F., Ashford, V.A., Xuong, N.-H., Taylor, S.S., and Sowadski, J.M. (1991). Science **253**, 407.
44. Narayana, N., Cox, S., Xuong, N., Ten Eyck, L.F., and Taylor, S.S. (1997). Structure **5**, 921.
45. Bossemeyer, D., Engh, R.A., Kinzel, V., Postingl, H., and Huber, R. (1993). EMBO J. **12**, 849.
46. Prade, L., Engh, R.A., Girod, A., Kinzel, V., Huber, R., and Bossemeyer, D. (1997). Structure **5**, 1627.
47. Engh, R.A., Girod, A., Kinzel, V., Huber, R., and Bossemeyer, D. (1996). J. Biol. Chem. **271**, 26157.
48. Narayana, N., Cox, S., Shaltiel, S., Taylor, S.S., and Xuong, N.-H. (1997). Biochemistry **36**, 4438.
49. Zhang, F., Strand, A., Robbins, D., Cobb, M.H., and Goldsmith, E.J. (1994). Nature **367**, 704.
50. Canagarajah, B.J., Khokhlatchev, A., Cobb, M.H., and Goldsmith, E.J. (1997). Cell **90**, 859.
51. Wilson, K.P., Fitzgibbon, M.J., Caron, P.R., Griffith, J.P., Chen, W., McCaffrey, P.G., Chambers, S.P., and Su, M.S.-S. (1996). J. Biol. Chem. **271**, 27696.
52. Wang, Z., Harkins, P.C., Ulevitch, R.J., Han, J., Cobb, M.H., and Goldsmith, E.J. (1997). Proc. Natl. Acad. Sci. USA **94**, 2327.
53. Tong, L., Pav, S., White, D.M., Rogers, S., Crane, K.M., Cywin, C.L., Brown, M.L., and Pargellis, C.A. (1997). Nature Struct. Biol. **4**, 311.
54. Owen, D.J., Noble, M.E.M., Garman, E.F., Papageorgiou, A.C., and Johnson, L.N. (1995). Structure **3**, 467.
55. Lowe, E.D., Noble, M.E.M., Skamnaki, V.K., Oikonomakos, N.G., Owen, D.J., and Johnson, L.N. (1997). EMBO J. **16**, 6646.
56. De Bondt, H.L., Rosenblatt, J., Jancarik, J., Jones, H.D., Morgan, D.O., and Kim, S.-H. (1993). Nature **363**, 595.
57. Lawrie, A.M., Noble, M.E.M., Tunnah, P., Brown, N.R., Johnson, L.N., and Endicott, J.A. (1997). Nature Struct. Biol. **4**, 796.
58. Russo, A.A., Jeffrey, P.D., and Pavletich, N.P. (1996). Nature Struct. Biol. **3**, 696.
59. Bourne, Y., Watson, M.H., Hickey, M.J., Holmes, W., Rocque, W., Reed, S.I., and Tainer, J.A. (1996). Cell **84**, 863.
60. Xu, R.-M., Carmel, G., Sweet, R.M., Kuret, J., and Cheng, J. (1995). EMBO J. **14**, 1015.
61. Xu, R.-M., Carmel, G., Kuret, J., and Cheng, X. (1996). Proc. Natl. Acad. Sci. USA **93**, 6308.
62. Longenecker, K.L., Roach, P.J., and Hurley, T.D. (1996). J. Mol. Biol. **257**, 618.
63. Niefind, K., Guerra, B., Pinna, L.A., Issinger, O.-G., and Schomburg, D. (1998). EMBO J. **9**, 2451.

64. Hu, S.-H., Parker, M.W., Lei, J.Y., Wilce, M.C.J., Benian, G.M., and Kemp, B.E. (1994). Nature **369**, 581.
65. Kobe, B., Heierhorst, J., Feil, S.C., Parker, M.W., Benian, G.M., Weiss, K.R., and Kemp, B.E. (1996). EMBO J. **15**, 6810.
66. Goldberg, J., Nairn, A.C., and Kuriyan, J. (1996). Cell **84**, 875.
67. Hubbard, S.R. (1997). EMBO J. **16**, 5572.
68. Mohammadi, M., Schlessinger, J., and Hubbard, S.R. (1996). Cell **86**, 577.
69. Yamaguchi, H. and Hendrickson, W.A. (1996). Nature **384**, 484.
70. Xu, W., Harrison, S.C., and Eck, M.J. (1997). Nature **385**, 595.
71. Williams, J.C., Weijland, A., Gonfloni, S., Thompson, A., Courtneidge, S.A., Superti-Furga, G., and Wierenga, R.K. (1997). J. Mol. Biol. **274**, 757.
72. Sicheri, F., Moarefi, I., and Kuriyan, J. (1997). Nature **385**, 602.
73. Goldberg, J., Huang, H., Kwon, Y., Greengard, P., Nairn, A.C., and Kuriyan, J. (1995). Nature **376**, 745.
74. Egloff, M.-P., Cohen, P.T.W., Reinemer, P., and Barford, D. (1995). J. Mol. Biol. **254**, 942.
75. Egloff, M.-P., Johnson, D.F., Moorhead, G., Cohen, P.T.W., Cohen, P., and Barford, D. (1997). EMBO J. **16**, 1876.
76. Griffith, J.P., Kim, J.L., Kim, E.E., Sintchak, M.D., Thomson, J.A., Fitzgibbon, M.J., Fleming, M.A., Caron, P.R., Hsiao, K., and Navia, M.A. (1995). Cell **82**, 507.
77. Kissinger, C.R., Parge, H.E., Knighton, D.R., Lewis, C.T., Pelletier, L.A., Tempczyk, A., Kalish, V.J., Tucker, K.D., Showalter, R.E., Moomaw, E.W., Gastinel, N.L., Habuka, N., Chen, X., Maldonado, F., Barker, J.E., Backuet, R., and Villafranca, J.E. (1995). Nature **378**, 641.
78. Das, A.K., Helps, N.R., Cohen, P.T.W., and Barford, D. (1996). EMBO J. **15**, 6798.
79. Jia, Z., Barford, D., Flint, A.J., and Tonks, N.K. (1995). Science **268**, 1754.
80. Barford, D., Flint, A.J., and Tonks, N.K. (1994). Science **263**, 1397.
81. Bilwes, A.M., den Hertog, J., Hunter, T., and Noel, J.P. (1996). Nature **382**, 555.
82. Hoffmann, K.M.V., Tonks, M.K., and Barford, D. (1997). J. Biol. Chem. **272**, 27505.
83. Hof, P., Pluskey, S., Dhe-Paganon, S., Eck, M.J., and Shoelson, S.E. (1998). Cell **92**, 441.
84. Stuckey, J.A., Schubert, H.L., Fauman, E.B., Zhang, Z.-Y., Dixon, J.E., and Saper, M.A. (1994). Nature **370**, 571.
85. Fauman, E.B., Yuvaniyama, C., Schubert, H.L., Stuckey, J.A., and Saper, M.A. (1996). J. Biol. Chem. **271**, 18780.
86. Yuvaniyama, J., Denu, J.M., Dixon, J.E., and Saper, M.A. (1996). Science **272**, 1328.
87. Fauman, E.B., Cogswell, J.P., Lovejoy, B., Rocque, W.J., Holmes, W., Montana, V.G., Piwnica-Worms, H., Rink, M.J., and Saper, M.A. (1998). Cell **93**, 617.
88. Su, X.-D., Taddel, N., Stefani, M., Ramponi, G., and Nordlund, P. (1994). Nature **370**, 575.
89. Zhang, M., Van Etten, R.L., and Stauffacher, C.V. (1994). Biochemistry **33**, 11097.
90. Zhang, M., Zhou, M., Van Etten, R.L., and Stauffacher, C.V. (1997). Biochemistry **36**, 15.

List of suppliers

Amersham

Amersham International plc., Lincoln Place, Green End, Aylesbury, Buckinghamshire HP20 2TP, UK.

Amersham Corporation, 2636 South Clearbrook Drive, Arlington Heights, IL 60005, USA.

Anderman

Anderman and Co. Ltd., 145 London Road, Kingston-Upon-Thames, Surrey KT17 7NH, UK.

Beckman Instruments

Beckman Instruments UK Ltd., Progress Road, Sands Industrial Estate, High Wycombe, Buckinghamshire HP12 4JL, UK.

Beckman Instruments Inc., PO Box 3100, 2500 Harbor Boulevard, Fullerton, CA 92634, USA.

Becton Dickinson

Becton Dickinson and Co., Between Towns Road, Cowley, Oxford OX4 3LY, UK.

Becton Dickinson and Co., 2 Bridgewater Lane, Lincoln Park, NJ 07035, USA.

Bio

Bio 101 Inc., c/o Statech Scientific Ltd, 61–63 Dudley Street, Luton, Bedfordshire LU2 0HP, UK.

Bio 101 Inc., PO Box 2284, La Jolla, CA 92038–2284, USA.

Bio-Rad Laboratories

Bio-Rad Laboratories Ltd., Bio-Rad House, Maylands Avenue, Hemel Hempstead HP2 7TD, UK.

Bio-Rad Laboratories, Division Headquarters, 3300 Regatta Boulevard, Richmond, CA 94804, USA.

Boehringer Mannheim

Boehringer Mannheim UK (Diagnostics and Biochemicals) Ltd., Bell Lane, Lewes, East Sussex BN17 1LG, UK.

Boehringer Mannheim Corporation, Biochemical Products, 9115 Hague Road, P.O. Box 504 Indianopolis, IN 46250–0414, USA.

Boehringer Mannheim Biochemica, GmbH, Sandhofer Str. 116, Postfach 310120 D-6800 Ma 31, Germany.

British Drug Houses (BDH) Ltd., Poole, Dorset, UK.

Difco Laboratories

Difco Laboratories Ltd., P.O. Box 14B, Central Avenue, West Molesey, Surrey KT8 2SE, UK.

Difco Laboratories, P.O. Box 331058, Detroit, MI 48232–7058, USA.

Du Pont

Dupont (UK) Ltd., Industrial Products Division, Wedgwood Way, Stevenage, Herts, SG1 4Q, UK.

Du Pont Co. (Biotechnology Systems Division), P.O. Box 80024, Wilmington, DE 19880–002, USA.

European Collection of Animal Cell Culture, Division of Biologics, PHLS Centre for Applied Microbiology and Research, Porton Down, Salisbury, Wilts. SP4 0JG, UK.

Falcon (Falcon is a registered trademark of Becton Dickinson and Co.)

Fisher Scientific Co., 711 Forbest Avenue, Pittsburgh, PA 15219–4785, USA.

Flow Laboratories, Woodcock Hill, Harefield Road, Rickmansworth, Herts. WD3 1PQ, UK.

Fluka

Fluka-Chemie AG, CH-9470, Buchs, Switzerland.

Fluka Chemicals Ltd., The Old Brickyard, New Road, Gillingham, Dorset SP8 4JL, UK.

Gibco BRL

Gibco BRL (Life Technologies Ltd.), Trident House, Renfrew Road, Paisley PA3 4EF, UK.

Gibco BRL (Life Technologies Inc.), 3175 Staler Road, Grand Island, NY 14072–0068, USA.

Arnold R. Horwell, 73 Maygrove Road, West Hampstead, London NW6 2BP, UK.

Hybaid

Hybaid Ltd., 111–113 Waldegrave Road, Teddington, Middlesex TW11 8LL, UK.

Hybaid, National Labnet Corporation, P.O. Box 841, Woodbridge, N.J. 07095, USA.

HyClone Laboratories 1725 South HyClone Road, Logan, UT 84321, USA.

International Biotechnologies Inc., 25 Science Park, New Haven, Connecticut 06535, USA.

Invitrogen Corporation

Invitrogen Corporation 3985 B Sorrenton Valley Building, San Diego, C.A. 92121, USA.

Invitrogen Corporation c/o British Biotechnology Products Ltd., 4–10 The Quadrant, Barton Lane, Abingdon, OX14 3YS, UK.

Kodak: Eastman Fine Chemicals 343 State Street, Rochester, NY, USA.

Life Technologies Inc., 8451 Helgerman Court, Gaithersburg, MN 20877, USA.

Merck

Merck Industries Inc., 5 Skyline Drive, Nawthorne, NY 10532, USA.

Merck, Frankfurter Strasse, 250, Postfach 4119, D-64293, Germany.

Millipore

Millipore (UK) Ltd., The Boulevard, Blackmoor Lane, Watford, Herts WD1 8YW, UK.

Millipore Corp./Biosearch, P.O. Box 255, 80 Ashby Road, Bedford, MA 01730, USA.

New England Biolabs (NBL),

New England Biolabs (NBL), 32 Tozer Road, Beverley, MA 01915–5510, USA.

New England Biolabs (NBL), c/o CP Labs Ltd., P.O. Box 22, Bishops Stortford, Herts CM23 3DH, UK.

Nikon Corporation, Fuji Building, 2–3 Marunouchi 3-chome, Chiyoda-ku, Tokyo, Japan.

Perkin-Elmer

Perkin-Elmer Ltd., Maxwell Road, Beaconsfield, Bucks. HP9 1QA, UK.

Perkin Elmer Ltd., Post Office Lane, Beaconsfield, Bucks. HP9 1QA, UK.

Perkin Elmer-Cetus (The Perkin-Elmer Corporation), 761 Main Avenue, Norwalk, CT 0689, USA.

Pharmacia Biotech Europe Procordia EuroCentre, Rue de la Fuse-e 62, B-1130 Brussels, Belgium.

Pharmacia Biosystems

Pharmacia Biosystems Ltd., (Biotechnology Division), Davy Avenue, Knowlhill, Milton Keynes MK5 8PH, UK.

Pharmacia LKB Biotechnology AB, Björngatan 30, S-75182 Uppsala, Sweden.

Promega

Promega Ltd., Delta House, Enterprise Road, Chilworth Research Centre, Southampton, UK.

Promega Corporation, 2800 Woods Hollow Road, Madison, WI 53711–5399, USA.

Qiagen

Qiagen Inc., c/o Hybaid, 111–113 Waldegrave Road, Teddington, Middlesex, TW11 8LL, UK.

Qiagen Inc., 9259 Eton Avenue, Chatsworth, C.A. 91311, USA.

Schleicher and Schuell

Schleicher and Schuell Inc., Keene, NH 03431A, USA.

Schleicher and Schuell Inc., D-3354 Dassel, Germany. Schleicher and Schuell Inc., c/o Andermann and Company Ltd.

Shandon Scientific Ltd., Chadwick Road, Astmoor, Runcorn, Cheshire WA7 1PR, UK.

Sigma Chemical Company

Sigma Chemical Company (UK), Fancy Road, Poole, Dorset BH17 7NH, UK.

Sigma Chemical Company, 3050 Spruce Street, P.O. Box 14508, St. Louis, MO 63178–9916.

Sorvall DuPont Company, Biotechnology Division, P.O. Box 80022, Wilmington, DE 19880–0022, USA.

Stratagene

Stratagene Ltd., Unit 140, Cambridge Innovation Centre, Milton Road, Cambridge CB4 4FG,UK.

Strategene Inc., 11011 North Torrey Pines Road, La Jolla, CA 92037, USA.

United States Biochemical, P.O. Box 22400, Cleveland, OH 44122, USA.

Wellcome Reagents, Langley Court, Beckenham, Kent BR3 3BS, UK.

Index